Control Algorithm Designs in Power Electronics

Control Algorithm Designs in Power Electronics

Edited by **Annie Kent**

CWILLFORD PRESS

New York

Published by Willford Press,
118-35 Queens Blvd., Suite 400,
Forest Hills, NY 11375, USA
www.willfordpress.com

Control Algorithm Designs in Power Electronics
Edited by Annie Kent

International Standard Book Number: 978-1-68285-096-1 (Hardback)

Printed in the United States of America.

Contents

Preface

The main aim of this book is to educate learners and enhance their research focus by presenting diverse topics covering this vast field. This is an advanced book which compiles significant studies by distinguished experts. This book addresses successive solutions to the challenges arising in the area of application, along with it; the book provides scope for future developments.

Power electronics is an important branch of electrical and electronic engineering and has diverse applications. This book unravels the recent studies pertaining to control algorithms in power electronics. The various studies that are constantly contributing towards advancing technologies and evolution of power electronics and control algorithms are examined in detail through topics like electromagnetic and thermal performance, power quality and utility, modeling, simulation, etc. This book includes contributions of experts and scientists which will provide innovative insights into this field. It is a vital tool for all researching and studying this field.

It was a great honour to edit this book, though there were challenges, as it involved a lot of communication and networking between me and the editorial team. However, the end result was this all-inclusive book covering diverse themes in the field.

Finally, it is important to acknowledge the efforts of the contributors for their excellent chapters, through which a wide variety of issues have been addressed. I would also like to thank my colleagues for their valuable feedback during the making of this book.

Editor

Improved Torque Control Performance in Direct Torque Control using Optimal Switching Vectors

Muhd Zharif Rifqi Zuber Ahmadi[1], Auzani Jidin[2], Maaspaliza Azri[3], Khairi Rahim[4], Tole Sutikno[5]

[1,2,3,4]Faculty of Electrical Engineering, Faculty of Electrical Engineering,Universiti Teknikal Malaysia Melaka
Hang Tuah Jaya,76100 Durian Tunggal, Melaka Malaysia
[5]Department of Electrical Engineering, Universitas Ahmad Dahlan, Yogyakarta, Indonesia

Keyword:

Cascaded H-Bridge
Direct Torque Control
Induction machine
Multilevel Inverter
Switching Vector

ABSTRACT

This paper presents the significant improvement of Direct Torque Control (DTC) of 3-phases induction machine using a Cascaded H-Bidge Multilevel Inverter (CHMI). The largest torque ripple and variable switching frequency are known as the major problem founded in DTC of induction motor. As a result, it can diminish the performance induction motor control. Therefore, the conventional 2-level inverter has been replaced with CHMI the in order to increase the performance of the motor either in dynamic or steady-state condition. By using the multilevel inverter, it can produce a more selection of the voltage vectors. Besides that, it can minimize the torque ripple output as well as increase the efficiency by reducing the switching frequency of the inverter. The simulation model of the proposed method has been developed and tested by using Matlab software. Its improvements were also verified via experimental results.

Corresponding Author:

Muhd Zharif Rifqi Zuber Ahmadi
Faculty of Electrical Engineering
Universiti Teknikal Malaysia Melaka
Hang Tuah Jaya, 76100 Durian Tunggal, Melaka Malaysia
Email: zharifrifqi@student.utem.edu.my

1. INTRODUCTION

In the middle 1980's, a simple control strategy to enhance performance of induction motor was proposed by Takahashi and Noguchi. The control strategy is popularly known as Direct Torque Control (DTC) [1]. This method gradually replacing the traditional method of Field Oriented Control (FOC) proposed by F.Blaskhe [2]. At early stages, the FOC was extensivey used to established the control of AC quantities of stator flux, currents and voltages by using vector control approach. However, this scheme is complicated due to the existence of frame transformation, current controller and requires knowledge of machine parameters. In DTC, the torque and flux are controlled independently, in which their demands are satisfied simultaneously by choosing suitable voltage vectors according to the digitized status produced from hysteresis controllers. Unlike the FOC, the torque and flux are controlled based on producing the current components (d-q axis component of stator current referring to excitation reference frame) which results in complex mathematical equations.

Despite the DTC simplicity, it is known to have two major problems, namely variable switching frequency and large torque ripple. These problems that have arisen due to the unpredictable torque and flux control behavior for various operating conditions in hysteresis operation. Obviously, many researchers have extensively proposed some/minor adjustments to minimize the problems. Space vector modulation technique is one of the popular methods to overcome the problem. This way is widely used by researchers in order to achieve greater performance motor as was reported in [3]. The major different between DTC hysteresis based and DTC-SVM is the how to generate the stator voltage reference. In DTC-SVM, the stator voltage reference can be produced by calculating within a sampling time[4, 5]. By doing so, it can produce the constant switching as opposed the DTC-hysteresis based. However, to generate the stator voltage references involve the complex calculation and burden the processor device. Another improvement used is a variable hysteresis band. Basically, when reduce the bandwidth hysteresis band, the torque ripple has also become minimize. Even so, the possibility to select the reverse voltage vector can be occurred whenever the torque changes rapidly at the extreme conditions (i.e. at very low speeds). This mean, overshoot and undershoot of torque to vary outside the hysteresis bands might be happened. As a result, the extreme torque ripple is produced due to the inappropriate selection voltage vector. To improve the switching frequency, the dithering method is used [6, 7]. This method was applied with injecting the high switching frequency of the error component for flux and torque. However, it still also not maintains the switching frequency. Furthermore, many kinds of technique were adopted in DTC drives in order to overcome the problem as well as enhance the excellent performance of motor drives such as [8, 9].

In recent years, the researches on DTC drives utilizing multilevel inverter become the hot topic for providing the more excellent and precision of selection voltage vector to improve DTC performance[10-12]. In general, multilevel inverter can be categorized in three layouts, namely, CHMI, neutral point capacitor multilevel inverter and flying capacitor multilevel inverter as was reported [13, 14] . The all kinds of multilevel have different configurations, number of switching devices, switching states/vectors and arrangements. Multilevel inverter can offer significant advantages to improve DTC performance, especially for medium and high-power voltage application. Furthermore, it also can operate at high voltage and produce lower harmonic (i.e. slope of voltage changed dv/dt)[15].

In this paper, the DTC performances, in terms of torque ripple, harmonics distortion and switching frequency were improved by applying appropriate selection of voltage vectors offered in CHMI topology. The selection of the appropriate vectors depends on the motor operating conditions which inherently determined by the output status of 7-level of torque hysteresis comparator. The application of simple DTC structure and fast instantaneous control with high control bandwidth offered in hysteresis based DTC can be retained. This paper is organized by section as followed; Section II described about the concept of DTC-hysteresis based, Section III presents the topology and switching vectors available in CHMI topology, Section IV discusses the proposed selection of vectors in DTC-CHMI; Section VI presents the simulation results to show the improvements offered and finally Section VII gives the conclusion.

2.   CONCEPT OF DTC-HYSTERESIS BASED

DTC has a simple structure configuration as shown in Figure 1, yet it is superior to enhance torque and flux control, in terms of fast dynamic and reliable control due to the hysteresis operation. By doing so, the appropriate selection of voltage vectors can independently control both torque and flux. Therefore, it can offer a faster instantaneous control of torque and flux based. Selection voltage vector or switching state can be obtained from the look-up table as was tabulated in Table I. Where, in the switching table contains three main components, namely the status of torque, flux and status flux orientation for selecting the appropriate switching state. The switching states are choosing based on the requirement of torque and stator flux, either to increase or decrease and also stator flux sector. In order to make the decision either to increase or decrease can be obtained from the 3-level and 2-level hysteresis of torque and stator flux, respectively. Besides that, the estimated value of flux and torque can be produce from the calculation of voltage and current component. In power circuits, the voltage source inverter is performed by IGBT device. The schematic diagram of 3-phase voltage source inverter is realized in Figure 2. According to these figures, the inverter has contained six switch modes to operate in the 3-phase induction machine. Therefore, it can generate eight voltage space vectors, as illustrated in Figure 3. Each voltage space vector, has a three switching state, [Sa,Sb,Sc]. Six active voltage vector ($\overline{v}_1$ to $\overline{v}_6$) and two non-active or zero voltage vector ($\overline{v}_0$ and $\overline{v}_7$) corresponding to [0 0

0] and [1 1 1], respectively. Each switching device must be complementing each other for (upper and lower switch) to avoid short circuit conditions.

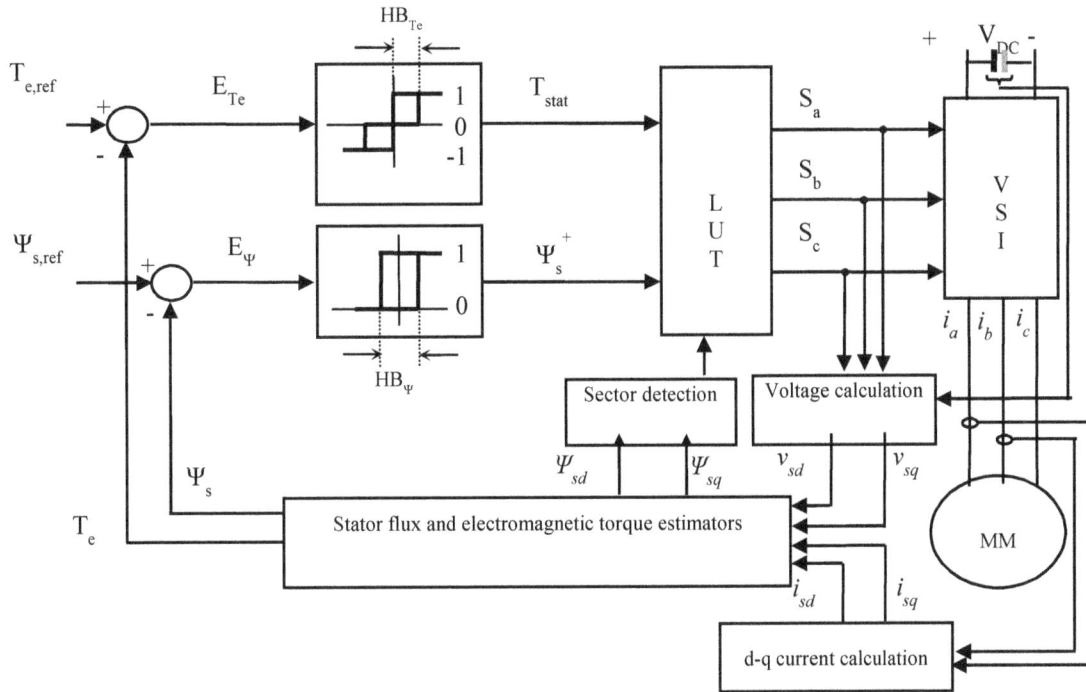

Figure 1. Complete structure of DTC-Hysteresis Based

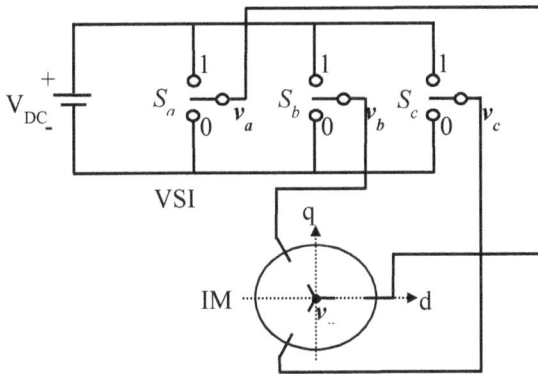

Figure 2. Schematic diagram of VSI

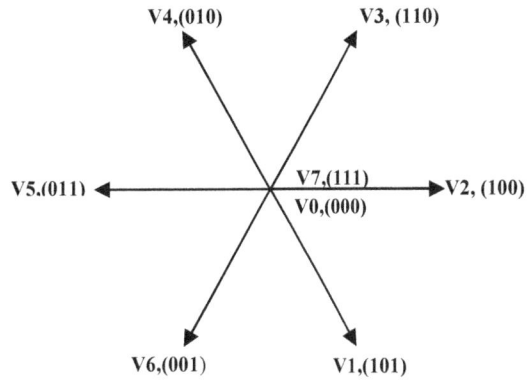

Figure 3. Voltage vectors are generated by VSI

Table 1. Look-up table

Stator flux error status, Ψ_s^+	Torque error status, T_{stat}	Sec I	Sec II	Sec III	Sec IV	Sec V	Sec VI
1	1	[100]	[110]	[010]	[011]	[001]	[101]
	0	[000]	[111]	[111]	[000]	[000]	[111]
	-1	[001]	[101]	[100]	[110]	[010]	[011]
0	1	[110]	[010]	[011]	[001]	[101]	[100]
	0	[111]	[000]	[000]	[111]	[111]	[000]
	-1	[011]	[001]	[101]	[100]	[110]	[101]

3. CASCADED H-BRIDGE MULTILEVEL INVERTER (CHMI)

CHMI is one of the popular power circuit topologies used in high-power medium voltage. These names were given because, it uses multiple units of power cells connected in a series to operate in medium or high voltage as well as to generate lower harmonic ripple. A few isolated DC sources are required for this inverter to synthesize an output voltage waveform. The structure of this inverter is shown in Figure 4, which each phase consists of two H-Bridge. Each cell has single DC-link source to fed the induction motor with connected individually. So, for three phases motor required three isolated DC-links. Four switches off the device (Sa+, Sa-, Sb+ and Sb-) are operating when they receive the signal from the gate drives. By four switches of the inverter to trigger, can produce three discrete output Vab with the level $+V_{dc}$ for (S1 and S4 switch is ON), $-V_{dc}$ for (S2 and S3 switch is ON) and 0V for (all switch OFF). The number of voltage level, L for CHMI can be defined by L=2m+1, where m, for numbers of H-bridge cell per phases. For a 3-level CHMI, the voltage vector can generate 3^3= 27 different voltage vectors and 3L (L - 1) +1= 19 voltage vectors practices used in CHMI topology. Figure 5 shows the voltage vector available as shown in CHMI on a d-q axis . It can be seen that, the outer hexagon which contains 12 voltage vectors with single switching state combination, while for inner hexagon which contains 6 voltage vectors are produced with two combination switching state. Therefore, when increase the level of multilevel inverter, more voltage vectors will be produced. So that, the voltage vectors can be categorized in three conditions (i.e: low speed, medium and high) according the speedy operation. As a result, the total number of switching state become increase, and it offered the many possibilities to improve control strategies of induction motor.

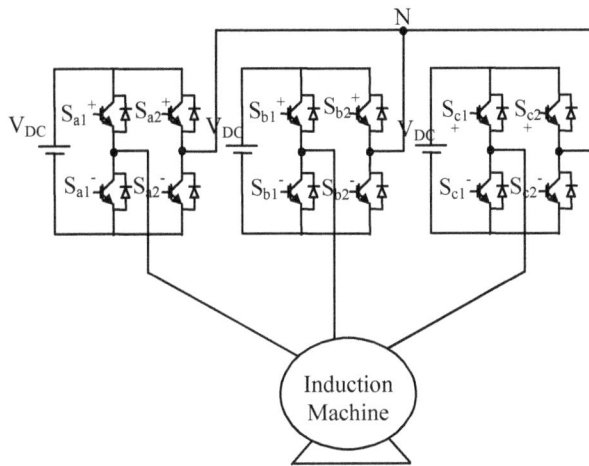

Figure 4. 3-Level CHMI connected to 3-phase induction machine.

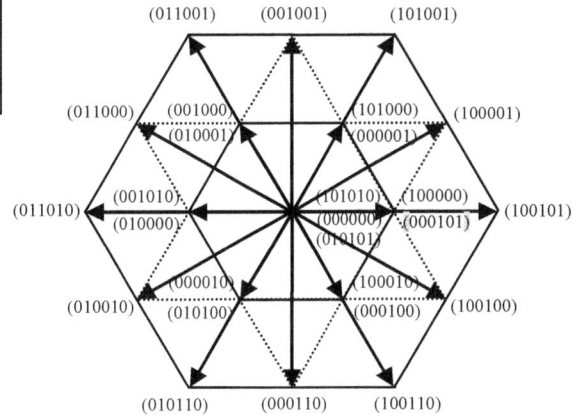

Figure 5. Voltage vector generated by the 3-level CHMI on d-q plane.

4. PROPOSED SWITCHING STRATEGY

In the proposed strategy, a new block modification of torque error status is introduced in the DTC structure by implementing the CHMI topology, is called "Optimum Status Detection or OSD". This block is responsible to modify the torque error status (T_{stat}), which produces new torque status ($T_{stat,new}$) for selecting the optimum switching vectors. Figure 6 shows the complete structure of the DTC-CHMI with inclusion of OSD block (gray color). From the figure, it can be noticed that, some different parts as compared with the DTC conventional. These include the definition of the stator flux plane, calculation of voltage phase for d and q component, and modified the look-up-table for DTC-CHMI. The following subsections discuss the functions or equations used to model the parts.

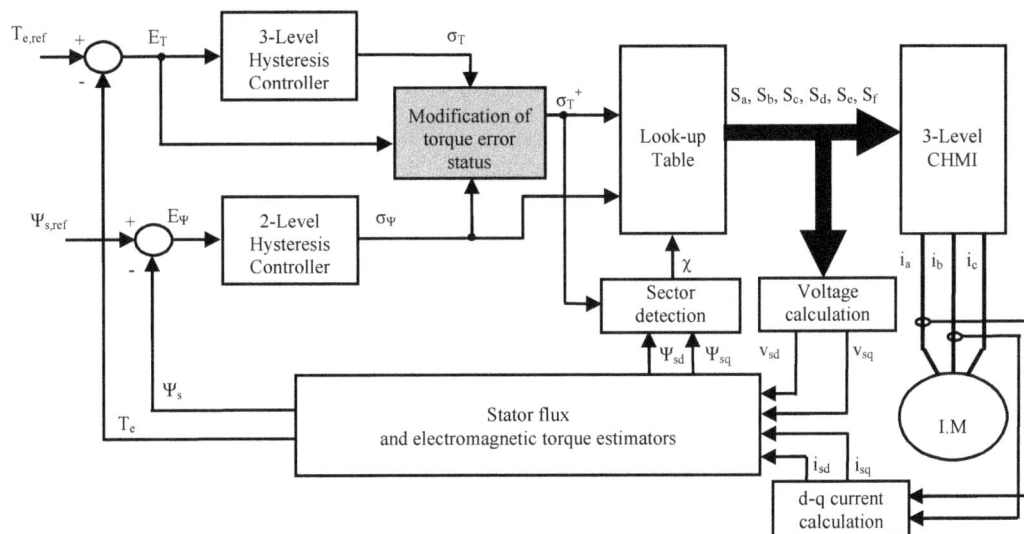

Figure 6 Structure of DTC-hysteresis based induction machine with the proposed modification of flux error status

(i) Definition of sector flux plane

A sector definition of stator flux plane in proposed method is split into two different sequences and angle between sectors is maintained to 60 degrees. It is because, in implementing the CHMI topology produces more voltage vector and the number of switching state also increases. Figure 7. is showing a two different definition of the stator flux plane to apply in this research. Figure 7(i) is shown the definition of stator flux plane for middle speed operation. In this case, the middle voltage vector amplitude is chosen to increase and decrease the flux. While for low and high speed, the sector definition of the stator flux plane as in Figure 7 (ii) is used either to increase or decrease flux. The threshold value for both of the definition stator flux plane can be determined using this equation;

$$\chi = \begin{cases} \vartheta, & \text{if } \sigma_T^+ = 0, 1 \text{ or } 3 \\ \vartheta', & \text{if } \sigma_T^+ = 2 \end{cases} \tag{1}$$

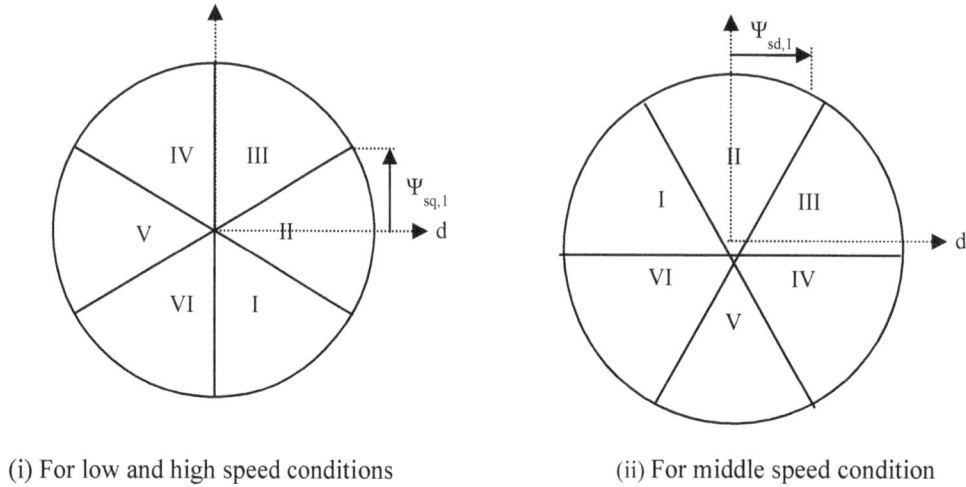

(i) For low and high speed conditions   (ii) For middle speed condition

Figure 7. Difference sector definition

(ii) Calculation of d and q component

By applying this topology; the voltage component can be obtained from the switching pattern of the 3-phase voltage source inverter as follows

$$V_d = \frac{V_{dc}}{3} (2S_{a1} - 2S_{a2} - S_{b1} + S_{b2} - S_{c1} + S_{c2}) \tag{2}$$

$$V_q = \frac{V_{dc}}{\sqrt{3}} (S_{b1} - S_{b2} - S_{c1} + S_{c2}) \tag{3}$$

(iii) Modified Look-Up-Table

As a shown the previous section, look up table is an important part in DTC drives. In the look-up-table have three conditions must be satisfied to choose the appropriate switching state, i.e. torque status, flux status and sector. In look-up table with a proposed structure consist of voltage vectors with three difference amplitudes, (i.e. Short, medium and long). The selection of voltage vectors depends on the torque error status. For example, when the high speed condition mode, the torque error status is selected 3 to indicate is high speed as well as is choosing the longest voltage vector. For the medium speed, the torque error is select the status 2 is shown the motor in middle speed mode. Finally, at low speed operation, the shortest amplitude voltage vector is selected. The labeling the voltage vector based on the speed operation as shown in Figure 9 and Table II shows the look-up table of DTC with proposed structure and strategy.

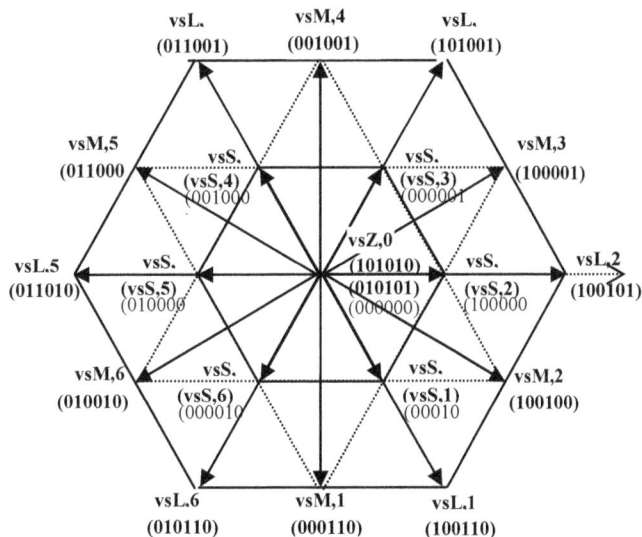

Figure 9. Voltage Vector available in CHMI

Table 2. The Look-Up Table For DTC drives by applied

Stator flux error status, Ψ_s^+	Torque error status, T_{stat}	Sector I	Sector II	Sector III	Sector IV	Sector V	Sector VI
	3	vsL,2	vsL,3	vsL,4	vsL,5	vsL,6	vsL,1
	2	vsM,6	vsM,5	vsM,4	vsM,3	vsM,2	vsM,1
	1	vsS,2	vsS,3	vsS,4	vsS,5	vsS,6	vsS,1
1	0	vsZ,0	vsZ,0	vsZ,0	vsZ,0	vsZ,0	vsZ,0
	-1	vsS,6	vsS,1	vsS,2	vsS,3	vsS,4	vsS,5
	-2	vsM,4	vsM,3	vsM,2	vsM,1	vsM,6	vsM,5
	-3	vsL,6	vsL,1	vsL,2	vsL,3	vsL,4	vsL,5
	3	vsL,3	vsL,4	vsL,5	vsL,6	vsL,1	vsL,2
	2	vsM,1	vsM,6	vsM,5	vsM,4	vsM,3	vsM,2
	1	vsS,3	vsS,4	vsS,5	vsS,6	vsS,1	vsS,2
0	0	vsZ,0	vsZ,0	vsZ,0	vsZ,0	vsZ,0	vsZ,0
	-1	vsS,5	vsS,6	vsS,1	vsS,2	vsS,3	vsS,4
	-2	vsM,3	vsM,2	vsM,1	vsM,6	vsM,5	vsM,4
	-3	vsL,5	vsL,6	vsL,1	vsL,2	vsL,3	vsL,4

A description of the optimal selection based on the operating speed is determined from the behavior of error status (flux and torque) and torque error as follows:

1) Selection of the longest voltage vector happened, when the switching frequency of torque status is less than the switching flux status (Tstat < Fstat) during the torque demands or high speed operation.

2) Selection of the medium of the vector is active when the switching frequency of torque status is slightly higher than that of flux status.

3) The shortest amplitude voltage vector is chosen when the switching frequency of torque status is higher that flux status (Tstat > Fstat). This case occurring during at low speed operation and negative torque demand.

5. EXPERIMENTAL SETUP

The feasibility of the DTC of implementing CHMI topology in reduced the torque ripple output and constant switching frequency has been realized with a complete drive system. In the experimental test, the proposed algorithm switching strategy has conducted by dSPACE ds1104. To interface in real time between hardware and software system, the Control Desk has been provided in order to easily control the parameter.

The control Desk application is very friendly user to configure the layout experiment to monitor the experimental result. For FPGA device, the algorithm blanking times for 3-level CHMI is constructed in proper for preventing the short circuit at power device circuit (IGBT). This device is responsible to achieve the proposed strategy in smooth and successful. The sampling period in this system is set 50us with same the simulation sampling time. Figure 10 shows the complete hardware setup has been done. Three units of the DC-Link is set 120V for testing DTC-CHMI and for DTC conventional inverter is set 240V. The motor parameters are determined based on the blocked rotor and no load test. The motor parameter as tabulated in Table III. In the experiment, induction motor is coupled to a DC motor as a load. To control the load is by applying the voltage supply at the armature winding in the DC motor. The DC motor was manufactured by LO RENZO and the power rating is 1.1kW.

Figure 10. Complete experimental setup

Table 3. Motor parameters

Induction motor parameters	
Rated power, P	1.1 kW
Rated voltage, V_s	380 V
Rated current, $i_{s,rated}$	2.7 A
Rated speed, ω_m	2800 rpm
Stator resistance, R_s	6.1 Ω
Rotor resistance, R_r	4.51 Ω
Stator self inductance, L_s	306.5 mH
Rotor self inductance, L_r	306.5 mH
Mutual inductance, L_m	291.9 mH
Combined inertia, J	0.0565 kg-m^2
Combined viscous friction, B	0.0245 N.m.s
Number of pole pairs, P	2

6. IMPLEMENTATION AND EXPERIMENTAL RESULTS

These sections present the experimental results of significant improvement by using CHMI configuration. Before to the algorithm conducted at hardware testing, the simulation has been simulated to make sure the proposed strategy can achieve the excellent performance of DTC drives via CHMI. Matlab R2011a version is used as a tool to confirm that a strategy. Then, the proposed switching strategy was verified through experimental test. Some test conditions were conducted on DTC-drives control schemes to evaluate the performances. The two different topologies of inverter and control scheme used in these evaluations are conducted as follows;

i.   Control of DTC-Hysteresis based on the 2-level conventional inverter.

ii. Control of DTC based on the 3-level CHMI using proposed switching strategy.

Each test was performed under the same condition in order to have fair comparisons for both of inverter topology as tabulated in Table 4.

Table 4. Comparison between the conventional and proposed method

Control Parameters		Inverter topology	
		(a) 2-Level Conventional Inverter	(c) 3-Level CHMI
Torque reference Change	Low to High	0.7 Nm – 2.5 Nm	
	High to Low	2.5 Nm – 0.7 Nm	
Sampling time, T_s		50 µs	
Torque hysteresis bandwidth, HB_T		0.36 Nm	
Flux hysteresis bandwidth, HB_i		0.02 Wb	
Torque limit, T_{limit}		4 Nm	

The torque, flux, voltage phases and current waveform result for both of inverter topology are shown in Figure 11. Two different references of torque have been applied for both of the inverter. The significant improvement in term torque ripple and switching pattern can be obtained by applying the proposed switching strategy. A step change of reference torque (T_{ref}) was applied from 0.7 Nm to 2.5 Nm at t=0.4s as shown in Figure 11(i) and (iii) and 2.5Nm to 0.7Nm as shown in Figure 11(ii) and (iv) for both of the inverter . In initially, the stator flux angular velocity as well as motor speed is slower. So, in this condition, the torque error status is selected between S_t=0 and S_t=1 in order to choose the lower amplitude of voltage vector. When the torque reference suddenly changed to 2.5Nm, it can be seen that, the torque error status S_t=2 momentarily for very short of time response during torque transient was occurring. This is show that, when have a new demand, the longest amplitude voltage vector is applied to produce the faster dynamic torque response to increase. Based on the observation of the experimental results, it can be seen that when the torque reference change to lower the magnitude, suddenly the negative longest reverse voltage is chosen. That means, the torque error status is selected S_t=-3 in order to quick response to reach the demand of the motor. Therefore, it shows the torque error status St gradually changes from S_T=3↔2 to S_T=2↔1 and finally to S_T=1↔0 for decreasing the output voltage in satisfying the torque demand as the stator flux angular velocity reduce. The effects of torque ripple and switching frequency by applying proposed switching strategy can be clearly seen in the magnified image of Figure 12. By observing the proposed switching strategy, it can be seen that, the waveform for voltage phase is to show the length of slope when torque reference is applied. This is indicated the longest voltage vector is chosen when suddenly torque reference changed either for forward or reverse condition operation.

(i)   **(a) Conventional inverter**   (ii)

(iii)   (iv)

(b) Proposed strategy

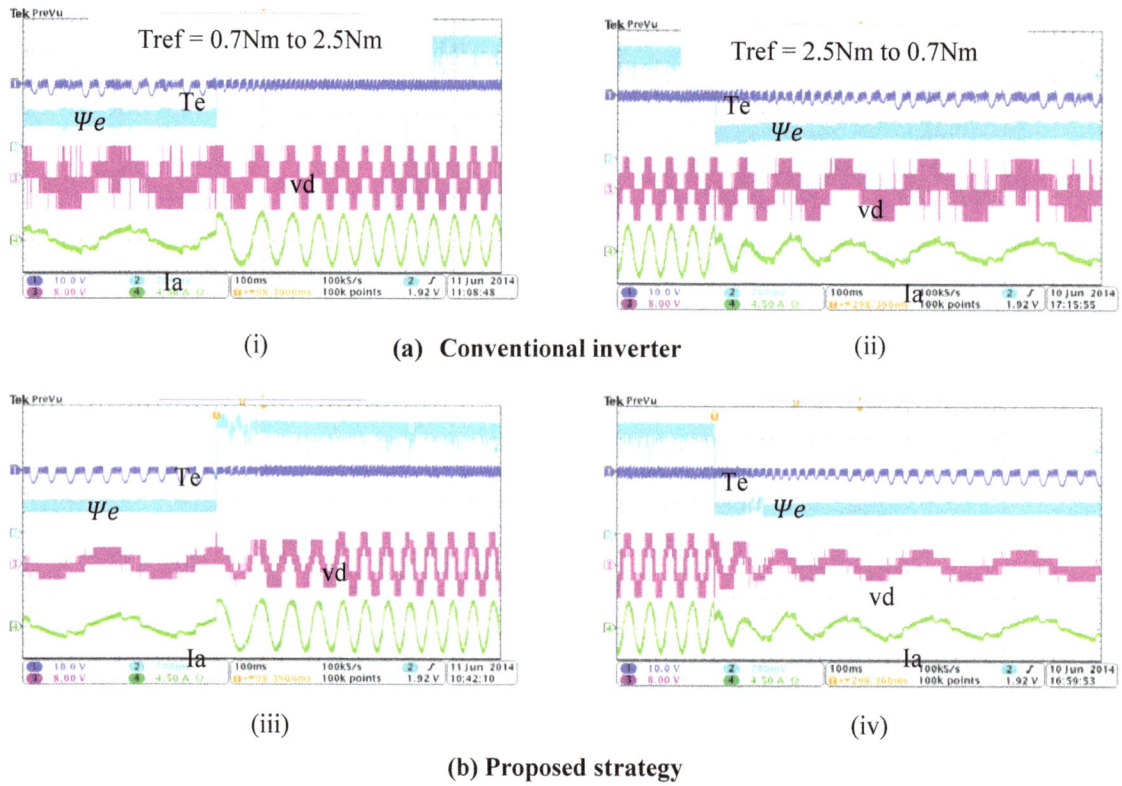

Figure 11. Experimental result for torque referecene was applied for condition

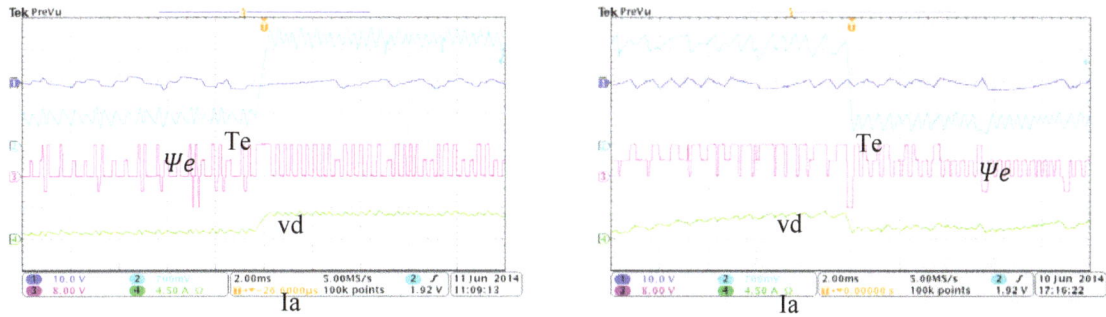

(a) **Conventional inverter**

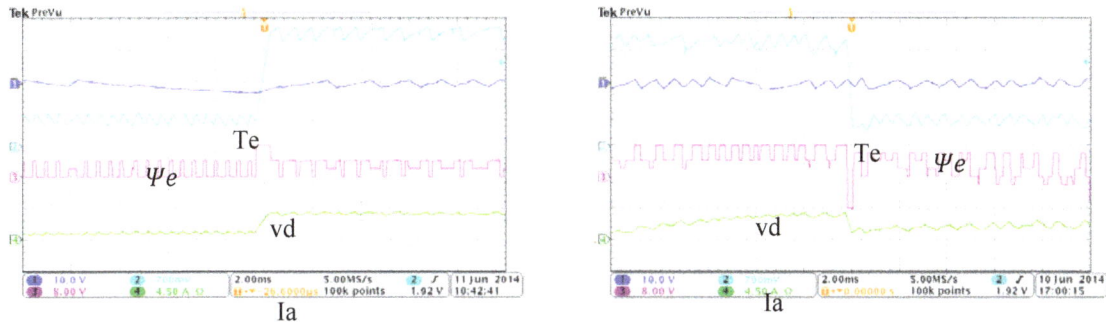

(b) **Proposed Strategy**

Figure12 Magnified image for all experimental results

7. CONCLUSION

This paper presented a simple implementation by introducing the block modification on DTC of induction motor fed by 3-level CHMI topology. This multilevel can generate more voltage vectors as well as increase the number of switching states. Besides that, the appropriate voltage vector was selected based on the status generated from the modified OSD block. By increasing the number of voltage vectors, it can improve the DTC performances in term torque ripples, switching frequency and dynamic response. Besides that, by employing the CHMI topology, the extreme torque slope can be prevented by applying the lower amplitude vector during at low speed as well as the rate change of voltage can also minimize (dv/dt). Whereas, in condition at high speed operation, the longest vector is chosen in order to improve torque control capability in satisfying its demand.

ACKNOWLEDGEMENTS

The authors would like thanks the Ministry of Education Malaysia (KPM) and Universiti Teknikal Malaysia Melaka (UTeM) for providing the research grant FRGS/2013/FKE/TK02/02/1/F00159 for this research.

REFERENCES

[1] Takahashi, I. and T. Noguchi, *A New Quick-Response and High-Efficiency Control Strategy of an Induction Motor.* Industry Applications, IEEE Transactions on, 1986. IA-22(5): p. 820-827.

[2] Ohtani, T., N. Takada, and K. Tanaka. *Vector control of induction motor without shaft encoder.* in *Industry Applications Society Annual Meeting, 1989., Conference Record of the 1989 IEEE.* 1989.

[3] Jun-Koo, K. and S. Seung-Ki. *Torque ripple minimization strategy for direct torque control of production motor.* in *Industry Applications Conference, 1998. Thirty-Third IAS Annual Meeting. The 1998 IEEE.* 1998.

[4] Casadei, D., G. Serra, and A. Tani. *Improvement of direct torque control performance by using a discrete SVM technique.* in *Power Electronics Specialists Conference, 1998. PESC 98 Record. 29th Annual IEEE.* 1998.

[5] Tripathi, A., A.M. Khambadkone, and S.K. Panda, *Torque ripple analysis and dynamic performance of a space vector modulation based control method for AC-drives.* Power Electronics, IEEE Transactions on, 2005. 20(2): p. 485-492.

[6]   Noguchi, T., et al., *Enlarging switching frequency in direct torque-controlled inverter by means of dithering.* Industry Applications, IEEE Transactions on, 1999. 35(6): p. 1358-1366.

[7]   Behera, R.K. and S.P. Das. *High performance induction motor drive: A dither injection technique.* in *Energy, Automation, and Signal (ICEAS), 2011 International Conference on.* 2011.

[8]   Bird, I.G. and H. Zelaya-De La Parra, *Fuzzy logic torque ripple reduction for DTC based AC drives.* Electronics Letters, 1997. 33(17): p. 1501-1502.

[9]   Sutikno, T., et al., *An Improved FPGA Implementation of Direct Torque Control for Induction Machines.* Industrial Informatics, IEEE Transactions on, 2013. 9(3): p. 1280-1290.

[10]   Ahmadi, M.Z.R.Z., et al. *Improved performance of Direct Torque Control of induction machine utilizing 3-level Cascade H-Bridge Multilevel Inverter.* in *Electrical Machines and Systems (ICEMS), 2013 International Conference on.* 2013.

[11]   Ahmadi, M.Z.R.Z., et al. *Minimization of torque ripple utilizing by 3-L CHMI in DTC.* in *Power Engineering and Optimization Conference (PEOCO), 2013 IEEE 7th International.* 2013.

[12]   Alloui, H., A. Berkani, and H. Rezine. *A three level NPC inverter with neutral point voltage balancing for induction motors Direct Torque Control.* in *Electrical Machines (ICEM), 2010 XIX International Conference on.* 2010.

[13]   Nabae, A., I. Takahashi, and H. Akagi, *A New Neutral-Point-Clamped PWM Inverter.* Industry Applications, IEEE Transactions on, 1981. IA-17(5): p. 518-523.

[14]   Malinowski, M., et al., *A Survey on Cascaded Multilevel Inverters.* Industrial Electronics, IEEE Transactions on, 2010. 57(7): p. 2197-2206.

[15]   Ahmadi, M.Z.R.Z., et al. *High efficiency of switching strategy utilizing cascaded H-bridge multilevel inverter for high-performance DTC of induction machine.* in *Energy Conversion (CENCON), 2014 IEEE Conference on.* 2014.

2

A Performance Comparison of DFIG using Power Transfer Matrix and Direct Power Control Techniques

K. Viswanadha S Murthy, M. Kirankumar, G.R.K. Murthy
Department of Electrical and Electronics Engineering, KL University, Andhra Pradesh, India

ABSTRACT

Keyword:

Doubly-fedInduction generator (DFIG)
Direct Power control (DPC)
Power Transfer Matrix
Pulse width modulation (PWM)
Wind energy

This paper presents a direct power control and power transfer matrix model for a doubly-fed induction generator (DFIG) wind energy system (WES). Control of DFIG wind turbine system is traditionally based on either stator-flux-oriented or stator-voltage-oriented vector control. The performance of Direct Power Control (DPC) and Power transfer Matrix control for the same wind speed are studied. The Power transfer matrix Control gave better results. The validity and performance of the proposed modelling and control approaches are investigated using a study system consisting of a grid connected DFIG WES. The performance of DFIG with Power Transfer Matrix and Direct Power Control (DPC) techniques are obtained through simulation. The time domain simulation of the study system using MATLAB Simulink is carried out. The results obtained in the two cases are compared.

Corresponding Author:

G.R.K. Murthy,
Departement of Electrical and Electronics Engineering
KL University
Greenfields, Vaddeswaram, Guntur District, Andhra Pradesh 522502, India
Email: drgrkmurthy@kluniversity.in

1. INTRODUCTION

Generation of electricity has been largely dominated by nuclear, hydro and fossil-fueled thermal plants. Generally this type of generation is considered as conventional power generation. The main drawback of most conventional power plants is the adverse impact on the environment. The gradual depletion of fossil-fuel (such as coal, gas) reserves is also a concern. The solution to these problems lies in adopting non-conventional methods such as wind, solar etc. in power generation. Wind is regarded as the best suitable renewable energy resource for production of power and the best alternative to the conventional energy resources mainly because of availability of large wind turbines [1].

For the last two decades, research is being carried out specifically on Wind Energy Systems to capture more power at fluctuating wind speeds. With the improvement in the power electronic technology constant speed constant frequency (CSCF) generators were replaced by variable speed constant frequency (VSCF) generators in WES. The Doubly Fed Induction Generator (DFIG) is currently the choice of generator for multi-MW wind turbines .The aerodynamic system must be capable of operating over a wide wind speed range in order to achieve optimum aerodynamic efficiency by tracking the optimum tip-speed ratio.Therefore, the generator's rotor must be able to operate at a variable rotational speed. TheDFIG system, therefore operates in both sub- synchronous and super-synchronous modes with a rotorspeed range around the synchronous speed. The stator circuit is directly connected to thegrid while the rotor winding is connected via slip-rings to a three-phase converter. Forvariable - speed systems where the speed range requirements are small, (for example ±30% ofsynchronous speed) the DFIG offers adequate performance and is sufficient for the speedrange required to exploit typical wind resources.

The doubly fed induction generator (DFIG) based wind turbine with variable speed and variable pitch control scheme is the most popular wind power generation system in the wind power industry. This machine can be operated either in grid connected mode or in standalone mode. This system has recently become very popular as generator for variable speed wind turbines. The major advantage of the doubly fed induction generator (DFIG), which has made it popular, is that the power electronic equipment has to handle only a fraction (20-30%) of the total system power [2], [3]. That means the losses in the power electronic equipment can be reduced in comparison to power electronic equipment that has to handle the total system power as for a direct-driven synchronous generator, apart from the cost saving of using smaller converters. Control of the DFIG is more complicated than the control of a standard induction machine. In order to control the DFIG rotor current is controlled by a power electronic converter. One common way of controlling the rotor current is by means of Field oriented (vector) control. Direct torque control (DTC) of induction machines, provides an alternative to vector control [5]. Based on the principles of DTC strategy, direct power control (DPC) was developed for three-phase pulse width modulation (PWM) converters.

Power transfer matrix is a control technique of DFIG which uses instantaneous real and reactive power instead of dq components of currents in a vector control scheme. The main features of the proposed model compared to conventional models in the dq frame of reference are [6].

a) Robustness: The waveforms of power components are independent of a reference frame; therefore, this approach is inherently robust against unaccounted dynamics such as PLL.

b) Simplicity of realization: The power components (state variables of a feedback control loop) can be directly obtained from phase voltage/current quantities, which simplify the Implementation of the control system.

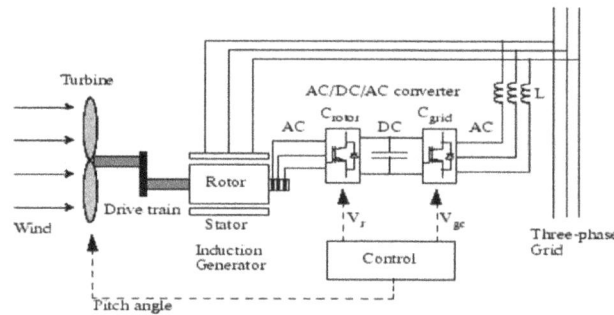

Figure 1. Structure of DFIG wind power generating system

2. WIND TURBINE MODEL

The wind turbine characteristics must be analyzed for getting optimum power curve (P_{opt}). The power output of Wind turbine is given by [4]:

$$P_0 = C_p * P_V = 0.5 \, \rho \, S_w \, V^3 C_p \tag{1}$$

Where 'ρ' is the air density; S_w is wind turbine blade swept area in the wind, V is wind speed. C_p represents the power conversion efficiency of the wind turbine. It is a function of λ (Tip-speed Ratio).

$$\lambda = \frac{2\pi RN}{V\infty} = \frac{R}{V} \, \omega \tag{2}$$

Where 'R' is the blade radius; ω is the angular velocity of the rotating blades; N is the rotational speed in revolutions per second, and V_∞ is the wind speed without the interruption of rotor. C_p can be calculated by using the formula:

$$C_p = 0.5716 * \left(\frac{1}{\lambda i}\right) (116 * - 0.4\beta - 5) * e\left[-21\left(\frac{1}{\lambda i}\right)\right] + 0.068 * \lambda \tag{3}$$

$$\frac{1}{\lambda i} = \frac{1}{\lambda + 0.08\beta} - \frac{0.035}{\beta^3 + 1}$$

Maximum power from the wind turbine is:

$$P_{max} = K \, \omega^3 \tag{4}$$

Where $k = 0.5 S_w \left(\frac{\lambda opt}{R} \right)^3 * C_p$

3. DIRECT POWER CONTROL OF DFIG

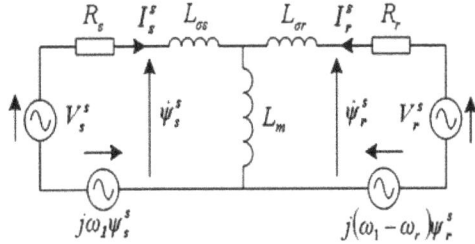

Figure 2. Equivalent circuit of DFIG in the synchronous d-q reference frame

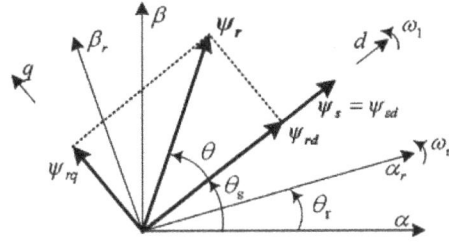

Figure 3. Stator and rotor flux vectors in synchronous d-q frame

The equivalent circuit of a DFIG in the synchronous d–q frame, rotating at the speed of ω_1, is shown in Figure 2. The d-axis of the synchronous frame is fixed to the stator flux, as shown in Figure 3. With reference to Figure 2, the stator voltage vector in the synchronous d–q reference frame is given as:

$$V_s^s = R_s I_s^s + (d\Psi_s^s/dt) + j\omega_1 \Psi_s^s \tag{5}$$

Under balanced ac voltage supply, the amplitude and rotating speed of the stator flux are constant. Therefore, in the synchronous d–q frame, the stator flux maintains a constant value [5]. Thus;

$$\Psi_s^s = \Psi_{sd}$$
$$(d\Psi_s^s/dt) = 0 \tag{6}$$

Considering Equation (5) and neglecting the voltage drop across the stator resistance, Equation (6) can be simplified as:

$$V_s^s = j\omega_1 \Psi_s^s = j\omega_1 \Psi_{sd} \tag{7}$$

The stator current in the synchronous d-q frame is given as:

$$I_s^s = \frac{L_r \psi_s^s - L_m \psi_r^s}{L_s L_r - L_m^2} = \frac{\psi_s^s}{\sigma L_s} - \frac{L_m \psi_r^s}{\sigma L_s L_r} \tag{8}$$

Thus the stator active and reactive power inputs can be calculated as:

$$P_s - jQ_s = 3/2\, j\omega_1\psi_{sd} \times \left(\frac{\psi_s^s}{\sigma} \Big/ L_s - \frac{L_m \psi_r^s}{\sigma} \Big/ L_s L_r \right)$$

$$P_s - jQ = 3/2\, j\omega_1\psi_{sd} \times \left(\frac{\psi_{sd}}{\sigma} \Big/ L_s - \frac{L_m (\psi_{rd} - j\psi_{rq})}{\sigma} \Big/ L_s L_r \right) \tag{9}$$

$$P_s - jQ_s = k\sigma\omega_1 \left[-\psi_{sd}\psi_{rq} + j\psi_{sd} \left(\frac{L_r\psi_{sd}}{L_m} - \psi_{rd} \right) \right]$$

Splitting Equation (9) into real and imaginary parts yields:

$$P_s = -k_\sigma\omega_1\psi_{sd}\psi_{rq}$$

$$Q_s = k_\sigma\omega_1\psi_{sd} \left(\psi_{rd} - \frac{L_r}{L_m}\psi_{sd} \right) \tag{10}$$

As the stator flux is constant, according to Equation (10), the active and reactive power changes over a constant period are given by :

$$\Delta P_s = -K_\sigma \omega_1 \psi_{sd} \Delta \psi_{rq}$$
$$\Delta Q_s = K_\sigma \omega_1 \psi_{sd} \Delta \psi_{rd}$$

(11)

Equation (10) indicates that the stator reactive and active power changes are determined by the changes of the rotor flux components on the d-q axis, i.e. , $\Delta \Psi$rd and ,$\Delta \Psi$rq, respectively.

4. ACTIVE AND REACTIVE POWER CONTROL

The active and reactive power control calculates the required rotor voltage that will reduce the active and reactive power errors to zero during a constant sampling time period of Ts. A PWM modulator is then used to generate the applied rotor voltage for the time period of Ts.

Within the time period of Ts, the rotor voltage required to eliminate the power errors in d-q reference frame are calculated as [7]-[9]:

$$V_{rd} = \frac{1}{T_s} \frac{\Delta Q_s}{k_\sigma \omega_1 \psi_{sd}} - \omega_s \psi_{rq}$$
$$V_{rq} = \frac{1}{T_s} \frac{-\Delta P_s}{k_\sigma \omega_1 \psi_{sd}} + \omega_s \psi_{rd}$$

(12)

However, its accuracy could be affected by the variation of Lm (Mutual inductance). An alternative method based on Equation (11) gives:

$$\psi_{rd} = \frac{Q_s}{k_\sigma \omega_1 \psi_{sd}} + \frac{L_r}{L_m} \psi_{sd}$$
$$\psi_{rq} = \frac{-P_s}{k_\sigma \omega_1 \psi_{sd}}$$

(13)

From the Equation (12) and (13) we get:

$$V_{rd} = \frac{1}{T_s} \frac{\Delta Q_s}{k_\sigma \omega_1 \psi_{sd}} + \omega_s \frac{P_s}{k_\sigma \omega_1 \psi_{sd}}$$
$$V_{rq} = \frac{1}{T_s} \frac{-\Delta P_s}{k_\sigma \omega_1 \psi_{sd}} + \omega_s \left(\frac{Q_s}{k_\sigma \omega_1 \psi_{sd}} + \frac{L_r}{L_m} \psi_{sd} \right)$$

(14)

The first terms on the right hand side reduce power errors while the second terms compensate the rotor slip that causes the different rotating speeds of the stator and rotor flux.

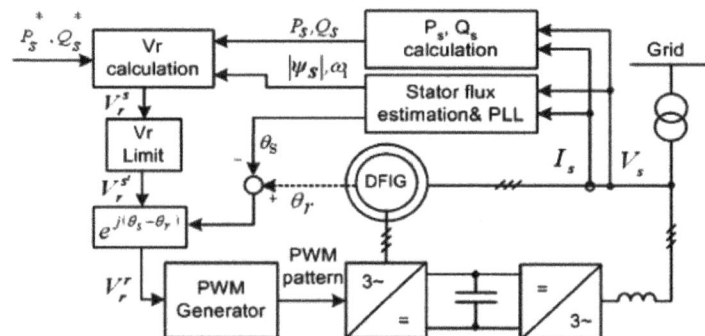

Figure 4. Schematic diagram of the DPC for a DFIG system

5. PRINCIPLES OF POWER TRANSFERMATRIX

The schematic diagram of a DFIG wind turbine generator is represented in Figure 1. The power Converter includes Rotor-side-converter (RSC) to control speed of the generator and Grid-side converter (GSC) to inject reactive power to the system. The instantaneous real and reactive power components of the grid side converter, $p_g(t)$and $q_g(t)$in the synchronous d-q frame of reference are [6]:

$$\begin{bmatrix} p_g(t) \\ q_g(t) \end{bmatrix} = \frac{3}{2} \begin{bmatrix} v_{sd} & v_{sq} \\ v_{sq} & -v_{sd} \end{bmatrix} \begin{bmatrix} i_{gd} \\ i_{gq} \end{bmatrix} \tag{15}$$

6. MODEL OF DFIG USING INSTANTANEOUS POWER COMPONENTS

The change in real power and reactive power can be expressed as [12]-[14]:

$$\frac{dp_s}{dt} = g_1 p_s - \omega_{sl} q_s - g_4 \psi_{sd} - g_5 \psi_{sq} + u_{rd} \tag{16}$$

$$\frac{dq_s}{dt} = \omega_{sl} p_s + g_1 q_s + g_4 \psi_{sq} - g_5 \psi_{sd} + u_{rd}$$

Where,

$$u_{rd} = g_2 v_{rd} + g_3 v_{rq} - 3\frac{v_s^2}{2L_s'}$$

$$u_{rq} = g_3 v_{rd} - g_2 v_{rq}$$

$$g_1 = -\frac{r_s L_r + L_s r_s}{L_s' L_r}; g_2 = \frac{3L_m v_{sd}}{2L_s' L_r}$$

$$g_3 = \frac{3L_m v_{sq}}{2L_s' L_r}$$

$$g_4 = \frac{3}{2}\left(\frac{r_r v_{sd} - L_r \omega_r v_{sq}}{L_s' L_r}\right) \tag{17}$$

$$g_5 = \frac{3}{2}\left(\frac{r_r v_{sd} + L_r \omega_r v_{sq}}{L_s' L_r}\right)$$

The electromechanical dynamic model of the machine is:

$$\frac{d\omega_r}{dt} = \frac{P}{J}\left(T_e - T_m\right) \tag{18}$$

Where P, J and Tm are the number of pole pairs, inertia of therotor, and mechanical torque of the machine, respectively. Theelectric torque is given by [10], [11]:

$$T_e = \frac{3}{2} P\left(\psi_{sd} i_{sq} - \psi_{sq} i_{sd}\right) \tag{19}$$

$$\frac{d\omega_r}{dt} = g_6 p_s + g_7 q_s - \frac{P}{J} T_m \tag{20}$$

Where,

$$g_6 = \frac{P^2}{J}\frac{\psi_{sq} v_{sd} - \psi_{sd} v_{sq}}{v_s^2} \tag{21}$$

$$g_7 = \frac{P^2}{J}\frac{\psi_{sd} v_{sd} + \psi_{sq} v_{sq}}{v_s^2}$$

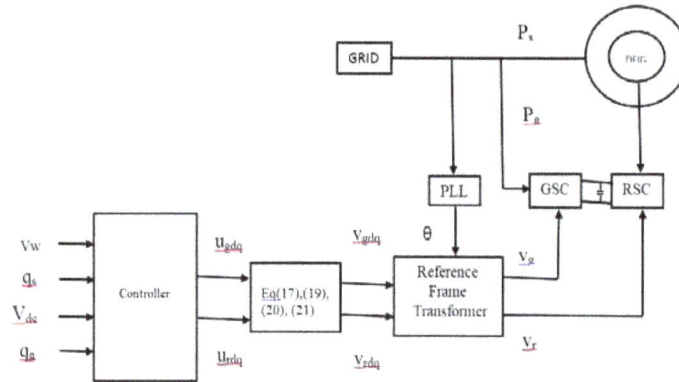

Figure 6. Schematic diagram of the study system of power transfer matrix

6. RESULTS AND COMPARISION
The following parameters are used to verify the real power, Reactive power and dc link voltages:

Parameters	Values	Units
Rated power(P)	1.5	MW
Rated voltage(V)	0.575	KV
Rated frequency(F)	60	Hz
Rated wind speed(V_w)	12	m/s
Stator resistance(R_s)	1.4	mΩ
Rotor resistance(R_r)	0.99	mΩ
Stator leakage inductance(L_S)	89.98	μH
Rotor leakage inductance(L_R)	82.08	μH
Magnetizationinductance(L_m)	1.526	mH
Stator/rotor turns ratio	1	-
Poles	6	-
Turbine rotor diameter	70	M
Lumped inertia constant	5.05	S

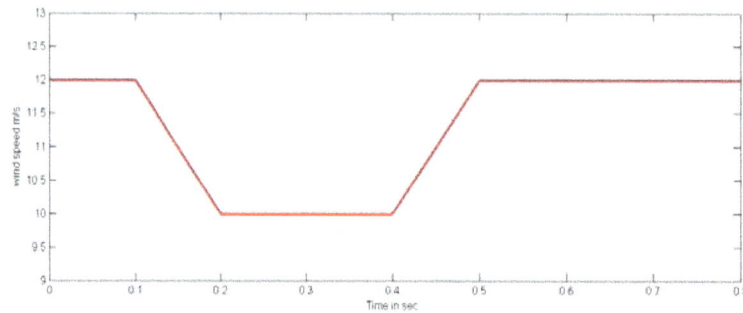

Figure 7. Trapezoidal pattern for wind speed

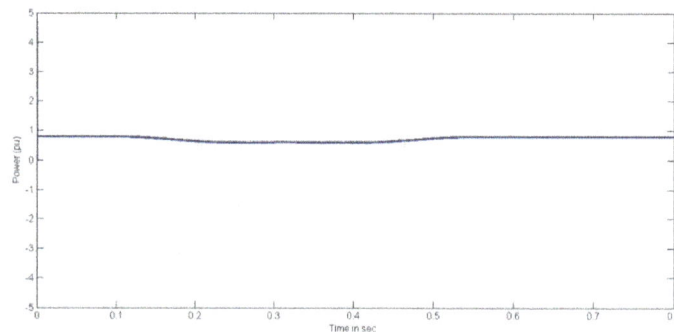

Figure 8(a). Real power from power transfer matrix control

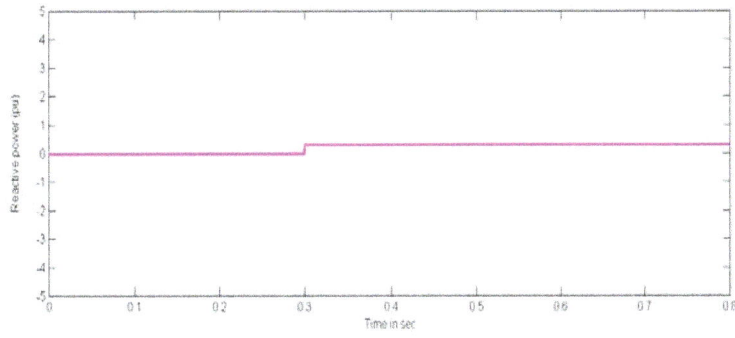

Figure 8(b). Reactive power from power transfer matrix control

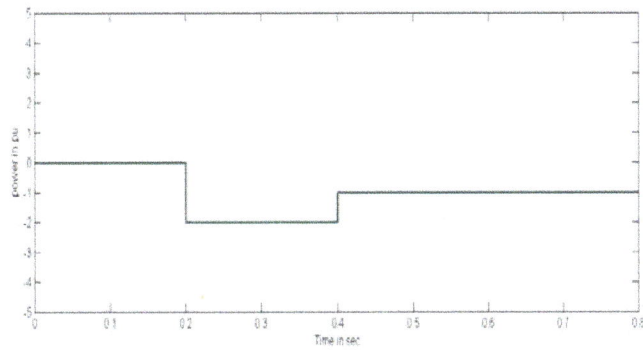

Figure 9(a). Real power from DPC

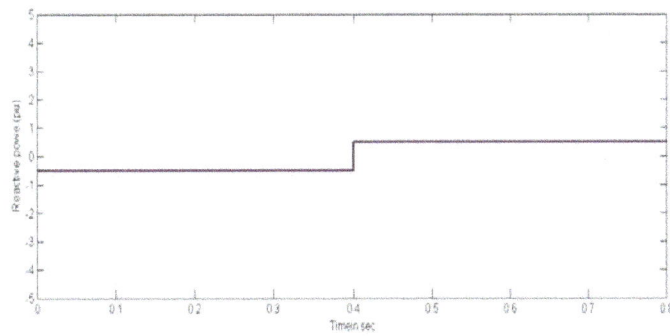

Figure 9(b). Reactive power from DPC

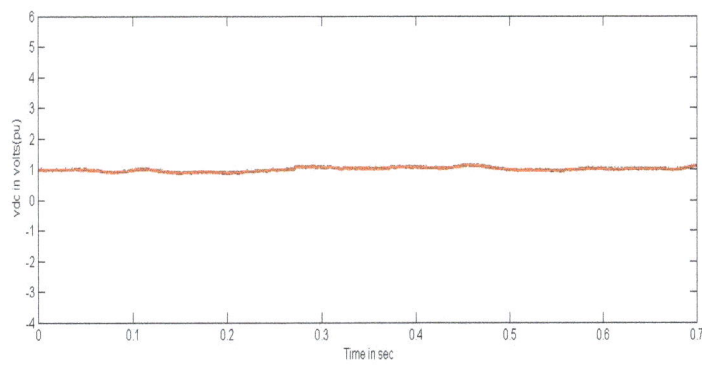

Figure 10(a). Vdc from Power transfer matrix

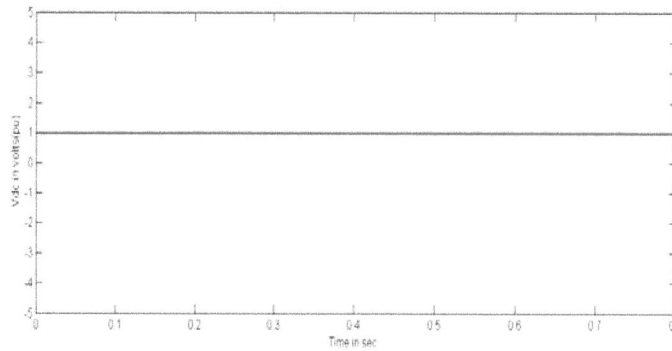

Figure 10(b). Vdc from DPC

The results are obtained at a wind speed of 12 m/sec. The trapezoidal wave form shown in Figure 7 shows the pattern of step change in the reactive reference which is applied to both the control techniques.The trapezoidal pattern wasselected to examine the system behavior following variation inthe wind speed with both negative and positive slopes. The selectedwind speed pattern spans an input mechanical wind powerfrom 0.7 to 1 p.u. (70 to 100% of the turbine-generator ratedpower).

Figures 8(a) and 8(b) show the Real and Reactive power tracking of DFIG against disturbances present in the given wind speed pattern. Because of coupling of all powers interlinked to each other, the coupling effect is obtained at t=0.3 sec.

Figure 9(a) shows the Real power tracking of DFIG against disturbances in the given wind speed pattern. Here the dip in the wave form shows the start of real power generation at t=0.2 sec. Figure 9(b) indicates the reactive power absorption for 0.4 sec.

Figures 10(a) and 10(b) show the dc link voltages of Power transfer matrix control and DPC.The change in wind speed leads to the fluctuations of the dc link voltage. Due to the coupling of all powers v_{dc} of power transfer matrix have some variations. Where as in DPC there is no coupling of the powers and the dc link voltage is constant. Change in wind speed does not affect dc link voltage

7. CONCLUSION

Upon examining the results of both Power transfer matrix and DPC techniques for the same disturbances the Real power generation is better in power transfer matrix control than with DPC. Also the generation of power starts in DPC with a delay of 0.2 sec. Hence power transfer matrix method is giving better results than the DPC method.

REFERENCES

[1] Wind technology 2011024/29. Global wind energy council (online).
[2] LH Hansen, L Helle, F Balaabjerg, E Ritchie, S munk-nielsen, H binder. *Conseptual Survey of Generators And Power Electronics For Wind Turbines.* Riso national laboratory.
[3] L Morel, H Godfroid. *Dfim Converter Optimisation And Field Oriented Control Without Positioning Sensor.* IEE proc.elctr.powerappl. 1998; 145(4): 360-368.
[4] SN Bhadra, D Kastha, S Banarjee. *Wind Electrical Systems.* Oxford University press
[5] Dawei Zhi, Lie Xu. *Direct Power Control of DFIG with Constant Swtching Frequency and Improved Transient Performance"IEEE Transactions on Energy Conversion.* 2007; 22(1).
[6] Esmaeil Rezaei, Ahmadreza Tabesh, Mohammad Ebrahimi. Dynamic Model and Control of DFIG Wind Energy Systems Based on Power Transfer Matrix. *IEEE Transactions on Power Delivery.* 2012; 27(3).
[7] P Krause, O Wasynczuk, S Sudhoff, IPE Society. *Analysis of Electric Machinery and Drive Systems.* Piscataway, NJ: IEEE. 2002.
[8] Abdelmalek Boulahia. Mentouri university of Constantine, Algeria; Khalil NABTI; Hocine BENALLA, Algeria Direct Power Control for Threelevel NPC Based PWM AC/DC/AC Converter in Doubly Fed Induction Generators Based Wind Turbine International Journal of Electrical and Computer Engineering (IJECE). 2012; 2(3): 425-436.
[9] NRN Idris, AHM Yatim.*Direct Torque Control of Induction Machines with Constant Switching Frequency and Reduced Torque Ripple. IEEE Trans. Ind. Electron.,* 2004; 51(4): 758–767.
[10] J Kang, S Sul. New Direct Torque Control of Induction Motor for Minimum Torque Ripple and Constant Switching Frequency. *IEEE Trans. Ind. Appl.,* 1999; 35(5): 1076–1082.

[11] T Noguchi, H Tomiki, S Kondo, I Takahashi. Direct Power Control Of Pwm Converter Without Power-Source Voltage Sensors. *IEEE Trans.Ind. Appl.*, 1998; 34(3): 473–479.

[12] G Escobar, AM Stankovic, JM Carrasco, E Galvan, R Ortega. Analysis and Design Of Direct Power Control (Dpc) For A Three Phase Synchronousrectifier Via Output Regulation Subspaces. *IEEE Trans. Power Electron.*, 2003; 18(3): 823–830.

[13] M Malinowski, MP Kazmierkowski, S Hansen, F Blaabjerg, GD Marques. VIRTUAL-FLUX-BASED DIRECT POWER CONTROL OF THREE-PHASEPWM RECTIFIERS. *IEEE Trans. Ind. Appl.*, 2001; 37(4): 1019–1027.

[14] KP Gokhale, DW Karraker, SJ Heikkila. *Controller for A Woundrotor Slip Ring Induction Machine.* U.S. Patent 6 448 735 B1. 2002.

[15] L Xu, P Cartwright. Direct active and reactive power control of DFIGfor wind energy generation. *IEEE Trans. Energy Convers.*, 2006; 21(3): 750–758.

[16] Wang Zezhong, Liu Qihui. Analysis of DFIG Wind Turbine during Steadystate and Transient Operation. *TELKOMNIKA Indonesian Journal of Electrical Engineering.* 2014; 12(6): 4148-4156.

3

Closed Loop Non Linear Control of Shunt Hybrid Power Filter for Harmonics Mitigation in Industrial Distribution System

A.Arivarasu, R.Balasubramanium
Department of Electrical and Electronics Engineering, SASTRA University, Thanjavur

Keyword:	**ABSTRACT**
d-q transformation Harmonics mitigation Non- linear function Reactive power compensation Shunt Hybrid Power Filter	In recent years, the amount of non-linear loads has increased considerably since there were improvements in power electronic equipment (such as adjustable speed drives or converter ac-dc, ac-ac, dc-ac and dc-dc) in industrial sectors which cause deterioration of the quality of the electric power supply through distortion of supply voltage and supply current. This has led to improvement of many stringent needs regarding generation of harmonic current, which are found in IEEE519 and IEC61000 standards. This paper proposes a non-linear function based closed loop control strategy (without load current extraction) for three-phase Shunt Active Power Line Conditioner and LC passive filter to compensate harmonics, power factor improvement and enhance the dynamic performance of Shunt Hybrid Power Filter (SHPF). By using a PI controller the DC bus voltage of the Shunt Active Power Filter is maintained constant. Results obtained from simulation shows the performance of expected hybrid filter in transient and steady state operation . This indicates that the controller is able to compensate even under severe load current imbalances.

Corresponding Author:

A.Arivarasu,
Departement of Electrical and Electronics Engineering,
SASTRA University,
Thirumalaisamuthiram, Vallam, Tanjavur 62102, Tamil Nadu, India.
Email: arivarasu.apa@gmail.com

1. INTRODUCTION

The increased level of nonlinear loads due to usage of more electronic equipment leads to deterioration of power quality in the power system. Whenever the nonlinear load draws harmonic current from a supply there occurs a distortion in the supply voltage waveform at the common coupling (PCC) point because of the source impedance. The distorted voltage and current may cause end-user equipment to malfunction and overheating of conductors. Due to this, the component connected at the PCC gets affected by reduction in efficiency and life period.

Usually, to reduce and avoid current harmonics, a passive LC power filter is used when connected parallel to the load. These types of passive power filter have some drawbacks, due to which it cannot provide a complete solution. These have the disadvantages of large size, resonance, and fixed compensation. In recent days, based on power electronic methods the harmonic suppression facilities have been improved. These facilities are known as active power filter which can suppress various order harmonic components simultaneously for loads which are nonlinear in nature. Depending on the compensation types, the active power filter is categorized into reactive power, harmonic, balancing of three-phase systems and multiple compensations.

This study of active power filter and its operating principle were introduced by H.Sasaki and T.machida in 1970 [1]. The current source converter type based active power filters were implemented with GTO thyristors for first time in the world in 1982 [2], Nowadays the IGBTs are been used for the real improvement in active power filter technology. Among the subjects related to the active filter's design techniques and applications, the technique used for extracting the harmonic load currents and determining the filter reference current has a major role. The accuracy and response speed of the SAPF are taken into consideration [3], [4]. The techniques of reference current generation are categorized as below: 1) time-domain and 2) frequency domain methods [5], [6]. Time-domain methods such as p–q transformation and$d$–q transformation etc. depend on the measurements and conversion of three-phase quantities. Fast response is the major advantage of the time-domain control techniques in comparison with the frequency-domain techniques depending on the fast Fourier transformation. On the other hand, frequency-domain control techniques detect the individual and multiple harmonic load current with more accuracy. The loop for control of operation of the controller can be categorized as open loop control and closed loop control. The control algorithm for closed loop control system is less complicated then in open loop method and requires minimal number of current sensors.

Many control techniques have been shown in the literature, such as instantaneous active & reactive power theory [7], synchronous reference frame [8], Fuzzy control [9], PI control [10], Sliding mode control [11],Neural Network approach [12], and Open loop Nonlinear control [13].

In this paper, a three-phase shunt hybrid power filter is modeled in the three-phase "abc" coordinates, and then, to avoid time dependence, the model is transformed to the rotating "dq" reference frame. On the other hand, A proportional–integral (PI) controller is used to control the SAPF dc bus voltage. The dynamic performances of the SAPF using the Non-linear function based closed loop control approach with shunt LC passive filter are obtained by simulation using Simulink.

2. RESEARCH METHOD

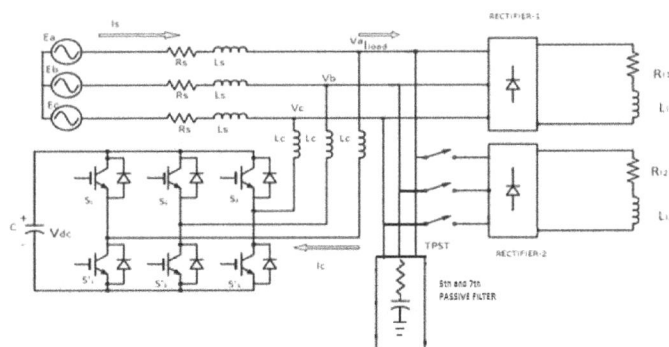

Figure 1. Configuration of Shunt Hybrid Power Filter

2.1. Estimation of Harmonics

Power quality measurements are done using a power quality analyzer. The power quality analyzer used in this work is YOKOGAWA CW240. It is capable of detecting the presence of voltage and current harmonics and measuring their characteristics (order, amplitude and phase).

Loads under study include nonlinear loads in the E & D building of Delphi-TVS, Mannur. Loads include power electronics equipment such as adjustable speed drives and DC drives. The readings are taken at the input distribution panel. The current THD is found to be 18.3% and voltage THD to be 2.251%.With onlypassive filter being implemented in the building, the current THD is found to be 8.2% and voltage THD to be 1.71%.

2.2. Hybrid Power Filter Configuration

The hybrid filter configuration is twofold, with a non-linear control of active filter and RC passive filter. An active power filter, APF, comprises a three phase pulse width modulation (PWM) voltage source inverter. The inverter has one 500µF capacitor in the DC side and is shunt connected with the electrical grid. Series passive element which consist of three 3mH inductors and 0.01Ω resistor. The passive filters are an important part of the Active Filter design. It must be dimensioned appropriately so that the switching frequency does not affect the source currents THDafter the compensation. During designing care must be taken to prevent the interference of the passive filters with the control of the Active Filter. In this paper an Active Filter is presented where inductor and resistor has been used as series passive filter, but the passive filter configuration can be a LCor RLC, or even more complex topologies. Each one of these topologies hasdrawbacks and advantages, which must be weighted according with the type of loads that will be compensated, the IGBT switching frequency, the control of the Active Filter and the final costThe shunt passive filter parameters are selected in such a manner that it is capable of eliminating 5th harmonics and 7th harmonics which are more prominent.

2.3. Modeling of Shunt Active Power Filter
2.3.1.   Modeling in Three Phase 'abc' Frame

The model of active filter is first developed in three-phase 'abc' frame. Kirchhoff's voltage and current laws are applied at the supply terminal, and it yields the following three differential equations in the stationary 'abc' frame [13].

$$E_1 = L_s \frac{di_{s1}}{dt} + R_s I_{s1} + V_1$$

$$E_2 = L_s \frac{di_{s2}}{dt} + R_s I_{s2} + V_2$$

$$E_3 = L_s \frac{di_{s3}}{dt} + R_s I_{s3} + V_3 \tag{1}$$

Where V_1, V_2, and V_3 indicate the line-to-ground voltages at the point of common coupling (PCC), E_1, $E_2$and E_3 indicate the line-to-ground voltage at the supply terminal.

Also,

$$V_1 = L_f \frac{di_{f1}}{dt} + V_{FM} + V_{1M}$$

$$V_2 = L_f \frac{di_{f2}}{dt} + V_{FM} + V_{2M}$$

$$V_3 = L_f \frac{di_{f3}}{dt} + V_{FM} + V_{3M}$$

The voltage drop across the inductor is small as compared to the Shunt Active Filter voltage, which gives us the relation,

$$L_f \frac{di_{f1}}{dt} \Box\ V_{FM} + V_{1M}$$

$$L_f \frac{di_{f2}}{dt} \Box\ V_{FM} + V_{2M}$$

$$L_f \frac{di_{f3}}{dt} \Box\ V_{FM} + V_{3M}$$

$$V_1 \Box\ V_{FM} + V_{1M}$$
$$V_2 \Box\ V_{FM} + V_{2M}$$
$$V_3 \Box\ V_{FM} + V_{3M}$$

By summing the three equations in (1), and with an assumption that the voltages of AC supply are balanced, and by neglecting the zero-sequence currents in the three wire systems, (i.e.,) using the following assumptions:

$$E_1 + E_2 + E_3 = 0\ ; V_1 + V_2 + V_3 = 0$$
$$I_{s1} + I_{s2} + I_{s3} = 0\ ; I_{f1} + I_{f2} + I_{f3} = 0$$

One can obtain:

$$V_{FM} = -\frac{1}{3} \sum_{f=1}^{3} V_{fM} \tag{2}$$

The switching function sw_k of the kth leg (for $k = 1, 2, 3$) of the converter can be defined as:

$$sw_k = \begin{cases} 1, if \ S_k \ is \ On \ and \ S'_k \ is \ Off \\ 0, if \ S_k \ is \ Off \ and \ S'_k \ is \ On \end{cases} \tag{3}$$

Therefore, $V_{kM} = sw_k V_{dc}$. The dynamic equation of phase-k filter's model is denoted by the equation given below:

$$E_k = L_s \frac{di_{sk}}{dt} + R_s I_{sk} + (sw_k - \frac{1}{3}\sum_{f=1}^{3} sw_f)V_{dc} \tag{4}$$

In addition, we may define a function ss_{nk} switching state function which is denoted as follows:

$$ss_{nk} = (sw_k - \frac{1}{3}\sum_{f=1}^{3} sw_f)_n \tag{5}$$

Equation (5) denotes that the value of ss_{nk} is dependent on the switching function sw_k of all three legs of the shunt active power filter. This shows that the three phases interacts with each other. Further, depending on (5) and from the eight allowable switching states of the active filter (n=1... 7), the conversion of $[sw_{nk}]$ to $[ss_{nk}]$ is given by the following:

$$\begin{pmatrix} ss_{n1} \\ ss_{n2} \\ ss_{n3} \end{pmatrix} = \begin{pmatrix} 2 & -1 & -1 \\ -1 & 2 & -1 \\ -1 & -1 & 2 \end{pmatrix}\begin{pmatrix} sw_1 \\ sw_2 \\ sw_3 \end{pmatrix} \tag{6}$$

Note that $[ss_{nk}]$ has no zero-sequence component $(ie., ss_{n1} + ss_{n2} + ss_{n3} = 0)$

On the other hand, Analysis of the dc component of the system gives:

$$\frac{dV_{dc}}{dt} = \frac{1}{C}i_{dc} = \frac{1}{C}\sum_{m=1}^{3} sw_k i_{sk} \tag{7}$$

It can be shown that,

$$\sum_{k=1}^{3} sw_k i_{sk} = \sum_{m=1}^{3} ss_{nk} i_{sk} \tag{8}$$

And it can be verified that:

$$\frac{dV_{dc}}{dt} = \frac{1}{C}\sum_{m=1}^{3} ss_{nk} i_{sk} \tag{9}$$

And, using $i_{s1} + i_{s2} + i_{s3} = 0$ and $[ss_{nk}]$ in the functions leads to the differential equation on the dc side as shown below:

$$\frac{dV_{dc}}{dt} = \frac{1}{C}(2ss_{n1} + ss_{n2})i_{s1} + \frac{1}{C}(ss_{n1} + 2ss_{n2})i_{s1} \tag{10}$$

From this result, active filter in the 'abc' referential obtains it complete model by using (4) for phases '1' and '2', and (10):

$$L_s \frac{di_{s1}}{dt} = -R_s I_{s1} - ss_{n1}V_{dc} + E_1$$

$$L_s \frac{di_{s2}}{dt} = -R_s I_{s2} - ss_{n2}V_{dc} + E_2$$

$$C \frac{dV_{dc}}{dt} = (2ss_{n1} + ss_{n2})i_{s1} + (ss_{n1} + 2ss_{n2})i_{s1}$$

(11)

The interaction between the three phases indicates the disadvantage of the 'abc' model. Therefore, for achieving control, this model can be converted to 'dq' reference frame. The positive-sequence components are made constant because of time-varying transformation, and there is no interaction effect between the phases at the switching state decision level.

2.3.2. The Model Transformed into the 'dq' Reference Frame

Using Park's transformation, the three-phase quantities are converted to a 'dq' reference frame. The general transformation matrix is:

$$PT_{dq}^{123} = \begin{bmatrix} \cos\theta & \cos(\theta - 2\pi/3) & \cos(\theta - 4\pi/3) \\ -\sin\theta & -\sin(\theta - 2\pi/3) & -\sin(\theta - 4\pi/3) \end{bmatrix}$$

(12)

Where $\theta = \omega t$ represents the actual phase angle of the line voltage space vector and $PT_{123}^{dq} = (PT_{dq}^{123})^{-1} = (PT_{dq}^{123})^T$ coordinate matrix transformation.

The synchronous 'dq' frame obtained from the transformed model is denoted as:

$$L_s \frac{di_d}{dt} = -R_s I_d - ss_{nd}V_{dc} + L_s\omega i_q + E_d$$

$$L_s \frac{di_q}{dt} = -R_s i_q - ss_{nq}V_{dc} - L_s\omega i_q + E_q$$

$$C \frac{dV_{dc}}{dt} = ss_{nd}i_d + ss_{nd}i_q$$

(13)

The resultant model from the synchronous orthogonal rotating frame is denoted as follows:

$$\frac{d}{dt}\begin{pmatrix} i_d \\ i_q \\ V_{dc} \end{pmatrix} = \begin{pmatrix} -\dfrac{R_s}{L_s} & \omega & -\dfrac{ss_{nd}}{L_s} \\ -\omega & -\dfrac{R_s}{L_s} & -\dfrac{ss_{nq}}{L_s} \\ ss_{nd} & ss_{nq} & 0 \end{pmatrix}\begin{pmatrix} i_d \\ i_q \\ V_{dc} \end{pmatrix} + \begin{pmatrix} E_d \\ E_q \\ 0 \end{pmatrix}$$

(14)

The model given in (14) has nonlinear nature because of the multiplication terms present between the state variables $\{i_d, i_q, V_{dc}\}$ and the inputs $\{ss_{nd}, ss_{nq}\}$. However, this model is independent of time for a given switching period. Here, these three variables should have an independent control. Therefore, the currents i_d and i_q should be made to follow a reference current $\{i_d, i_q\}$ of varying nature. For maintaining the performance of the active filter in a compensatory manner, the DC voltage level V_{dc} is adjusted to a set point when there are dynamic variations.

2.3.3. Current Controller

In the current loop, one has the following expressions for switching functions ss_{nd} and ss_{nq} as:

$$ss_{nd} = \frac{1}{V_{dc}}\left[-L_s \frac{di_d}{dt} - R_s i_d + \omega L_s i_q + E_d \right]$$

$$ss_{nq} = \frac{1}{V_{dc}}\left[-L_s \frac{di_q}{dt} - R_s i_q - \omega L_s i_d + E_q \right]$$

(15)

Let,

$$\begin{bmatrix} \dfrac{di_d}{dt} \\ \dfrac{di_q}{dt} \end{bmatrix} = \begin{bmatrix} u_d \\ u_q \end{bmatrix}$$

u_d and u_q can be used to control the currents i_d and i_q. An integrator is added for attenuating to track the steady-state error which is used as a tracking controller. It is designed by using the following expressions [13]:

$$u_d = \frac{di_d}{dt} = \frac{di_d^*}{dt} + k_p \, \tilde{i}_d + k_i \int \tilde{i}_d \, dt$$

$$u_q = \frac{di_q}{dt} = \frac{di_q^*}{dt} + k_p \, \tilde{i}_q + k_i \int \tilde{i}_q \, dt$$

Where $\tilde{i}_d = i_d^* - i_d$ and $\tilde{i}_q = i_q^* - i_q$ are current errors and $\{i_d^*, i_q^*\}$ are the references of $\{i_d, i_q\}$ correspondingly. The proportional (k_{pc}) and integral (k_{ic}) gains are obtained as follows:

$$k_{pc} = 2\xi\omega_n$$
$$k_{ic} = 2\xi\omega_n^2$$

Where ξ is the damping factor, and ω_n is the current loop natural angular frequency.

2.3.4. DC Voltage Regulation

To maintain considerable level of V_{dc} across the SAPF dc capacitor, the losses through the active power filter's resistive–inductive branches can be managed by working on the source current. Ideally, it must work on the active component of current i_d. For this purpose, an outer control loop is designed by using a PI regulator:

$$I_{dc} = k_p (V_{dc}^* - V_{dc}) + k_i \int (V_{dc}^* - V_{dc}) dt \tag{16}$$

The closed-loop transfer function of the outer loop is given as follows:

$$k_{pv} = 2\xi\omega_v C_{dc} .$$
$$k_{iv} = 2\xi\omega_v^2 C_{dc}$$

Where ξ is the damping factor, and ω_v is the outer loop natural angular frequency.

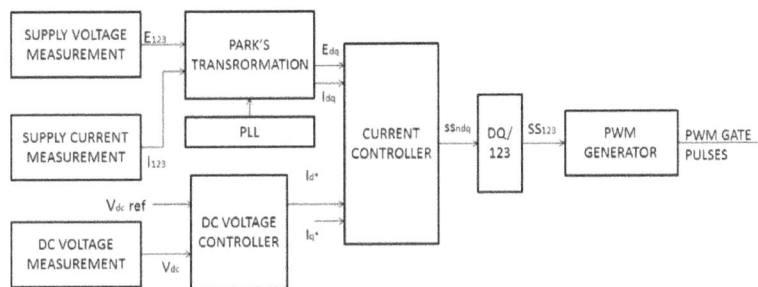

Figure 2. Control strategy of shunt active power filter

2.4. Modeling of shunt passive power filter

The passive power filter is tuned to eliminate 5th and 7th harmonics.The parameters are selected based on the following expression.

$$h = \sqrt{X_l / X_c}$$

Where h is the harmonics number, X_l and X_c are reactance of passive element.

3.   RESULTS AND ANALYSIS

The proposed control strategy has been simulated under MATLAB-Simulink environment and its performance is verified. The nonlinear load consists of two three-phase rectifier, so that the effectiveness of the control scheme to compensate for unbalanced load was tested. The rectifiers are feeding *R–L*-type circuits.For variation in loads; the THD is obtained by analyzing the source current waveforms determined from the results of simulation. The main objective of the simulation is made to analyze different aspects such as: reactive power compensation and harmonic load currents compensation; for variations in load the corresponding dynamic response of the SHPF.Some results are presented to demonstrate the performance of non-linear function-based control scheme. The simulation results are shown in Figure 3-6. The parameters taken in these simulations are shown below in Table 1.

Table 1. Shunt Hybrid Filter Parameters

PARAMETERS	VALUE
Line voltage and frequency	V_s=230V(rms),f_s=50Hz
Active filter parameters	R=0.01Ω,L=0.1mH
	C_{dc}=750uF,V_{dc}=700V
Non Linear Load	R_1=10 Ω,L_1=10mH
	R_2=10 Ω,L_2=10mH
Regulator	K_{pc}=7K_{ic}=800
	K_{pv}=4.5K_{iv}=30
Series elements	R=0.01 Ω ,L=3mH
Passive elements	L_5=13mH C_5=30uF
	L_7=6.5mH C_7= 30uF

Figure 3 shows the system performance without hybrid filter. The THD level of voltage and current before compensation are shown in Figure 4. The current THD levels are observed to be 18.2% .voltage THD were 2.25%.The current THD levels are observed to be 8.71%, voltage THD were 1.71% with only passive filter being installed. The waveform and THD level are shown in Figure 5 and 6 respectively. The current THD level after compensation reduced to 1.32% and voltage THD level to 0.4% which can be seen with waveform and THD level in Figure 7 and 8. The dynamic performance of filter is seen in Figure 7. It can be observed that there is smooth changeover from one load value to another value,The DC bus voltage of SAPF settles to its steady-state value within two cycle of sine wave. From these results, it can be concluded that SHPF offers a very good dynamic performance for a stepped load current.

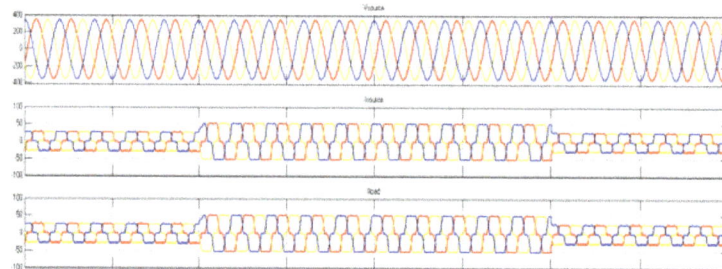

Figure 3. Voltage and current waveform before compensation

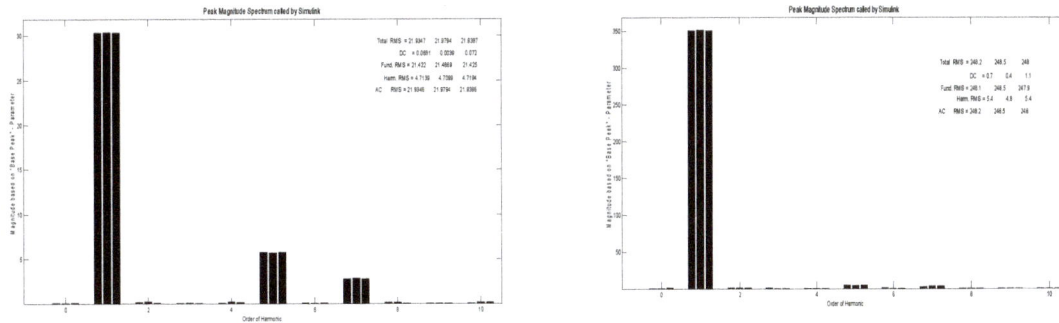

Figure 4. Current and Voltage THD before compenstion

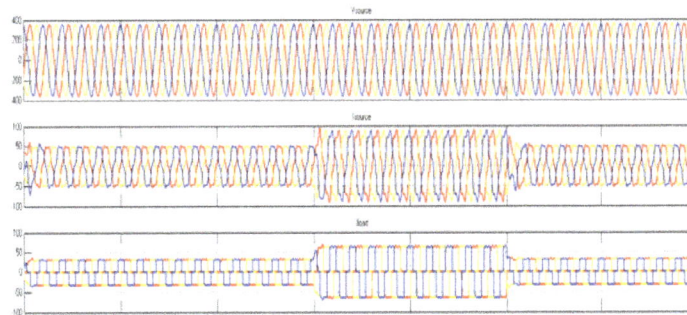

Figure 5. Voltage and current waveform with passive filter

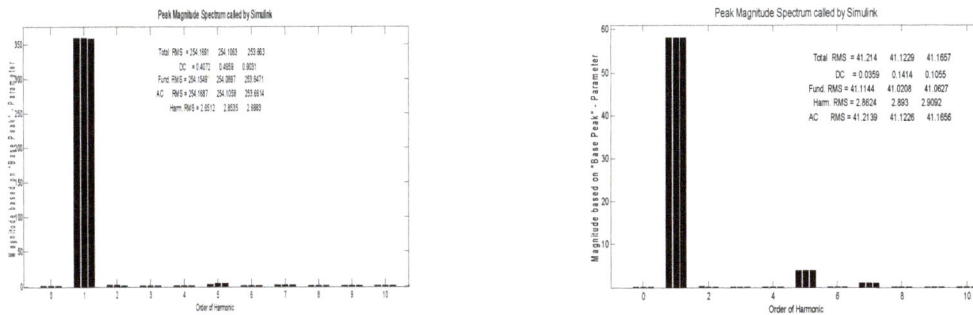

Figure 6. Current and Voltage THD with passive filter

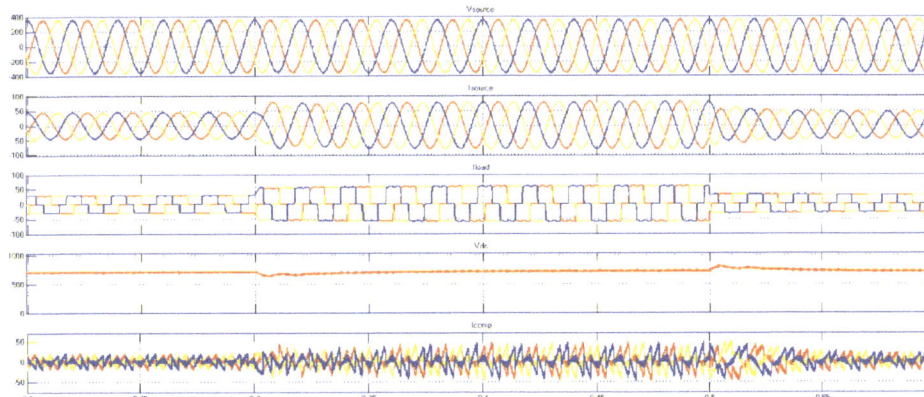

Figure 7. Voltage and current waveform after compensation

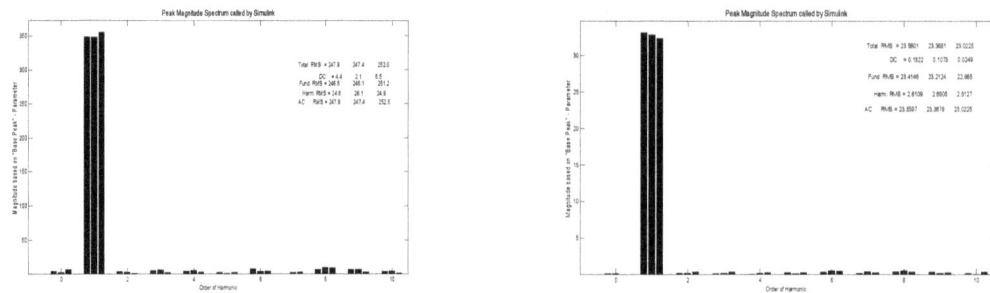

Figure 8. Current THD before and after compensation

4.   CONCLUSION

The Shunt Hybrid Power Filter based on non-linear function control has been proposed and simulated under MATLAB environment to evaluate the dynamic performance for varying voltage-source type of nonlinear load conditions. From simulated results, it has also been shown that the control strategy has a fast dynamic response during large load variations and is capable of maintaining the THD of the voltage at PCC and the supply currents well below the mark of 5% specified in the IEEE-519 standard.

ACKNOWLEDGEMENTS

The authors thank Delphi-TVS,Mannur to carry out the project work on power quality.

REFERENCES

[1]   H Sasaki, T Machida. A New Method To Eliminate AC Harmonic Currents By Magnetic Flux Compensation Consideration Basic Design. *IEEE Trans. Power App. Syst.,* 1971; 9(5): 2009-2019.

[2]   H Kawahira, T Nakamura, S Nakazawa. Active Power Filters. *IEEJ IPEC - Tokyo.* 1983; 981-992.

[3]   Akagi H. Trends in active power line conditioners. *IEEE Trans. Power Electron.,* 1994; 9(3): 263–268

[4]   Singh B, Al-Haddad K, Chandra A. A review of active filters for power quality improvement. *IEEE Trans. Ind. Electron.,* 1999; 46(5): 960–971.

[5]   BN Singh, B Singh, A Chandra, P Rastgoufard, K Al-Haddad. An improved control algorithm for active filters. *IEEE Trans. PowerDel.,* 2007; 22(2): 1009–1020.

[6]   Singh, Bhim, Kamal Al-Haddad, Ambrish Chandra. A review of active filters for power quality improvement" *Industrial Electronics, IEEE Transactions on.* 1999; 46(5): 960-971

[7]   RS Herrera, P Salmeron, H Kim. Instantaneous reactive power theory applied to active power filter compensation: Different approaches, assessment, and experimental results. *IEEE Trans. Ind. Electron.,* 2008; 55(1): 184–196.

[8]   Mendalek, Nassar, Kamal Al-Haddad. Modeling and nonlinear control of shunt active power filter in the synchronous reference frame. *Harmonics and Quality of Power, 2000. Proceedings. Ninth International Conference on.* IEEE. 2000; 1.

[9]   Kumar, Parmod, AlkaMahajan. Soft computing techniques for the control of an active power filter." *Power Delivery, IEEE Transactions on.* 2009; 24(1): 452-461.

[10] A Hamadi, K AI-Haddad, S Rahmani, H Kanaan. Comparison of Fuzzy logic and Proportional Integral Controller of Voltage Source Active Filter CompensatingCurrent Harmonics and Power Factor. *International Conference on industrial Technology, IEEE ICIT04.* 2004; 645-650.

[11] Ghamri A, MT Benchouia, A Golea. Sliding-mode control based three-phase shunt active power filter: Simulation and experimention. *Electric Power Components and Systems.* 2012; 40(4): 383-398.

[12] Bhattacharya, Avik, Chandan Chakraborty. A shunt active power filter with enhanced performance using ANN-based predictive and adaptive controllers. *Industrial Electronics, IEEE Transactions on.* 2011; 58(2): 421-428.

[13] N Mendalek, K Al-Haddad, F Fnaiech, LA Dessaint. Nonlinear control technique to enhance dynamic performance of a shunt active power filter. *IEE Proceedings-Electric Power Applications.* 2003; 150(4): 373-379.

4

Speed Sensorless Direct Rotor Field-Oriented Control of Single-Phase Induction Motor Using Extended Kalman Filter

Mohammad Jannati, Seyed Hesam Asgari, Nik Rumzi Nik Idris, Mohd Junaidi Abdul Aziz
UTM-PROTON Future Drive Laboratory, Faculty of Electrical Engineering, Universiti Teknologi Malaysia, Johor Bahru, MALAYSIA

ABSTRACT

Keyword:

DRFOC

EKF

Speed sensorless

SPIM

Nowadays, Field-Oriented Control (FOC) strategies broadly used as a vector based controller for Single-Phase Induction Motors (SPIMs). This paper is focused on Direct Rotor FOC (DRFOC) of SPIM. In the proposed technique, transformation matrices are applied in order to control the motor by converting the unbalanced SPIM equations to the balanced equations (in this paper the SPIM with two different stator windings is considered). Besides this control technique, a method for speed estimation of SPIM based on Extended Kalman Filter (EKF) to achieve the higher performance of SPIM drive system is presented. Simulation results are provided to demonstrate the high performance of the presented techniques.

Corresponding Author:

Seyed Hesam Asgari,
Faculty of Electrical Engineering,
Universiti Teknologi Malaysia,
UTM Skudai, 81310 Johor, Malaysia.
Email: seyedhesamasgari@gmail.com

1. INTRODUCTION

Single-Phase Induction Motors (SPIMs) are used in both domestic and industrial purposes. They can be employed in air conditioners, fans, refrigerators, compressors, dryers, washing machines and other applications. Generally, the stator of SPIMs has two windings which are orthogonal in space. They are different in terms of resistances and impedances. Since the main and auxiliary stator windings are unbalanced, therefore SPIM will be encountered the torque pulsations [1]. Consequently, studying about SPIMs has been increased dramatically. It has been recommended by the researchers, various studies, which focused on improving the performance and efficiency of the SPIMs. These researches, presented design and optimization, study on the power factor, research on improved modeling and analysis, progress on improvement of torque performance and researching on influence of harmonic [2]-[4].Variable Frequency Control (VFC) techniques which are used in the Variable Frequency Drives (VFDs) have advantages in terms of saving of energy and high performance applications of IMs [5]-[7]. VFC methods are categorized into scalar and vector based control. The scalar methods will not be able to fulfill the requirement of dynamic drives and has a slow reaction to transient but, vector control is an excellent control method to handle transient and satisfy the requirement of dynamic drives. Generally vector control is classified into Direct Torque Control (DTC) and Field-Oriented Control (FOC). FOC method is proposed broadly as a vector based controller for IMs and is classified into Stator FOC (SFOC) and Rotor FOC (RFOC). Another classification of this method is also performed based on the calculation of rotor flux position which includes Direct FOC (DFOC) and Indirect FOC (IFOC) [8].

From a review of literature, there are many papers which have been suggested for vector control of SPIMs based on FOC. In 2000, Correa et al. investigated IRFOC technique for SPIM. In the proposed technique, they eliminated the asymmetry terms of the SPIM equations by selecting appropriate transformation for the stator variables [9]. It was suggested in [9], to use hysteresis current controller. In [10], Correa et al. proposed vector control of SPIM based on ISFOC. In the proposed technique, for reduction of electromagnetic torque oscillations in SPIM, they designed a double-sequence current controller to control stator current [10]. In [11], decoupling vector control of SPIM has been proposed by Vaez-Zadeh and Reicy. In the proposed vector control method in paper [12], the maximum potential operation of SPIM is obtained according to maximum torque per ampere. In [13], [14], the authors proposed and implemented ISFOC and IRFOC for SPIM respectively. In fact they used the same variable changing based on [9], to eliminate the asymmetrical terms in SPIM. To have high performance vector control techniques such as [9]-[14], it is necessary to employ feedback speed control. For this purpose, an encoder is normally used to provide this information. Using sensor causes more instrumentation, increasing cost and size, decrease robustness and reliability of the drive system. Therefore, instead of implementation of sensor, it is better to apply speed estimation techniques. Generally, speed estimation is categorized into two main parts, speed estimation based on motor model and speed estimation through signal injection [8]. It is proposed speed estimation methods in IMs by different authors. In [15], a study has been proposed for sensorless IRFOC of SPIM. The applied method for estimation of speed in [15] is based on SPIM model. In paper [16], a method for ISFOC of SPIM with estimation of rotor speed based on the motor currents and reference q-axis current has been proposed. In [17], Model Reference Adaptive System (MRAS) strategy has been used for speed sensorless IRFOC of SPIM. The MRAS speed sensorless vector control of IMs is sensitive to variations of resistance [18]. For this, in [19], MRAS strategy by an online stator resistance estimator and in [20], a Recursive Least Square (RLS) algorithm is employed to calculate the SPIM parameters in sensorless vector control of this motor. Using Extended Kalman Filter (EKF) is another technique to estimate the rotor speed. Since the nonlinearities and uncertainties of IM are well-suited to the EKF, therefore it would be able to estimate the parameters simultaneously at the short interval of time [21], [22]. Moreover, in this method the measurement and system noises which are not normally considered in the previously presented techniques for SPIMs such as [15]-[20] are regarded. In this paper, a novel technique for DRFOC of SPIM (unbalanced two-phase IM) with estimation of mechanical speed using EKF is discussed and verified using MATLAB/SIMULINK. The presented EKF in this paper is the conventional EKF which has been developed of SPIM. Besides the removing of mechanical speed sensor such as tachogenerator and encoder, the proposed DRFOC in this work, eliminates the pure integration which is used in IFOC. Using integration operator in the vector control of IM suffers the well-known difficulties of integration effect especially at the low frequencies [23]. The results of this research show that the proposed speed sensorless control for SPIM has reasonably good torque and speed response dynamics and satisfactory tracking capability.

2. SPIM MODEL

The mathematical model of squirrel cage SPIM can be shown in a stationary reference frame as follows [9]:

$$
\begin{bmatrix} v_{ds}^s \\ v_{qs}^s \\ 0 \\ 0 \end{bmatrix} = \begin{bmatrix} R_{ds} + L_{ds}\dfrac{d}{dt} & 0 & M_{ds}\dfrac{d}{dt} & 0 \\ 0 & R_{qs} + L_{qs}\dfrac{d}{dt} & 0 & M_{qs}\dfrac{d}{dt} \\ M_{ds}\dfrac{d}{dt} & \omega_r M_{qs} & R_r + L_r\dfrac{d}{dt} & \omega_r L_r \\ -\omega_r M_{ds} & M_{qs}\dfrac{d}{dt} & -\omega_r L_r & R_r + L_r\dfrac{d}{dt} \end{bmatrix} \begin{bmatrix} i_{ds}^s \\ i_{qs}^s \\ i_{dr}^s \\ i_{qr}^s \end{bmatrix} \tag{1}
$$

$$
\begin{bmatrix} \lambda_{ds}^s \\ \lambda_{qs}^s \\ \lambda_{dr}^s \\ \lambda_{qr}^s \end{bmatrix} = \begin{bmatrix} L_{ds} & 0 & M_{ds} & 0 \\ 0 & L_{qs} & 0 & M_{qs} \\ M_{ds} & 0 & L_r & 0 \\ 0 & M_{qs} & 0 & L_r \end{bmatrix} \begin{bmatrix} i_{ds}^s \\ i_{qs}^s \\ i_{dr}^s \\ i_{qr}^s \end{bmatrix} \tag{2}
$$

$$
\tau_e = \frac{Pole}{2}(M_{qs} i_{qs}^s i_{dr}^s - M_{ds} i_{ds}^s i_{qr}^s) \tag{3}
$$

$$
\frac{Pole}{2}(\tau_e - \tau_l) = J\frac{d\omega_r}{dt} + F\omega_r \tag{4}
$$

Where, v^s_{ds}, v^s_{qs}, i^s_{ds}, i^s_{qs}, i^s_{dr}, i^s_{qr}, λ^s_{ds}, λ^s_{qs}, λ^s_{dr} and λ^s_{qr} are the d and q axes voltages, currents and fluxes of the stator and rotor, R_{ds}, R_{qs}, R_r, L_{ds}, L_{qs}, L_r, M_{ds} and M_{qs} denote the d and q axes resistances, self and mutual inductances of the stator and rotor. Moreover, ω_r, τ_e, τ_l, J and F are the motor speed, electromagnetic torque, load torque, inertia and viscous friction coefficient, respectively. Based on Equation (1)-(4) it is assumed that the main and auxiliary stator windings have different values ($R_{ds}{\neq}R_{qs}$, $L_{ds}{\neq}L_{qs}$ and $M_{ds}{\neq}M_{qs}$). To compensate the asymmetry in unbalanced IMs in [24]-[29], Jannati et al. proposed the use of unbalanced transformation matrices for stator voltage and current variables. In this work, similar to [25]-[29], these transformation matrices are employed to compensate the asymmetry between the main and auxiliary stator windings in SPIM (in [27]-[29], the transformation matrices have been used for IRFOC of three-phase IM under open-phase fault). These matrices are as follows:

Transformation matrix for stator voltage variables:

$$
\begin{bmatrix} v^e_{ds} \\ v^e_{qs} \end{bmatrix} = \begin{bmatrix} T^e_{vs} \end{bmatrix} \begin{bmatrix} v^s_{ds} \\ v^s_{qs} \end{bmatrix} = \begin{bmatrix} \dfrac{M_{qs}}{M_{ds}}\cos\theta_e & \sin\theta_e \\ -\dfrac{M_{qs}}{M_{ds}}\sin\theta_e & \cos\theta_e \end{bmatrix} \begin{bmatrix} v^s_{ds} \\ v^s_{qs} \end{bmatrix}
\tag{5}
$$

Transformation matrix for stator current variables:

$$
\begin{bmatrix} i^e_{ds} \\ i^e_{qs} \end{bmatrix} = \begin{bmatrix} T^e_{is} \end{bmatrix} \begin{bmatrix} i^s_{ds} \\ i^s_{qs} \end{bmatrix} = \begin{bmatrix} \dfrac{M_{ds}}{M_{qs}}\cos\theta_e & \sin\theta_e \\ -\dfrac{M_{ds}}{M_{qs}}\sin\theta_e & \cos\theta_e \end{bmatrix} \begin{bmatrix} i^s_{ds} \\ i^s_{qs} \end{bmatrix}
\tag{6}
$$

Where, θe is the angle between the stationary reference frame and the rotor flux-oriented reference frame (in this paper superscript "e" indicates that the variables are in the rotating reference frame and superscript "s" indicates that the variables are in the stationary reference frame. Ii is shown by using these transformation matrices, the stator and rotor variables of the main and auxiliary windings are transformed into equations that have similar structure to balanced IM equations. The stator and rotor voltage equations, rotor flux equations and electromagnetic torque equation after applying Equation (5) and Equation (6) are given by (7)-(11).

Stator voltage equations:

$$
\begin{bmatrix} v^e_{ds} \\ v^e_{qs} \end{bmatrix} = \begin{bmatrix} R_{ds} + L_{qs}\dfrac{d}{dt} & -\omega_e L_{qs} \\ \omega_e L_{qs} & R_{qs} + L_{qs}\dfrac{d}{dt} \end{bmatrix} \begin{bmatrix} i^e_{ds} \\ i^e_{qs} \end{bmatrix} + \begin{bmatrix} M_{qs}\dfrac{d}{dt} & -\omega_e M_{qs} \\ \omega_e M_{qs} & M_{qs}\dfrac{d}{dt} \end{bmatrix} \begin{bmatrix} i^e_{dr} \\ i^e_{qr} \end{bmatrix}
$$
$$
+ \begin{bmatrix} \left(\dfrac{M^2_{qs}}{M^2_{ds}}R_{ds} - R_{qs}\right) + \left(\dfrac{M^2_{qs}}{M^2_{ds}}L_{ds} - L_{qs}\right)\dfrac{d}{dt} & -\omega_e\left(\dfrac{M^2_{qs}}{M^2_{ds}}L_{ds} - L_{qs}\right) \\ \omega_e\left(\dfrac{M^2_{qs}}{M^2_{ds}}L_{ds} - L_{qs}\right) & \left(\dfrac{M^2_{qs}}{M^2_{ds}}R_{ds} - R_{qs}\right) + \left(\dfrac{M^2_{qs}}{M^2_{ds}}L_{ds} - L_{qs}\right)\dfrac{d}{dt} \end{bmatrix} \begin{bmatrix} i^{-e}_{ds} \\ i^{-e}_{qs} \end{bmatrix}
\tag{7}
$$

Where,

$$
\begin{bmatrix} i^{-e}_{ds} \\ i^{-e}_{qs} \end{bmatrix} = \begin{bmatrix} \cos^2\theta_e & -\sin\theta_e\cos\theta_e \\ -\sin\theta_e\cos\theta_e & \sin^2\theta_e \end{bmatrix} \begin{bmatrix} i^e_{ds} \\ i^e_{qs} \end{bmatrix}
\tag{8}
$$

Rotor voltage equations:

$$
\begin{bmatrix} 0 \\ 0 \end{bmatrix} = \begin{bmatrix} M_{qs}\dfrac{d}{dt} & -(\omega_e - \omega_r)M_{qs} \\ (\omega_e - \omega_r)M_{qs} & M_{qs}\dfrac{d}{dt} \end{bmatrix} \begin{bmatrix} i_{ds}^e \\ i_{qs}^e \end{bmatrix}
$$
$$
+ \begin{bmatrix} R_r + L_r\dfrac{d}{dt} & -(\omega_e - \omega_r)L_r \\ (\omega_e - \omega_r)L_r & R_r + L_r\dfrac{d}{dt} \end{bmatrix} \begin{bmatrix} i_{dr}^e \\ i_{qr}^e \end{bmatrix}
\tag{9}
$$

Rotor flux equations:

$$
\begin{bmatrix} \lambda_{dr}^e \\ \lambda_{qr}^e \end{bmatrix} = \begin{bmatrix} M_{qs} & 0 \\ 0 & M_{qs} \end{bmatrix} \begin{bmatrix} i_{ds}^e \\ i_{qs}^e \end{bmatrix} + \begin{bmatrix} L_r & 0 \\ 0 & L_r \end{bmatrix} \begin{bmatrix} i_{dr}^e \\ i_{qr}^e \end{bmatrix}
\tag{10}
$$

Electromagnetic torque equation:

$$
\tau_e = \frac{Pole}{2} M_{qs} (i_{qs}^e i_{dr}^e - i_{ds}^e i_{qr}^e)
\tag{11}
$$

In (7), ω_e is the angle between the stationary reference frame and the rotor flux reference frame. As can be seen from Equation (7)-(11), using Equation (5) and Equation (6), the asymmetrical equations of SPIM changed into symmetrical equations. Thus, the FOC principles can be applied.

3.　DRFOC OF SPIM

In this study, the DRFOC technique for vector control of SPIM was used. Based on (7)-(11) and after simplifying of equations, the equations of the RFOC technique for a SPIM are obtained as following equations (in this method the rotor flux vector is aligned with d-axis):

$$
\omega_e = \omega_r + \frac{M_{qs} i_{qs}^e}{T_r |\lambda_r|}
\tag{12}
$$

$$
\tau_e = \frac{Pole}{2} \frac{M_{qs}}{L_r} |\lambda_r| i_{qs}^e
\tag{13}
$$

$$
|\lambda_r| = \frac{M_{qs} i_{ds}^e}{1 + T_r \dfrac{d}{dt}}
\tag{14}
$$

$$
v_{ds}^e = v_{ds}^d + v_{ds}^{ref} + v_{ds}^{-e} \quad , \quad v_{qs}^e = v_{qs}^d + v_{qs}^{ref} + v_{qs}^{-e}
\tag{15}
$$

Where,

$$
v_{ds}^d = -\omega_e i_{qs}^e (L_{qs} - \frac{M_{qs}^2}{L_r}) + \frac{M_{qs}}{L_r}(\frac{M_{qs} i_{ds}^e - |\lambda_r|}{T_r})
\tag{16}
$$

$$
v_{qs}^d = \omega_e i_{qs}^e (L_{qs} - \frac{M_{qs}^2}{L_r}) + \omega_e M_{qs} \frac{|\lambda_r|}{L_r}
\tag{17}
$$

$$
v_{ds}^{ref} = (\frac{R_{ds} M_{qs}^2 + R_{qs} M_{ds}^2}{2 M_{ds}^2})i_{ds}^e + (L_{qs} - \frac{M_{qs}^2}{L_r})\frac{di_{ds}^e}{dt}
\tag{18}
$$

$$
v_{qs}^{ref} = (\frac{R_{ds} M_{qs}^2 + R_{qs} M_{ds}^2}{2 M_{ds}^2})i_{qs}^e + (L_{qs} - \frac{M_{qs}^2}{L_r})\frac{di_{qs}^e}{dt}
\tag{19}
$$

$$
\begin{bmatrix} v_{ds}^{-e} \\ v_{qs}^{-e} \end{bmatrix} = \left(\frac{R_{ds} M_{qs}^2 - R_{qs} M_{ds}^2}{2 M_{ds}^2}\right) \begin{bmatrix} \cos 2\theta_e & -\sin 2\theta_e \\ -\sin 2\theta_e & -\cos 2\theta_e \end{bmatrix} \begin{bmatrix} i_{ds}^e \\ i_{qs}^e \end{bmatrix}
\tag{20}
$$

Where, $T_r=L_r/R_r$ is the rotor time constant. In (14) and (15), $v_{ds}{}^d$ and $v_{qs}{}^d$ are generated using Decoupling Circuit and $v_{ds}{}^{ref}$ and $v_{qs}{}^{ref}$ are generated using current PI controllers in the RFOC block diagram of the SPIM (see Figure 2). Moreover, the values of ω_r, λ_r and θ_e in (12)-(20) are calculated using estimated values of rotor d and q axis fluxes as follows (in this paper, the motor speed and rotor d and q axis fluxes are estimated using EKF):

$$\hat{\lambda}_r = \sqrt{\hat{\lambda}_{dr}^2 + \hat{\lambda}_{qr}^2} \tag{21}$$

$$\hat{\theta}_r = \tan^{-1}\left(\frac{\hat{\lambda}_{dr}}{\hat{\lambda}_{qr}}\right) \tag{22}$$

4. EKF FOR ROTOR SPEED ESTIMATION IN SPIM

In this paper, an Extended Kalman Filter is used to estimate the mechanical speed and rotor fluxes. The state space model of SPIM is shown by Equation (23):

$$\dot{x} = Ax + Bu + w(t) \quad , \quad y = Cx + v(t) \tag{23}$$

Where, A_n, B_n and C_n are the input and output matrixes of system and x, y and u are the system state matrix, system output matrix and system input matrix respectively. The covariance matrices of $w(t)$ and $v(t)$ are defined as follows ($w(t)$: system noise; $v(t)$: measurement noise):

$$Q = \text{cov}\,(w) = E\{ww'\} \quad , \quad R = \text{cov}\,(v) = E\{vv'\} \tag{24}$$

In this filter, the state matrix (x_n) is the stator d and q axis currents, rotor d and q axis fluxes and rotor speed, the input matrix (u_n) is stator d and q axis voltages and the output matrix (y_n) is stator d and q axis currents.

$$x_n = \begin{bmatrix} i_{ds}^{(n)} & i_{qs}^{(n)} & \lambda_{ds}^{(n)} & \lambda_{qs}^{(n)} & \omega_r^{(n)} & \tau_l^{(n)} \end{bmatrix}^T \tag{25}$$

$$u_n = \begin{bmatrix} v_{ds}^{(n)} & v_{qs}^{(n)} \end{bmatrix}^T \tag{26}$$

$$y_n = \begin{bmatrix} i_{ds}^{(n)} & i_{qs}^{(n)} \end{bmatrix}^T \tag{27}$$

Based on d-q model of SPIM (Equation (1)-(4)) and Equation (25)-(27), the matrixes A_n, B_n and C_n are obtained as Equation (28).

$$A_n = \begin{bmatrix} 1 - \frac{1}{k_1}\left(R_{ds} + \frac{M_{ds}^2 R_r}{L_r^2}\right)dt & 0 & \frac{M_{ds}R_r}{k_1 L_r^2}dt & \frac{M_{ds}R_r}{k_1 L_r}dt & 0 & 0 \\ 0 & 1 - \frac{1}{k_2}\left(R_{qs} + \frac{M_{qs}^2 R_r}{L_r^2}\right)dt & \frac{-M_{qs}R_r}{k_2 L_r}dt & \frac{M_{qs}R_r}{k_2 L_r^2}dt & 0 & 0 \\ \frac{M_{ds}R_r}{L_r}dt & 0 & 1 - \frac{R_r}{L_r}dt & -R_r dt & 0 & 0 \\ 0 & \frac{M_{qs}R_r}{L_r}dt & R_r dt & 1 - \frac{R_r}{L_r}dt & 0 & 0 \\ -\frac{1.5(Pole)^2 M_{ds}\lambda_{qr}^s}{2JL_r}dt & \frac{1.5(Pole)^2 M_{qs}\lambda_{dr}^s}{2JL_r}dt & 0 & 0 & 1 & -\frac{1}{J} \\ 0 & 0 & 0 & 0 & 0 & 1 \end{bmatrix}$$

$$B_n = \begin{bmatrix} \frac{1}{k_1}dt & 0 \\ 0 & \frac{1}{k_2}dt \\ 0 & 0 \\ 0 & 0 \\ 0 & 0 \\ 0 & 0 \end{bmatrix}$$

$$C_n = \begin{bmatrix} 1 & 0 & 0 & 0 & 0 & 0 \\ 0 & 1 & 0 & 0 & 0 & 0 \end{bmatrix}$$

$$\tag{28}$$

Where:

$$k_1 = L_{ds} - \frac{M_{ds}^2}{L_r} \quad , \quad k_2 = L_{qs} - \frac{M_{qs}^2}{L_r} \tag{29}$$

Using (28), (29) and EKF algorithm (Equations (30)-(36)), the rotor speed and rotor fluxes can be estimated (in (30)-(36), H is the matrix of output prediction, P_n is error covariance matrix and Φ is the matrix of state prediction).

EKF Algorithm:

Prediction of State:

$$x_{n+1\ n} = \Phi\left(n+1, n, x_{n\ n-1}, u_n\right) \tag{30}$$

Where,

$$\Phi\left(n+1, n, x_{n\ n-1}, u_n\right) = A_n\left(x_{n\ n}\right)x_{n\ n} + B_n\left(x_{n\ n}\right)u_n \tag{31}$$

Estimation of Error Covariance Matrix:

$$P_{n+1\ n} = \frac{d\Phi}{dx}\bigg|_{x=x_{n\ n}} P_{n\ n}\frac{d\Phi^T}{dx}\bigg|_{x=x_{n\ n}} + Q \tag{32}$$

Computation of Kalman Filter Gain:

$$K_n = P_{n\ n-1}\frac{\partial H^T}{\partial x}\bigg|_{x=x_{n\ n-1}}\left(\frac{\partial H}{\partial x}\bigg|_{x=x_{n\ n-1}} P_{n\ n-1}\frac{\partial H^T}{\partial x}\bigg|_{x=x_{n\ n-1}} + R\right)^{-1} \tag{33}$$

Where,

$$H\left(x_{n\ n-1}, n\right) = C_n\left(x_{n\ n-1}\right)x_{n\ n-1} \tag{34}$$

State Estimation:

$$x_{n\ n} = x_{n\ n-1} + K_n\left(y_n - H\left(x_{n\ n-1}, n\right)\right) \tag{35}$$

Update of the Error Covariance Matrix:

$$P_{n\ n} = P_{n\ n-1} - K_n\frac{\partial H}{\partial x}\bigg|_{x=x_{n\ n-1}} P_{n\ n-1} \tag{36}$$

Based on Equation (30)-(36), the block diagram of EKF can be shown as Figure 1.

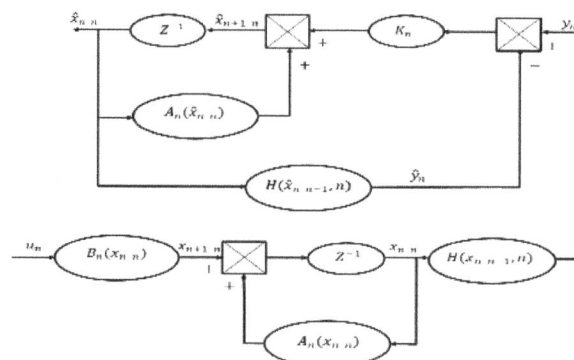

Figure 1. Block diagram of EKF

5. SIMULATIONS AND RESULTS

In this section, MATLAB simulation results were obtained for a SPIM. The simulated SPIM parameters are:

Voltage:110V, f=60Hz, No. of poles=4, R_{ds}=7.14Ω, R_{qs}=2.02Ω, R_r=4.12Ω, L_{ds}=0.1885H, L_{qs}=0.1844H, L_r=0.1826H, M_{qs}=0.1772H, J=0.0146kg.m^2

The simulated drive system is presented in Figure 2. As shown in this figure, the SPIM was fed by two-leg Voltage Source Inverter (VSI). In the simulation test, the motor speed varied from zero to ±400rpm (a trapezoidal reference speed) as shown in Figure 3(a), and the control drive system was feedback with the EKF.

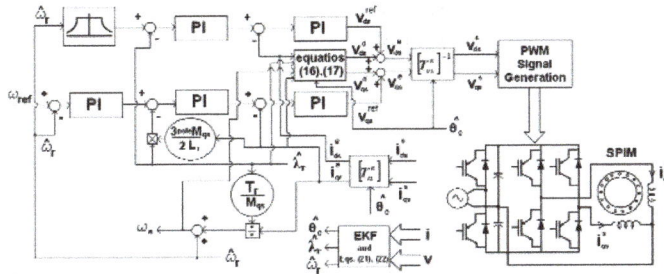

Figure 2. Block diagram of the proposed speed sensorless DRFOC for SPIM

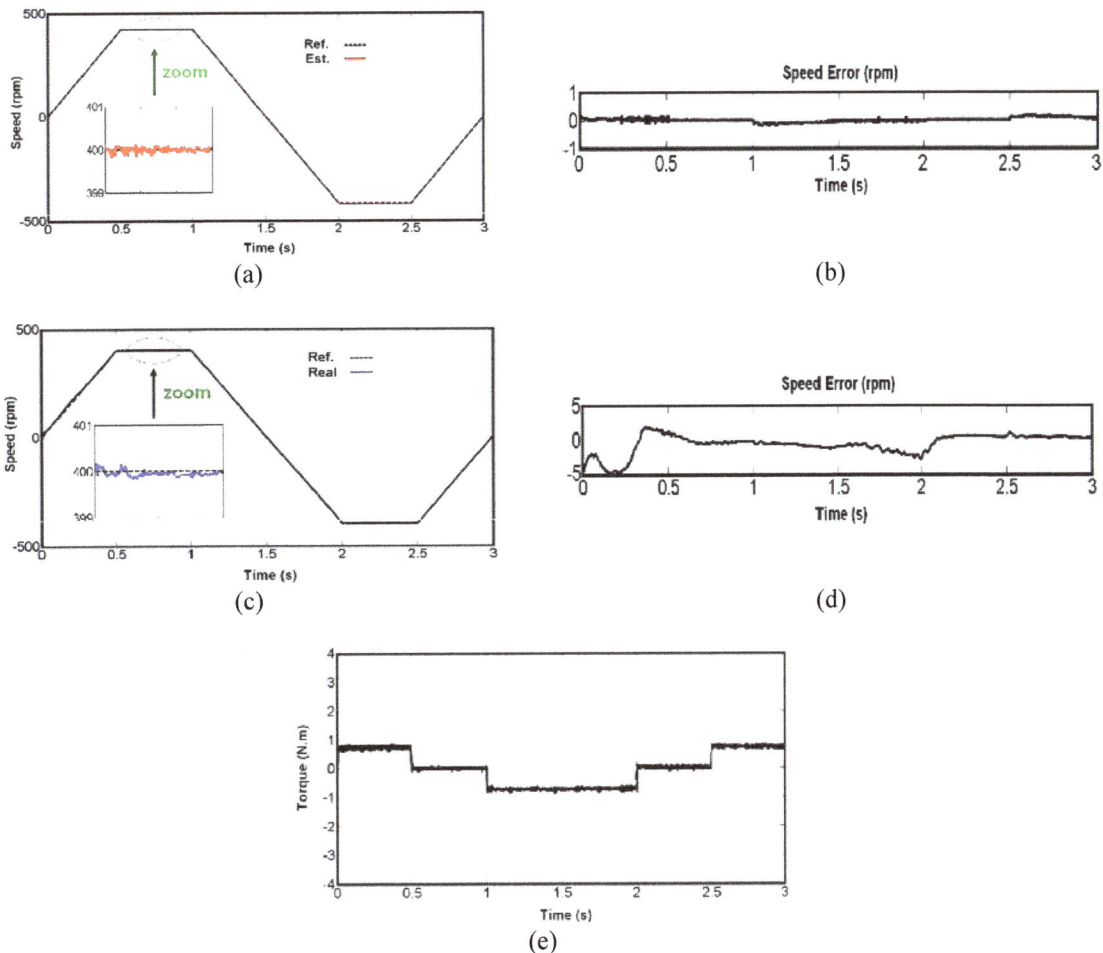

(a)

(b)

(c)

(d)

(e)

Figure 3. Simulation results of the speed sensorless DRFOC for a trapezoidal reference speed

Figure 3(a) presents the reference speed and estimated speed, while Figure 3(b) shows the error between reference speed and estimated speed. Figures 3(c) and 3(d) show the reference speed and motor speed and the error between reference speed and actual speed respectively. Figure 3(e) shows the simulated electromagnetic torque of SPIM. The simulation results have been shown the good speed and torque identification performance of the proposed drive system (e.g., the oscillations of electromagnetic torque is about 0.2N.m).

Figure 4 shows the good performance of the proposed drive system for speed sensorless DRFOC of SPIM at zero and low speed operation (ω_{ref}=0 and ω_{ref}=50rpm). It can be seen from Figure 4 that the dynamic performance of the proposed drive system for speed sensorless of SPIM at zero and low speed is extremely acceptable.

Figure 5 shows the simulation results of the proposed controller under load (step load). From t=0s to t=1.2s, the value of the load is 0N.m and from t=1.2s to t=1.5s, the value of the load is 1N.m. Results show that the proposed controller for vector control of SPIM is also robust to the load torque variations and produced good results (in this case, as can be seen in Figure 5(b), by using proposed controller, the torque oscillation after applying load torque and phase cut-off and at steady state is ~ 0.1N.m at load torque of 1N.m).

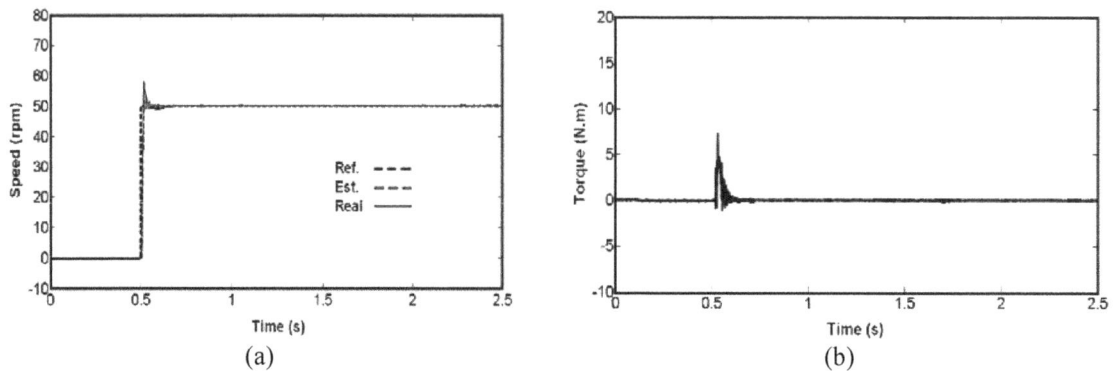

Figure 4. Simulation results of the speed sensorless DRFOC at zero and low speed; (a) Speed, (b) Torque

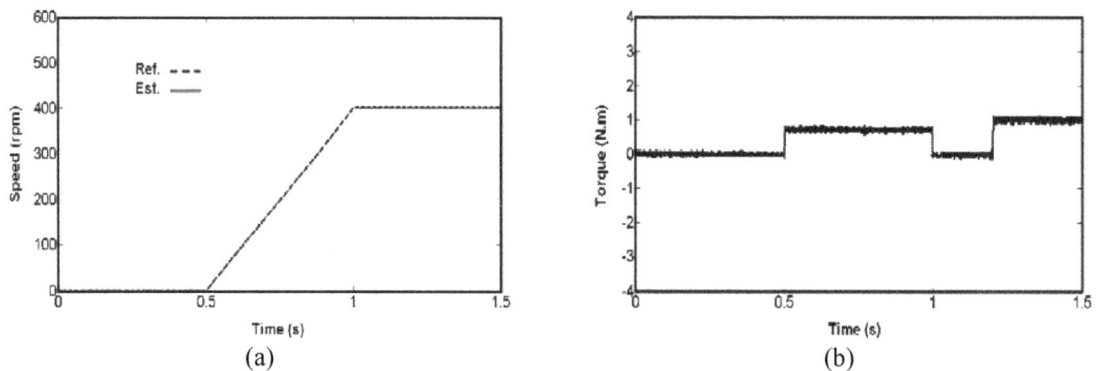

Figure 5. Simulation results of the speed sensorless DRFOC under load; (a) Speed, (b) Torque

6. CONCLUSION

This paper made a contribution to the speed sensorless DRFOC of SPIM. First, by applying transformation matrices to the SPIM equations, a novel DRFOC for SPIM is presented. Secondly, in order to get higher performance of SPIM drive, a speed estimation method based on EKF is proposed. The simulation results in this paper demonstrate the good performance of the suggested methods, in both controlling and speed estimation strategies.

REFERENCES
[1] HW Beatty, JL Kirtley. Electric motor handbook. New York, McGraw Hill. 1998.

[2] X Wang, H Zhong, Y Yang, X Mu. Study of a Novel Energy Efficient Single-Phase Induction Motor With Three Series-Connected Windings and Two Capacitors. *IEEE Trans. Energy Convers.*, 2010; 25(2): 433–440,.

[3] V Debusschere, B Multon, H Ben Ahmed, PE Cavarec. Life cycle design of a single-phase induction motor. *IET Electr. Power Appl.*, 2010; 4(5): 348–356.

[4] C Mademlis, I Kioskeridis, T Theodoulidis, A Member. Optimization of Single-Phase Induction Motors-Part I: Maximum Energy Efficiency Control. *IEEE Trans. Energy Convers.*, vol. 20, no. 1, pp. 187–195, 2005.

[5] K. Satyanarayana, P. Surekha, and P. Vijaya Prasuna, "A New FOC Approach of Induction Motor Drive Using DTC Strategy for the Minimization of CMV," *International Journal of Power Electronics and Drive System (IJPEDS)*, vol. 3, no. 2, pp. 241–250, 2013.

[6] B. Bossoufi, M. Karim, A. Lagrioui, and M. Taoussi, "FPGA-Based Implementation Nonlinear Backstepping Control of a PMSM Drive," *International Journal of Power Electronics and Drive System (IJPEDS)*, vol. 4, no. 1, pp. 12–23, 2014.

[7] R. Ramasamy, and N. Devarajan, "Dynamically Reconfigurable Control Struture for Three Phase Induction Motor Drives," *International Journal of Power Electronics and Drive System (IJPEDS)*, vol. 2, no. 1, pp. 43–50, 2011.

[8] G. Kohlrusz, and D. Fodor, "Comparison of Scalar and Vector Control Strategies of Induction Motors," *Hungarian J. Ind. Chem.* Veszprem, vol. 39, no. 2, pp. 265–270, 2011.

[9] M. B. de R. Correa, C. B. Jacobina, A. M. N. Lima, and E. R. C. Silva, "Rotor-Flux-Oriented Control of a Single-Phase Induction Motor Drive," *IEEE Trans. Ind. Electron.*, vol. 47, no. 4, pp. 832–841, 2000.

[10] M. B. de R. Correa, C. B. Jacobina, E. R. C. da Silva, and A. M. N. Lima, "Vector Control Strategies for Single-Phase Induction Motor Drive Systems," *IEEE Trans. Ind. Electron.*, vol. 51, no. 5, pp. 1073–1080, 2004.

[11] S. Vaez-Zadeh, and S. Reicy Harooni, "Decoupling Vector Control of Single-Phase Induction Motor Drives," *in Power Electronics Specialists Conference*, 2005, no. 1, pp. 733–738.

[12] S. Reicy, and S. Vaez-Zadeh, "Vector Control of Single-Phase Induction Machine with Maximum Torque Operation," *in IEEE ISIE*, 2005, pp. 923–928.

[13] H. Ben Azza, M. Jemli, and M. Gossa, "Full-Digital Implementation of ISFOC for Single-Phase Induction Motor Drive Using dSpace DS 1104 Control Board," *Int. Rev. Electr. Eng.*, vol. 3, no. 4, pp. 721–729, 2008.

[14] M. Jemli, H. Ben Azza, and M. Gossa, "Real-Time Implementation of IRFOC for Single-Phase Induction Motor Drive Using dSpace DS 1104 Control Board," *Simul. Model. Pract. Theory*, vol. 17, no. 6, pp. 1071–1080, 2009.

[15] M. B. de R. Correa, C. B. Jacobina, P. M. dos Santos, E. C. dos Santos, and A. M. N. Lima, "Sensorless IFOC for Single-Phase Induction Motor Drive System," *in Electric Machines and Drives*, 2005, pp. 162–166.

[16] M. Jemli, H. Ben Azza, M. Boussak, and M. Gossa, "Sensorless Indirect Stator Field Orientation Speed Control for Single-Phase Induction Motor Drive," *IEEE Trans. Power Electron.*, vol. 24, no. 6, pp. 1618–1627, 2009.

[17] R. P. Vieira, and H. A. Grundling, "Sensorless Speed Control with a MRAS Speed Estimator for Single-Phase Induction Motors Drives," *in Power Electronics and Applications, 2009. EPE '09. 13th European Conference on*, 2009, pp. 1–10.

[18] S. Bolognani, L. Peretti, and M. Zigliotto, "Parameter Sensitivity Analysis of an Improved Open-Loop Speed Estimate for," *IEEE Trans. Power Electron.*, vol. 23, no. 4, pp. 2127–2135, 2008.

[19] H. Ben Azza, M. Jemli, M. Boussak, and M. Gossa, "High performance sensorless speed vector control of SPIM Drives with on-line stator resistance estimation," *Simul. Model. Pract. Theory*, vol. 19, no. 1, pp. 271–282, 2011.

[20] R. Z. Azzolin, T. A. Bernardes, R. P. Vieira, C. C. Gastaldinit, and H. A. Griindling, "Decoupling and Sensorless Vector Control Scheme for Single-Phase Induction Motor Drives," *in IECON 2012 - 38th Annual Conference on IEEE Industrial Electronics Society*, 2012, no. 1, pp. 1713–1719.

[21] L. Salvatore, and S. Stasi, "A New EKF-Based Algorithm for Flux Estimation in Induction Machines," *IEEE Trans. Ind. Electron.*, vol. 40, no. 5, pp. 496–504, 1993.

[22] O. S. Bogosyan, M. Gokasan, and C. Hajiyev, "An Application of EKF for the Position Control of a Single Link Arm," *in 27th Annual Conference of the IEEE Industrial Electronics Society, IECON'01*, vol. 1, 2001, pp. 564–569.

[23] P. Vas, "*Sensorless Vector and Direct Torque Control*," UK, Oxford University Press Oxford, 1998.

[24] M. Jannati, and E. Fallah, "Modeling and Vector Control of Unbalanced induction motors (faulty three phase or single phase induction motors)," *1st. Conference on Power Electronic & Drive Systems & Technologies (PEDSTC)*, 2010, pp. 208–211.

[25] M. Jannati, N. R. N. Idris, M. J. A. Aziz, A. Monadi, and A. A. Faudzi, "A Novel Scheme for Reduction of Torque and Speed Ripple in Rotor Field Oriented Control of Single Phase Induction Motor Based on Rotational Transformations," *Research Journal of Applied Sciences, Engineering and Technology*, vol. 7, no. 16, pp. 3405–3409, 2014.

[26] M. Jannati, A. Monadi, S. A. Anbaran, N. R. N. Idris, and M. J. A. Aziz, "An Exact Model for Rotor Field-Oriented Control of Single-Phase Induction Motors," *TELKOMNIKA Indonesian Journal of Electrical Engineering*, vol. 12, no. 7, pp. 5110–5120, 2014.

[27] M. Jannati, N. R. N. Idris, and Z. Salam, "A New Method for Modeling and Vector Control of Unbalanced Induction Motors," *Energy Conversion Congress and Exposition (ECCE)*, 2012, pp. 3625–3632.

[28] M. Jannati, N. R. N. Idris, and M. J. A. Aziz, "A New Method for RFOC of Induction Motor Under Open-Phase Fault," *in Industrial Electronics Society, IECON 2013*, 2013, pp. 2530–2535.

[29] M. Jannati, A. Monadi, N. R. N. Idris, M. J. A. Aziz, and A. A. Faudzi, "Vector Control of Faulty Three-Phase Induction Motor with an Adaptive Sliding Mode Control," *Przeglad Elektrotechniczny*, vol. 87, no. 12, pp. 116–120, 2013.

5

A Novel Direct Torque Control for Induction Machine Drive System with Low Torque and Flux Ripples using XSG

Souha Boukadida, Soufien Gdaim, Abdellatif Mtibaa
Laboratory EµE of the FSM, University of Monastir, Tunisia

Keyword:

DTC-SVM
FPGA
Induction machine
Matlab/Simulink
XSG

ABSTRACT

The conventional Direct Torque Control (DTC) is known to produce a quick and robust response in AC drives. However, during steady state, stator flux and electromagnetic torque which results in incorrect speed estimations and acoustical noise. A modified Direct Torque Control (DTC) by using Space Vector Modulation (DTC-SVM) for induction machine is proposed in this paper. Using this control strategy, the ripples introduced in torque and flux are reduced. This paper presents a novel approach to design and implementation of a high perfromane torque control (DTC-SVM) of induction machine using Field Programmable gate array (FPGA). The performance of the proposed control scheme is evaluated through digital simulation using Matlab\Simulink and Xilinx System Generator. The simulation results are used to verify the effectiveness of the proposed control strategy.

Corresponding Author:

Souha Boukadida,
Laboratory EµE of the FSM
University of Monastir, Tunisia
Email: boukadidasouha@yahoo.fr

1. INTRODUCTION

Since its inception, the Direct Torque Control has gained popularity for induction machine drives. Indeed, the control variables that are the stator flux and torque are calculated from the quantities related to the stator without the intervention of mechanical sensor. The response of the DTC is fast, however it has some drawbacks such as notable torque and flux ripples and the variable commutation frequency behavior of the inverter. Many papers presented different approaches to minimize the flux and torque ripples [1]-[4]. In [1] and [3], electromagnetic torque and flux are controlled directly by the selection of a switching vector from a table selection. Nevertheless, the selected vector is not always the best one because only the sector is considered, where the flux space vector lies without considering its location.

To overcome the several disadvantages of DTC a new control technique called Direct Torque Control – Space Vector Modulated (DTC-SVM) [5]-[6] is developped. In this new method, the disadvantages of the DTC are eliminated. The DTC-SVM strategies are based on the same fundamentals as classical DTC; it provides dynamic behavior comparable with classical DTC.

In practice, the vector control algorithm for an induction machine is implemented utilizing digital signal processor (DSP). The DSP control procedure is performed sequentially; this may result in a slower cycling period if complex algorithms are involved. Employing field programmable gate array (FPGA) in implementing vector control strategies provides advantages such as simpler hardware and software design, rapid prototyping, hence fast switching frequency and high speed computation [7]-[8].

The paper devotes to a comparative study between the performances of two approaches: (i) Classical DTC (ii) DTC-SVM. These strategies are designed using Xilinx System Generator (XSG) and Matlab/Simulink software packages and implemented on FPGA controller.

2.   BASIC PRINCIPLE OF DTC

The main idea of DTC is to recover the reduction of the ripples of torque and flux, and to have superior dynamic performances. Figure 1 present a possible schematic of Direct Torque Control. There are two different loops corresponding to the magnitudes of the stator flux and torque. The error between the estimated stator flux magnitude φ_s and the reference stator flux magnitude $\varphi_s*$is the input of a two level hysteresis comparator whereas the error between the estimated torque T_e and the reference torque T_e* is the input of a three level hysteresis comparator. The outputs of the stator flux error and torque error hysteresis blocks, together with the position of the stator flux are used as inputs of the switching table.

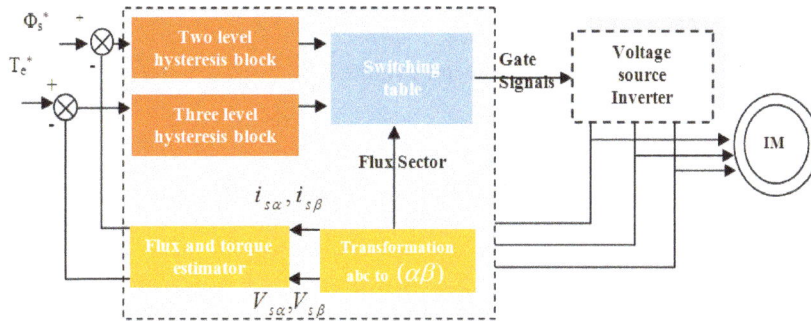

Figure 1. Block diagram of DTC

The selection vector is based on the hysteresis control of the torque and the stator flux. In the basic form the stator flux φ_s is estimated with:

$$\varphi_s = \int_0^t (V_s - R i_s)dt \tag{1}$$

The stator voltage and stator current are calculated from the state of three phase (Sa ,Sb ,Sc) and measured currents (ia, ib, ic).

$$V_s(S_a,S_b,S_c) = \sqrt{\frac{2}{3}} E_0(S_a + S_b e^{j\frac{2\pi}{3}} + S_c e^{j\frac{4\pi}{3}}) \tag{2}$$

$$i_s(i_a,i_b,i_c) = \frac{2}{3}(i_a + i_b e^{j\frac{2\pi}{3}} + i_c e^{j\frac{4\pi}{3}})$$

Phase angle and stator flux amplitude are calculated in expression (3).

$$\theta_s = arctg\left(\frac{\varphi_{s\beta}}{\varphi_{s\alpha}}\right)$$

$$\varphi_s = \sqrt{\varphi_{s\alpha}^2 + \varphi_{s\beta}^2} \tag{3}$$

The developed electromagnetic torque T_e of the machine can be evaluated by Equation (4):

$$T_e = \frac{3}{2} p (i_{s\beta} \phi_{s\alpha} - i_{s\alpha} \phi_{s\beta}) \tag{4}$$

The stator flux vector is moving along a straight axis colinear to that of the voltage vector required by the inverter:

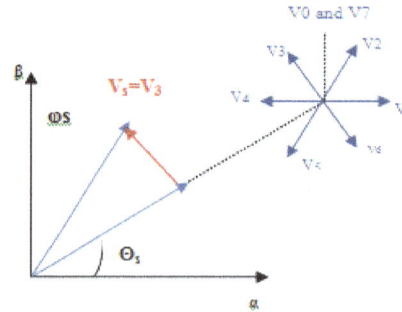

Figure 2. Stator flux vector evolution in the first sector

3. DTC SPACE VECTOR MODULATION

The DTC algorithm is based on the instantaneous values and directly calculated the gate signals for the inverter. The control algorithm in DTC-SVM is based on average values whereas the switching signals (Sa, Sb and Sc) for the inverter are calculated by space vector modulator [9]-[11].

3.1. Principle of Vector MLI

For each period of modulation of the inverter, the three phase voltages provided by the control algorithm can be expressed in a fixed reference linked to the stator, through their projections $V_{s\alpha}$ and $V_{s\beta}$.

The inverter has six switching cells, giving eight possible switching configurations. These eight switching configurations can be expressed in the plane (α, β) by 8 vectors tensions.

Knowing that in the graduation phase voltages (Va, Vb, Vc) are represented in the plane by a vector V_s. The principle of vector MLI is to project the desired stator voltage vector V_s on the two adjacent vectors corresponding to two switching states of the inverter. The values of these projections provide the desired commutation times.

3.2. General Structure of the Control DTC-SVM

Most existing blocks in the control DTC-SVM are identical to those of control DTC as shown in the following figure (3). The new blocks will be discussed below.

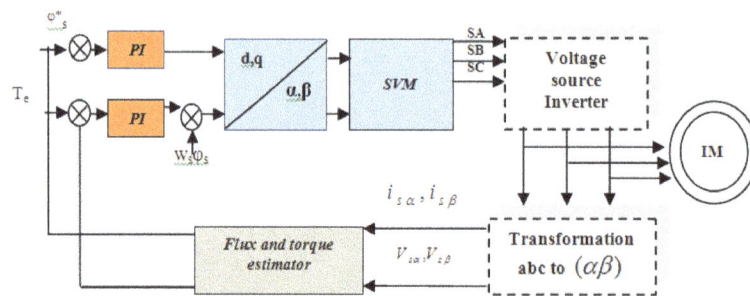

Figure 3. Block diagram of DTC-SVM

3.3. Calculation of time of application of the status of the inverter

Each modulation period T_{mod} of the inverter, the projected vector V_s on the two adjacent vectors assures the switching time of calculation.

The key step of the SVM technique is the determination of T_i and T_{i+1} during every modulation period T_{mod}. To illustrate the methodology we consider the case where V_s can be compounded by the active voltage vectors V_1 and V_2. The projection of the reference voltage vector on V_1 and V_2 is illustrated in the following figure:

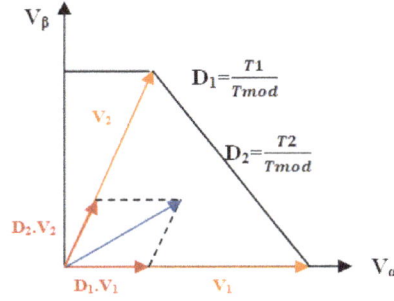

Figure 4. Projection of the reference voltage vector on V_1 and V_2

The active voltage vectors V1 and V2 are given as follow:

$$\begin{cases} \overrightarrow{V_1} = \sqrt{\dfrac{2}{3}}.E.e^{j0} \\ \overrightarrow{V_2} = \sqrt{\dfrac{2}{3}}.E.e^{j\frac{\pi}{3}} \end{cases} \tag{5}$$

Expressing the voltage vector Vs in the graduation (α, β) we have:

$$\overrightarrow{V_s} = V_{s\alpha} + jV_{s\beta} = \frac{T_1}{T_{mod}}\overrightarrow{V_1} + \frac{T_2}{T_{mod}}\overrightarrow{V_2} \tag{6}$$

Expanding this equation it is possible to express the time T_1 and T_2 in terms of $V_{s\alpha}$ and $V_{s\beta}$. The conduction time will be expressed as follows:

$$\begin{cases} T_1 = (\sqrt{\dfrac{3}{2}}V_{s\alpha} - \sqrt{\dfrac{1}{2}}V_{s\beta}).\dfrac{T_{mod}}{E} \\ T_2 = \sqrt{2}V_{s\beta}.\dfrac{T_{mod}}{E} \end{cases} \tag{7}$$

To facilitate the calculations, we normalize the voltages Vsα and Vsβ by posing:

$$\begin{cases} \hat{V}_{s\alpha} = \dfrac{V_{s\alpha}}{E}\sqrt{2} \\ \hat{V}_{s\beta} = \dfrac{V_{s\beta}}{E}\sqrt{2} \end{cases} \tag{8}$$

Consequently, the duties expressions are given as follows:

$$D_1 = \frac{\sqrt{3}}{2}.\hat{V}_{s\alpha} - \frac{1}{2}.\hat{V}_{s\beta}$$
$$D_2 = \hat{V}_{s\beta}$$
$$D_0 = 1 - D_1 - D_2 \tag{9}$$

The space vector in sector 1 is shown in figure (5).The time duration of zero vectors is divided equally into (V0, V1, V2, V7, V2, V1, V0), whereas the time duration of each nonzero vector is distributed into two parts. This sequence can ensure that is one phase switches when the switching pattern switches, thus can reduce the harmonic component of the output current and the loss of switching devices.

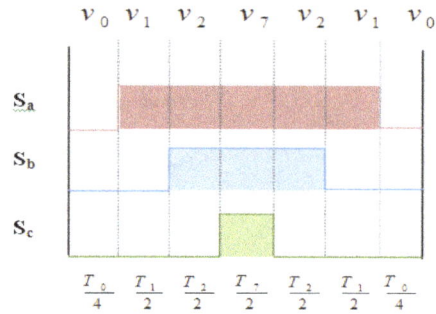

Figure 5. Sequences of the switches states in sector N1

The duties of each phase of the inverter are presented as follows:

$$S_a = 0.5(1 + D_1 + D_2)$$
$$S_b = 0.5(1 - D_1 + D_2)$$
$$S_c = 0.5(1 - D_1 - D_2)$$

(10)

4.   SIMULATION AND RESULT

The DTC and DTC-SVM scheme for induction machine are simulated using Matlab/Simulink and Xilinx System Generator and their results have been compared. The machine parameters used for simulation are given in this table.

Table 1. Induction Machine parameters

Voltage	220/380 v
Stator resistance R_s	5.717 Ω
Rotor resistance R_r	4.282 Ω
Stator inductance L_s	0.464 H
Rotor inductance L_r	0.464 H
Mutual inductance M	0.441 H
Moment of inertia J	0.0049 Kg.m^2

4.1. Simulink Model of Direct Torque Control

The simulation of DTC was conducted using Simulink\MATLAB. The inverter switching pulses are obtained from the switching table which decides the pulses from the error signals of torque and flux. The overall DTC model is shown in Figure 6.

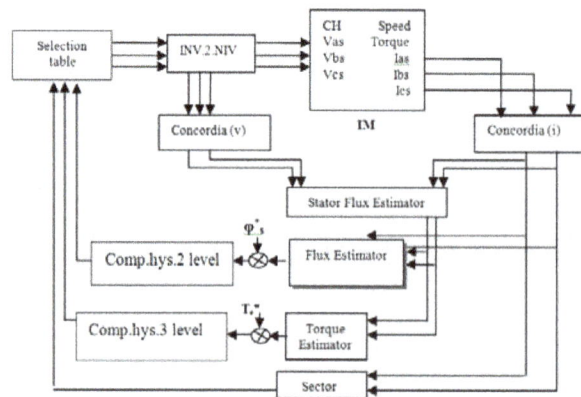

Figure 6. Simulink Model of DTC

4.2. Simulink Model of Space Vector Modulated Direct Torque Control obtained with Matlab\Simulink

Figure 7 illustrate the simulation block of the DTC-SVM control. The system is composed of the machine, PI controllers, three phase voltage source inverter, reference frame transformation blocks Concordia and Park. The Insulated-gate bipolar transistor IGBT switches are controlled using space vector modulation technique.

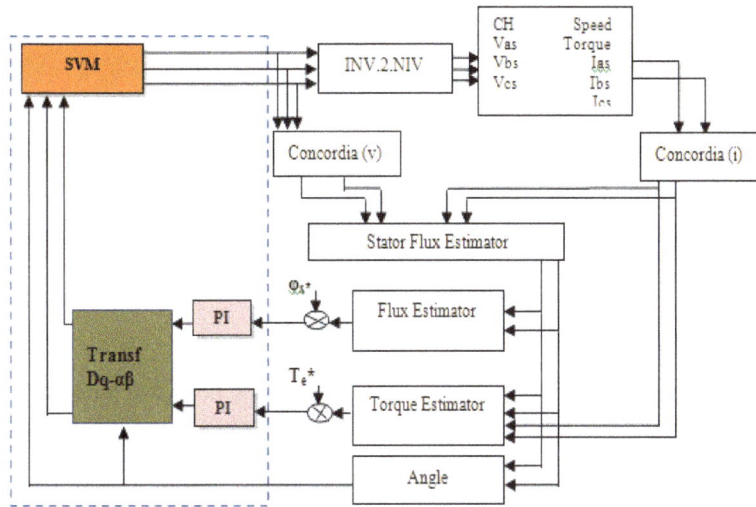

Figure 7. Simulink Model of DTC-SVM

The simulation of this technique is made through the following model:

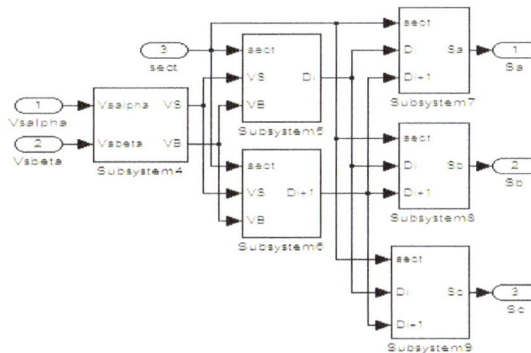

Figure 8. Simulink Model of bloc SVM

4.3. Simulink Model of Space Vector Modulated Direct Torque Control obtained with Xilinx System Generator

Initially, an algorithm is designed and simulated at the system level with the floating-point Simulink blocksets. A hardware representation of FPGA implementation is then derived using XSG. The XSG provides a bit-accurate model of FPGA circuits and automatically generates a synthesizable VHDL code for implementation in Xilinx FPGA. For DTC-SVM modeling, the blocks used are mostly multipliers, adders, Cordic sin cos, etc. The detailed steps are shown in the following diagram in Figure 9. The XSG design of proposed DTC-SVM is shown in Figure 10. The block Calcul_Vsalpha_Vsbeta is used to project the three-phase voltages in the repository (α, β) by performing the processing Clarke as shown in Figure 10(a). The block SVM generates a series of pulses to be used subsequently to carry out the control signals used in the model of the inverter as shown in Figure 10(b) and 10(c). The XSG design of torque and flux estimator is shown in Figure 11-12.

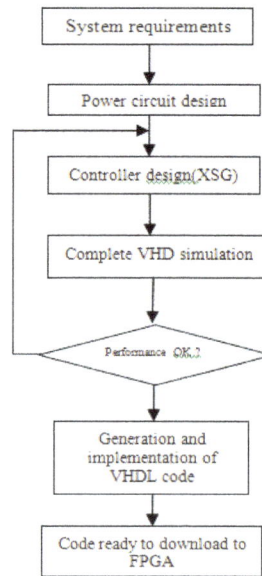

Figure 9. Induction machine drive controller design and implementation process

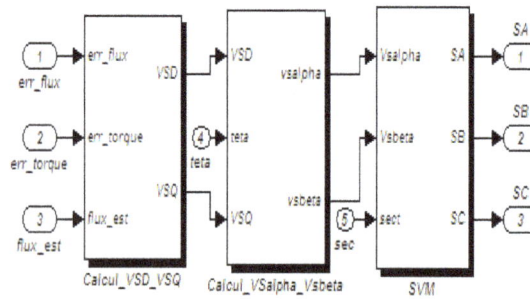

Figure 10. Xilinx Model of SVM

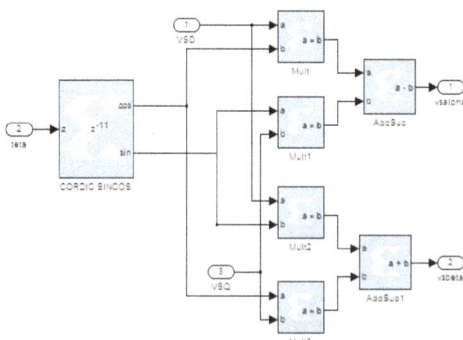

Figure 10(a). Calcul Vsalpha Vsbeta

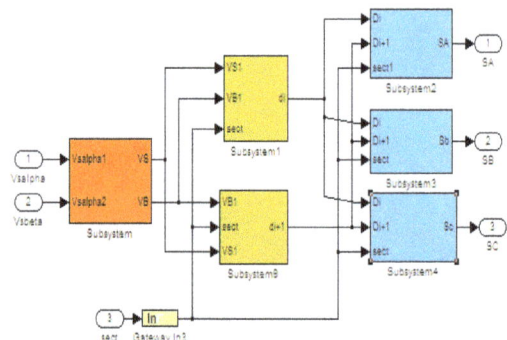

Figure 10(b). VM bloc in XSG

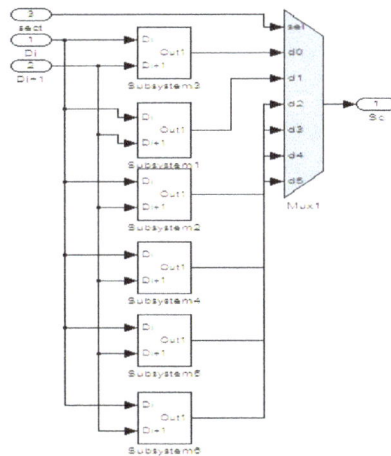

Figure 10(c). Calcul Sc

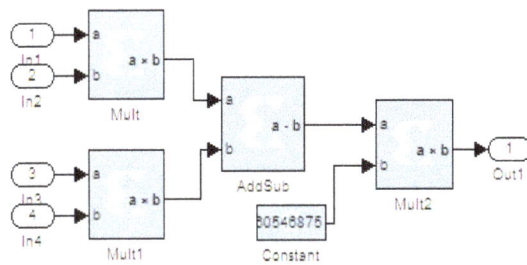

Figure 11. Model of electromagnetic torque

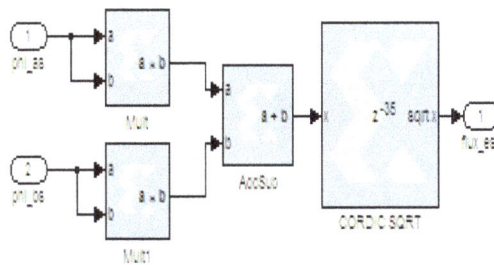

Figure 12. Model of flux estimator

4.4. SIMULATION RESULTS

The performance of the induction machine under different operating conditions was also investigated in order to verify the robustness of the proposed control scheme. The steady state behavior of induction machine with the conventional DTC and DTC-SVM are illustrated in Figure 13-15.

It is possible to see in Figure 13(a), (b), (c) an appreciable reduction of electromagnetic torque ripple has been obtained using the DTC-SVM. For the DTC, torque variation of the hysteresis band equal to 1.1. The high ripple observed in the DTC is reduced when we use the DTC-SVM, because in SVM, many vectors (IGBT states) are selected to adjust the flux and torque ripple in each sample time, whereas in DTC just one vector is selected to adjust ripple inside hysteresis bands of flux. Using SVM control provides the system with minimum ripple for flux as shown in Figure 14, where the flux ripple percentage is about 0.92%.

The DTC-SVM of induction machine presents the advanced performance to achieve tracking of the desired smooth circular trajectory of stator flux locus shown in Figure 15.

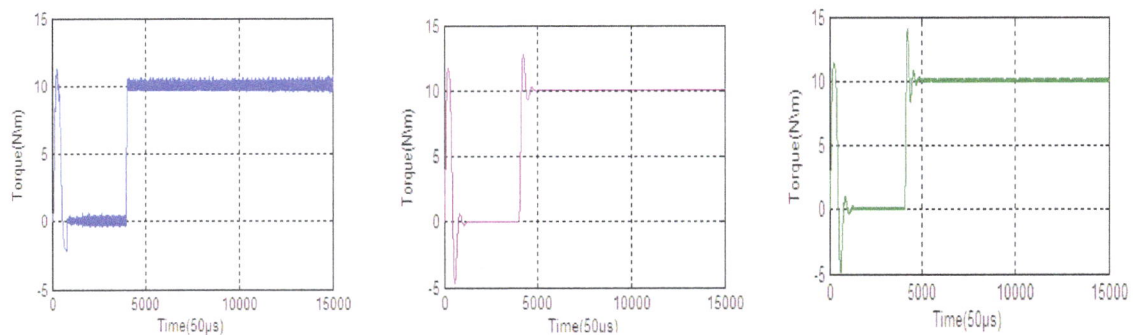

Figure 13. Electromagnetic torque, (a): DTC using MATLAB, (b) DTC-SVM using MATLAB, (c): DTC-SVM using XSG

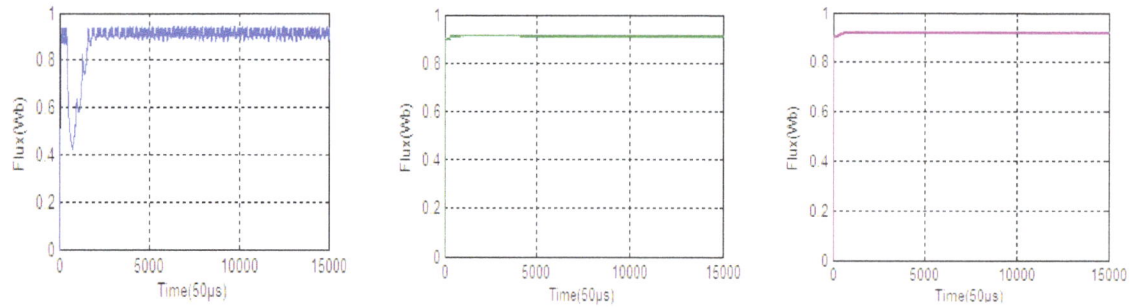

Figure 134. Stator flux DTC using MATLAB, (b): DTC-SVM using MATLAB, (c): DTC-SVM using XSG

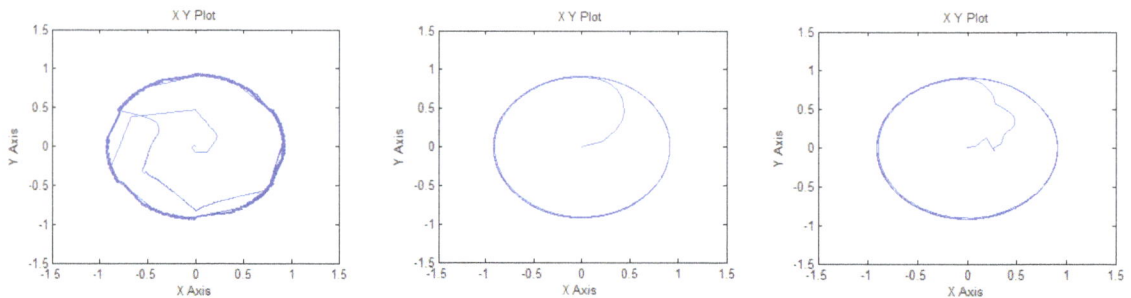

Figure 15. Trajectory of stator flux: DTC using MATLAB, (b): DTC-SVM using MATLAB, (c): DTC-SVM using XSG

Table 2. The percentage flux and torque error for DTC and DTC-SVM

Control strategies	Flux ripple (%)	Torque ripple (%)
DTC using Matlab	5.52	11
DTC-SVM using Matlab	0.92	1
DTC-SVM using XSG	1.84	2

The best results are given by DTC-SVM using MATLAB\SIMULINK, this is due to the arbitrary choice of the number of bits at XSG.

5.   FPGA SIMULATION RESULTS OF DTC-SVM

The above designed model is implemented using FPGA Editor. FPGA Editor reads the NCD file generated by the Map or Place & Route process, which contains the logic and routing of the design mapped to components, such as CLBs and IOBs.The internal structure of FPGA is shown in Figure 16.

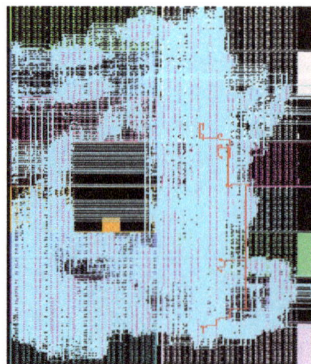

Figure 146. Internal structure of FPGA

The result of the resources used is shown in the following table:

Table 3. The result of the resources

Slice logic utilization	Used	Available	Utilization
Number of slices LUTs	10,511	44,800	23%
Number used as Logic	9,869	44,800	22%
Number of DSP48Es	109	128	85%
Number of slice registers	655	44,800	1%

6.   CONCLUSION

This paper has been devoted to the comparison between the performances of the DTC and DTC-SVM strategy. The steady state features of the induction machine as well as the transient behavior under both approaches have been commented and compared. The simulation result clearly indicates the high performance of DTC-SVM. The proposed high performance scheme is designed using XSG and Matlab/Simulink blocksets and implemented on Xilinx Virtex 5 FPGA. Numerical simulations have been carried out showing the advantages of the DTC-SVM with respect to the DTC. This work is the first step towards implemetation on FPGA of DTC-SVM. Future work will extend this experimental validation to the study.

REFERENCES

[1]   Z Li, L Wang, S Zhang, C Zhang, J Ahn. Torque Ripple Reduction in Direct Torque Controlled Brushless DC Motor. *IEEE Trans. Electrical Machines.* 2011; 1-4.

[2]   Y Cho, D Kim, K Lee, Y Lee, J Song. Torque Ripple Reduction and Fast Torque Response Strategy of Direct Torque Control for Permanent- Magnet Synchronous Motor. *IEEE Trans.Ind.Electronics.* 2013; 1-6.

[3]   J Beerten, J Verveckken, J Driesen. Predictive Direct Torque Control for Flux and Torque Ripple Reduction. *IEEE Trans.Ind.Electronics.* 2010; 57(1).

[4]   T Sutikno, N Rumzi, N Idris, A Jidin, N Cirstea. An Improved FPGA Implementation of Direct Torque Control for Induction Machines. *IEEE Trans.Ind.Electronics.* 2013; 1280-1290.

[5]   KN Achari, B Gururaj, DV Ashok Kumar, M Vijaya Kumar. A Novel MATLAB/Simulink Model of PMSM Drive using Direct Torque Control with SVM. *IEEE Multimedia Computing and Systems.* 2012; 1069-1075.

[6]   B Metidji, F Tazrart, A Azib, N Taib, T Rekioua. A New Fuzzy Direct Torque Control Strategy for Induction Machine Based on Indirect Matrix Converter. *International Journal of Research and Reviews in Computing Engineering.* 2011; 1(1).

[7]   MW Naouar, E Monmasson, AA Naassani. FPGA-based current controllers for AC machine drives-A review. *IEEE Transactions on Industrial Electronics.* 2007; 54(4): 1907- 1925.

[8]   JJ Rodriguez-Andina, MJ Moure, MD Valdes. Features, design tools, and application domains of FPGAs. *IEEE Transactions on Industrial Electroncis.* 2007; 54(4): 1810-1823.

[9]   Habetler TG, Profumo F, Pastorelli M. Direct torque control of induction machines over a wide speed range. *Proceedings of IEEE-IAS Conference.* 1992; 600-606.

[10]   Casadei D, Serra G, Tani A. Implementation of a direct control algorithm for induction motors based on discrete space vector modulation. *IEEE Transactions on Power Electronics.* 2000; 15: 769-777.

[11]   Tsung-Po Chen, Yen-Shin Lai, Chang-Huan Liu. A new space vector modulation technique for inverter control. *Power Electronics Specialist Conference.* 1999; 2: 777-782.

6

Fuzzy Bang-Bang Control Scheme of USSC for Voltage Sag Mitigation due to Short Circuits and Induction Motor Starting in Distribution System

M.Mohammadi, A.Mohammadi Rozbahani, S.Abasi Garavand, M.Montazeri, H.Memarinezhad
Department of Electrical Engineering, College of Engineering, Borujerd Branch, Islamic Azad University, Borujerd, Iran

ABSTRACT

Keyword:

Fuzzy bang-bang control
Power quality
Shunt-compensator
Unified series
Voltage sag

Unified series shunt compensator (USSC) has been widely used to mitigate various power quality disturbances in distribution network. The USSC is almost similar to the UPFC, but the only differences are that the UPFC inverters are in shunt series connection and used in transmission systems whereas the USSC inverters are in series-shunt connection and used in distribution systems. USSC, it is possible to compensate a different power quality problem as compared to DSTATCOM and DVR. It is noted that, mitigated load voltage by the DVR is lower than mitigated value obtained by USSC. In other words the USSC can mitigate voltage sag better in compared to DVR and D-STATCOM. Also in case of voltage flicker, unbalance and harmonics elimination it is much effective. Similarly, D-STATCOM is unable to control power flow. It is seen that the proposed USSC can mitigate variety of power quality (PQ) problems. Hence due to multi capability of USSC in power quality improvement, this paper presents the scheme based on fuzzy bang-bang control for USSC. Using Fuzzy Logic Control (FLC) based on bang-bang control; the USSC will contribute to improve voltage sag without deteriorating the effect of the other compensating devices.

Corresponding Author:

Ashkan Mohammadi Rozbahani
Department of Electrical Engineering, College of Engineering, Borujerd Branch,
Islamic Azad University,
Borujerd, Iran,
Email: mr.ashkan@iaub.ac.ir

1. INTRODUCTION

Voltage sag can cause loss of production in automated process since voltage sag can trip a motor or cause its controller to malfunction. Various methods have been applied to reduce or mitigate voltage sags. The conventional methods are by using capacitor banks, introduction of new parallel feeders and by installing uninterruptible power supplies (UPS) [1]. Coil hold-in devices are one of traditional mitigation method. These devices are connected between the AC supply and the contactor and can generally allow a contactor to remain energized [2]. A ferroresonant transformer, also known as a constant voltage transformer (CVT), is a transformer that operates in the saturation region of the transformer B-H curve. Voltage sags down to 30 % retained voltage can be mitigated through the use of ferroresonant transformers [3]. Flywheel systems use the energy stored in the inertia of a rotating flywheel to mitigate voltage sags.

In the most basic system, a flywheel is coupled in series with a motor and a generator which in turn is connected in series with the load. The flywheel is accelerated to a very high speed and when voltage sag occurs, the rotational energy of the decelerating flywheel is utilized to supply the load [4]. To compensate the voltage sag in a power distribution system, appropriate devices need to be installed at suitable locations. These devices are typically placed at the point of common coupling [PCC] which is defined as the point of the network changes. A SVC is a shunt connected power electronics based device which works by injecting reactive current into the load, thereby supporting the voltage and mitigating the voltage sag [5]. The DVR is one of the custom power devices which can improve power quality, especially, voltage sags [6]. The DVR injects three single-phase voltages in series with incoming supply voltages.

The magnitude and phase angle of injected voltage are variables which result in variable real and reactive power exchange between the DVR and the sensitive load or the distribution system Others have investigated new methods to improve power quality [7]. Usually the control voltage of the DVR in mitigating voltage sag is derived by comparing the supply voltage against a reference waveform [8]. There are many solutions in mitigating the power quality problems at a distribution system such as using surge arresters, active power filters, isolation transformer, uninterruptible power supply and static VAR compensator are some of new methods. In [9] authors proposed a new D-STATCOM control algorithm which enables separate control of positive and negative sequence currents and decoupled control of d- and q-axes current components. In [10] the mitigation of voltage flicker and reduction in THD by using STATCOM has been investigated.

Reference [11] use real time digital simulation of power electronic system which is a heavily computer intensive operation, and based on VSC D-STATCOM power system. From the studies, it is shown that all these equipments are capable in solving power quality problems.

The best equipment to solve this problem at distribution systems at minimum cost is by using Custom Power family of D-STATCOM. By using a unified approach of series-shunt compensators it is possible to compensate for a variety of power-quality problems in a distribution system including sag compensation, flicker reduction, unbalance voltage mitigation, and power-flow control [12]. Since this device is able to mitigate several of power quality disturbances, therefore this paper focuses on this device and presents a new control strategy based fuzzy logic bang-bang control to mitigate voltage sag.

2. VOLTAGE SAG CONCEPTS

Voltage sag is reduction in supply voltage magnitude followed by voltage recovery after a short period of time. In the IEEE Standard 1159-1995, the term "sag" is defined as a decrease in rms voltage to values between 0.1 to 0.9 p.u, for durations of 0.5 cycles to 1 min [8-10]. The two main causes of voltage sags are network faults and the starting of equipment which draw large currents, particularly direct-on-line motors. Voltage sag is characterized in terms of the following parameters, magnitude of sag and duration of sag and phase-angle jump. Depending on the type of fault, sag can be balanced or unbalanced. Naturally for the Three phase to ground (ABC-G) fault the sag is symmetrical (balanced) in all three phases as shown in Figure 1.

Whereas for unbalanced faults like A-G, B-C, BC-G the sag is unsymmetrical in all three phases, as shown in Figure 2.

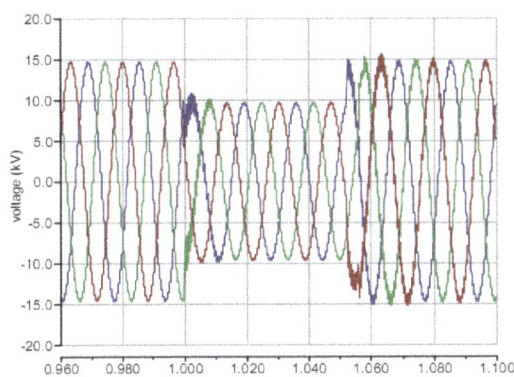

Figure 1. Balanced voltage sag in three phases   Figure 2. Balanced voltage sag in three phases

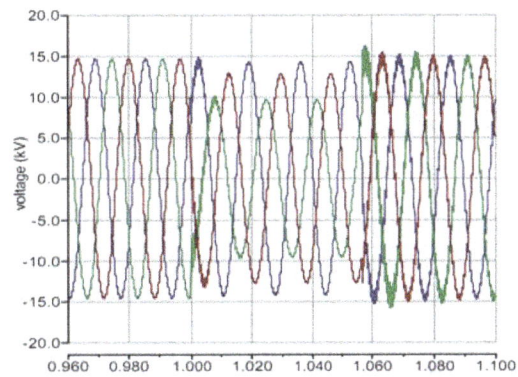

Voltage sags are measured using specialized power quality monitoring instrumentation. The instrumentation must be configured with a sag threshold voltage. That is, a voltage level that will trigger a sag capture when the rms voltage falls below it. Figure 3 shows a graphical representation of a voltage sag including the sag threshold and the parameters (duration, retained voltage) used to report the sag.

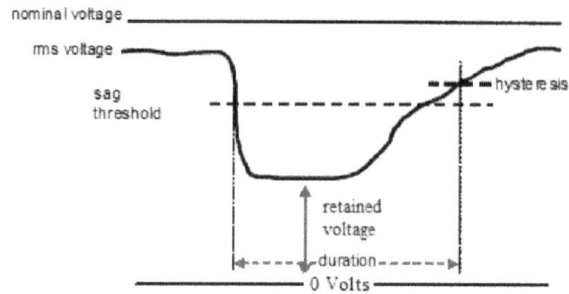

Figure 3. Graphical representation of voltage sag

3. USSC MODELING

The Unified Series Shunt Compensator is a combination of series and shunt voltage source inverters as shown in Figure 4. The basic components of the USSC are two 12-pulse voltage source inverters composed of forced commutated power semiconductor switches, typically Gate Turn Off thyristor valves. One voltage source inverter is connected in series with the line through a set of series injection transformers, while the other is connected in shunt with the line through a set of shunt transformers.

The dc terminals of the two inverters are connected together and their common dc voltage is supported by a capacitor bank [13]. The USSC is almost similar to the UPFC, but the only differences are that the UPFC inverters are in shunt series connection and used in transmission systems whereas the USSC inverters are in series-shunt connection and used in distribution systems [14].

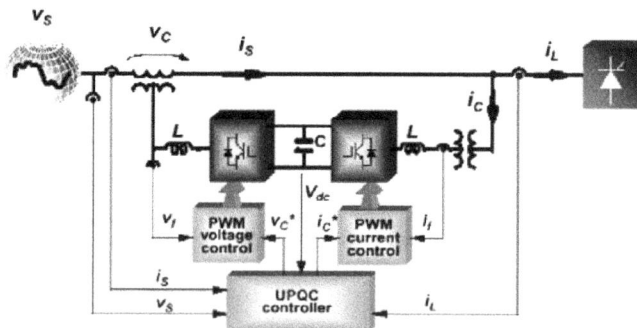

Figure 4. General Configuration of Unified Series Shunt Compensator-USSC

4. CAPABILITIES OF USSC VERSUS D-STATCOM AND DVR

Since the introduction of FACTS and custom power concept [15], devices such as unified power-flow controller (UPFC), synchronous static compensator (STATCOM), dynamic voltage restorer (DVR), solid-state transfer switch, and solid-state fault current limiter are developed for improving power quality and reliability of a system [16], [17]. Advanced control and improved semiconductor switching of these devices have achieved a new area for power-quality mitigation. Investigations have been carried out to study the effectiveness of these devices in power-quality mitigation such as sag compensation, harmonics elimination, unbalance compensation, reactive power compensation, power-flow control, power factor correction and flicker reduction [18-19]. These devices have been developed for mitigating specified power-quality problems. By using a unified approach of series-shunt compensators it is possible to compensate for a variety of power-quality problems in a distribution system including sag compensation, flicker reduction, unbalance voltage mitigation, and power-flow control [11]. Usually individual custom power devices such as DSTATCOM and DVR focus on solving specific power quality problems in a distribution system. However, by using USSC, it is possible to compensate a different power quality problem as compared to DSTATCOM and DVR as indicated in Table 1 [20].

Table 1. Power quality mitigation using USSC versus others custom power devices

Power Quality Mitigation	DVR	D-STATCOM	USSC
Voltage Flicker	YES	Limited	YES
Voltage Sag Compensation	NO	YES	YES
Unbalance	NO	YES	YES
UPS Mode	YES	YES	YES
Power Flow Control	NO	NO	YES
Harmonic Elimination	NO	YES	YES

It is noted that, mitigated load voltage by the DVR is a steady state value but this value is lower than mitigated value obtained by USSC. In other words the USSC can mitigate voltage sag better in compared to DVR and D-STATCOM. Also in case of voltage flicker, unbalance and harmonics elimination it is much effective. Similarly, D-STATCOM is unable to control power flow. It is seen that the proposed USSC can mitigate variety of PQ problems [21].

5. USSC INSTALLATION IN DISTRIBUTION SYSTEM

Before modeling the USSC, all distribution system components, i.e., lines and cables, loads, transformers, large motors and generators have to be converted into equivalent reactance (X) and resistance (R) on common bases. The main system component models are used in the formulation of impedance matrix for voltage sag calculation [22]. In steady state analysis, the series and shunt inverters of the USSC are presented by two voltage sources V_{dq} and V_{sh} respectively as shown in Figure 5.

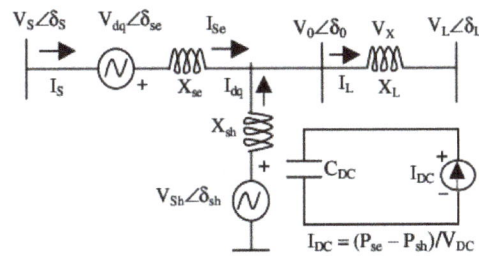

Figure 5. Equivalent circuit of USSC

X_{sc} and X_{sh} represents the reactance of the transformers associated with the series and shunt voltage source inverters, respectively. Therefore, voltage equation of series and shunt inverters can be expressed as follows:

$$V_s = -V_{dq} + I_{se}(jX_{se}) + V_0 \tag{1}$$

$$V_s + V_{dq} - I_{se}(jX_{se}) = V_{sh} + I_{dq}(X_{sh}) \tag{2}$$

$$I_s = I_{se} = I_{dq} + I_L = \frac{V_{sh} - V_0}{X_{sh}} + I_L \tag{3}$$

Where I_{sc} and I_{dq} are the series and shunt inverter currents, respectively.
The voltage across the distribution line reactance, X_L is:

$$V_X = V_s + V_{dq} - I_{se}(jX_{se}) - V_L =$$
$$V_0 - V_L = X_L . I_L \tag{4}$$

Where, I_L is distribution line current.
The voltage, V_X, across the distribution line can be changed by changing the inserted voltage, V_{dq},

which is in series with the distribution line. If we consider $V_{dq}=0$, the distribution line sending end voltage, V_S, leads the load voltage by an angle δ i.e $\delta_S - \delta_L$.

The resulting real and reactive power flows at the load side are P and Q, which are given as follows:

$$P_{ussc} = \frac{V_0.V_L}{X_L}\sin\delta \tag{5}$$

$$Q = \frac{V_0.V_L}{X_L}(1 - \cos\delta) \tag{6}$$

With an injection of V_{dq}, the distribution line voltage V_0 will lead the load voltage V_L, and $\delta_0 > \delta_L$, thus the resulting line current and amount of flow Will be changed. With a larger amount of V_{dq} injection, V_0 now lags the load voltage V_L, and $\delta_0 < \delta_L$.

Consequently, the line current and power flow will be reversed.

6. CONTROL STRATEGY OF VOLTAGE SAG MITIGATION

Series converter provides the main function the USSC by injecting a voltage Vdq with controllable magnitude Vdq and phase angle δ_{se} in series with the line via an insertion transformer. This injected voltage acts essentially as a synchronous ac voltage source. The feeder current flows through this voltage source resulting in reactive and real power exchange between it and its ac system. The reactive power exchanged at the ac terminal (ie. at the terminal of series injection transformer) is generated internally by the converter. The real power exchanged at the ac terminal is converted into dc power, which appears at the dc link as a positive or negative real power demand.

According to the theoretical concepts, the rotation of series voltage phasor Vdq with angle δ_{se} cause variation of both the transmitted real power 'P' and the reactive power 'Q' with δ_{se} in a sinusoidal manner. For validating the proposed circuit model of USSC, the magnitude of series injected voltage is kept constant at 2KV and its angle is varied from 0o to 360o. The variation in real and reactive power is investigated and it is observed that the variation of real and reactive power is sinusoidal with variation in angle, thus coinciding with theoretical concepts. It can be seen that the transmitted real power is maximum at angle 90o, minimum at angle 270o and medium at angle 0o. Hence, these values are selected in the switching function. The target of damping control is to conduct proper switching of C0, C1 or C2 at strategic times as to quickly mitigate voltage sag.

The output of series converter can be bang-bang controlled to three different values:

$$\text{Vdq} = \begin{cases} |V|\angle 0 & When \quad switch \quad C0 \quad is \quad closed. \\ |V|\angle 90 & When \quad switch \quad C1 \quad is \quad closed. \\ |V|\angle 270 & When \quad switch \quad C2 \quad is \quad closed. \end{cases} \tag{7}$$

Where Vdq is the voltage injected by the USSC; is the maximum magnitude of voltage that can be injected by the USSC.

Fuzzy logic controller is an intelligent technique which has been implanted in the control of facts devices on power system. Mridul Jha. and S.P. Dubey in [23] investigated the Neuro-Fuzzy based controller for a three phase four wire shunt active power filter . Also some authors have utilized the fuzzy approach in the control of renewable energies. By [24] the implementation of fuzzy logic controller in photovoltaic power generation using boost converter and boost inverter has been analyzed. The ultimate objective of this work is to implement fuzzy logic controller at the line in which USSC is connected. The inputs to fuzzy logic controller are V and δ measured at USSC terminals. For the output, the fuzzy logic controller will choose one of the three switch states from C0, C1 and C2 through competition. A simple fuzzy logic scheme comprises three functioning blocks, namely fuzzification, implication and inference, and selection of control. Input data are processed through these three blocks sequentially.

Fuzzification: Crisp input data need to be converted into membership grades to which they belong to each of the associated linguistic levels. These levels are represented by fuzzy sets. Fuzzification serves as data preprocessor for implications of linguistic rules in a later stage. There are 10 distinct linguistic levels, namely A1-10, for input V and 5 distinct linguistic levels, namely B1-5, for δ. Membership functions for the corresponding fuzzy sets are distinct and triangular. A heuristic trial-and-error procedure is needed to find the appropriate fuzzy partitioning by comparing the present and desired response for fuzzy logic control.

Implication and inferencing: Various fuzzified inputs are fed into a fuzzy rule base for implication and inferencing. Linguistic control rules are constructed based on observations of dynamic behaviors and switching curves.

With the use of two state inputs (V and δ), we obtain a two-dimensional rule base with 10x5 linguistic levels as in Table 2.

Table 2. Two-dimensional fuzzy control rules

δ	A_1	A_2	A_3	A_4	A_5	A_6	A_7	A_8	A_9	A_{10}
B_1	C_1	C_1	C_1	C_1	C_1	C_1	C_1	C_1	C_1	C_1
B_2	C_1	C_1	C_1	C_1	C_2	C_1	C_1	C_1	C_1	C_1
B_3	C_1	C_1	C_1	C_2	C_2	C_2	C_1	C_1	C_1	C_1
B_4	C_1	C_1	C_2	C_2	C_2	C_2	C_0	C_0	C_1	C_1
B_5	C_2	C_2	C_2	C_2	C_2	C_2	C_0	C_0	C_1	C_1

The rule base is a collection of fuzzy conditional statements in the form of 'if-then' rules. Each rule carries a weight α_i (called firing strength), which is a measure of the contribution of ith rule to the overall fuzzy control action. The firing strength α_i is defined as:

$$\alpha_i = \mu_A(x_0) \Lambda \mu_V(y_0)$$

(8)

Where $A \in V, B \in \delta$ A; μ denotes grade of membership defined for input state (V and δ), xo and yo are the input variables used at a particular time instant; and Λ is the fuzzy 'AND' operator.

The membership value of each possible switching state C0, C1 and C2 for the FLC is obtained as:

$$\mu_i(C_0) = \frac{\sum \alpha_i}{4} \quad i = 40,41,50,51$$

(9)

$$\mu_i(C_1) = \frac{\sum \alpha_i}{32} \quad i = 1,2,3,...$$

(10)

$$\mu_i(C_2) = \frac{\sum \alpha_i}{14} \quad i = 15,24,25,26,...$$

(11)

The main purpose of selection of control is to choose a non-fuzzy discrete control that best responds to current system oscillations. The final discrete FLC output indicates the final switching state chosen from C0, C1 and C2. The choice is competitive and only one switching state with highest membership μi among C0, C1 and C2 is chosen.

7. SIMULATION AND RESULT

The single line diagram of the network to study the voltage sag mitigation is shown in Figure 6.

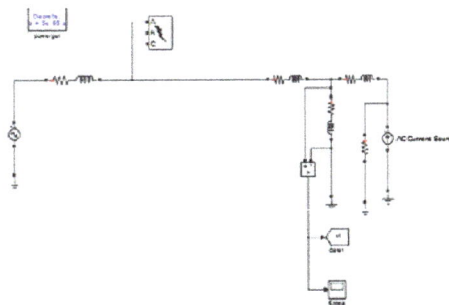

Figure 6. Single line diagram of the network to study the voltage sag mitigation

Voltage sag at PCC without USSC due to short circuit fault is shown in Figure 7.

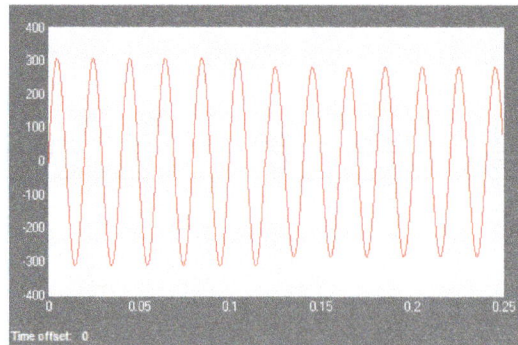

Figure 7. Voltage sag at PCC due to short circuit fault

The simulated system with MATLAB/SIMULINK software to study the fuzzy bang-bang controller on voltage sag mitigation using USSC is shown in Figure 8.

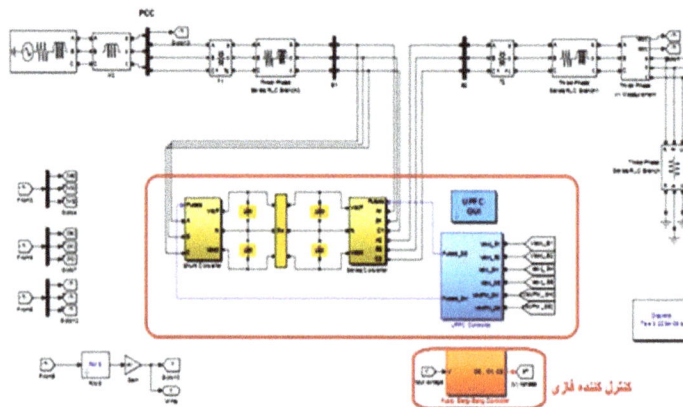

Figure 8. Simulated system in MATLAB/SIMULINK

The control structure of USSC used to illustrate the proposed fuzzy bang-bang controller is shown in Figure 9.

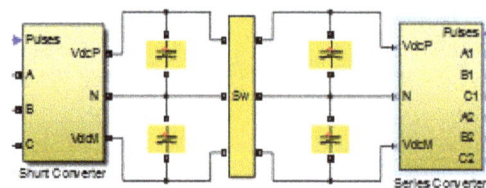

Figure 9. Series and shunt converters of USSC in MATLAB/SIMULINK

The control structure of USSC used to illustrate the proposed fuzzy bang-bang controller is shown in Figure 10. Te shunt converter can be controlled for maintaining constant voltage in dc bus and so it is controlled only to maintain dc bus voltage at *th* desired level.

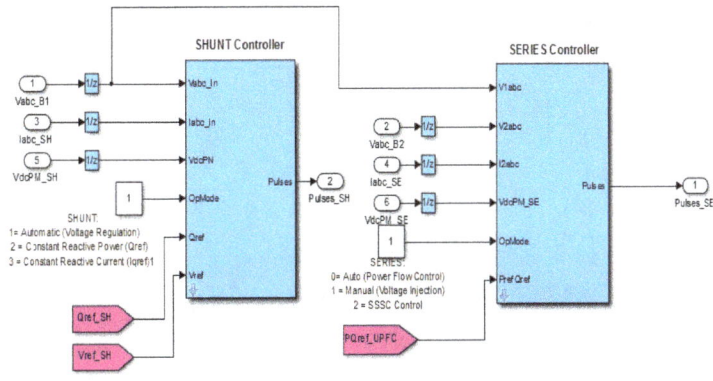

Figure 10. Fuzzy bang-bang controller designed for USSC

Changing state of switches C0, C1 or C2 as shown in Figure 11 can regulate the voltage injected by the series controller.

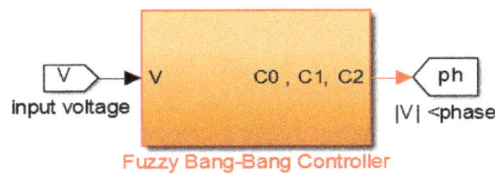

Figure 11. Changing state of switches C0, C1 or C2

The block diagram of the system control for reference voltage generation is shown in Figure 12.

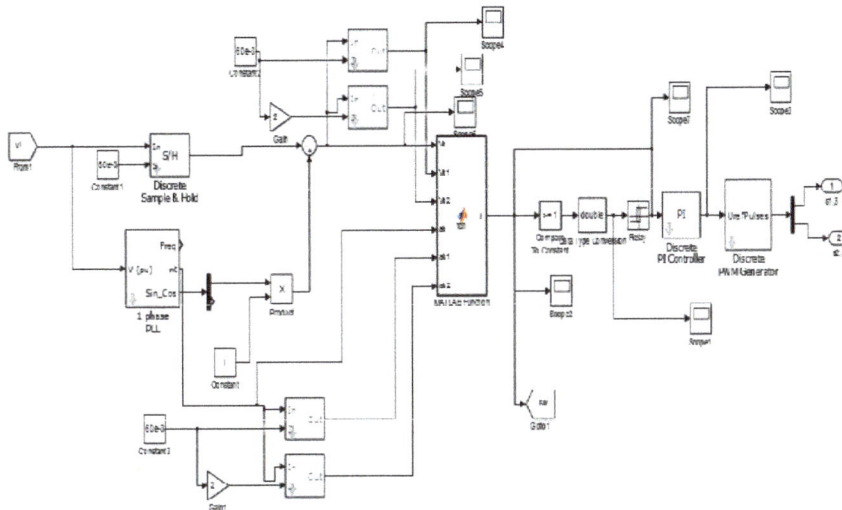

Figure 12. System control for reference voltage generation

The inhected voltage by USSC through series converter and its refrence is presented in Figure 13 and Figure 14. Figure 15 ahows the voltage sag compensated by USSC using fuzzy bang-bang based controller.

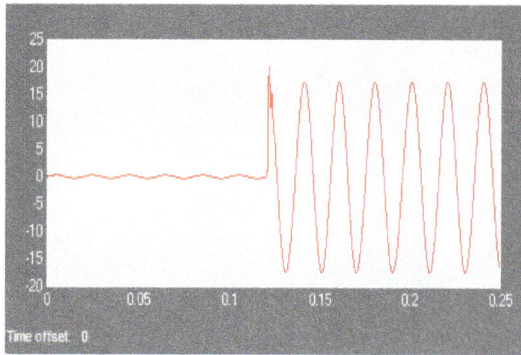

Figure 13. Injected volatge by series converter of USSC to mitigate volatge sag

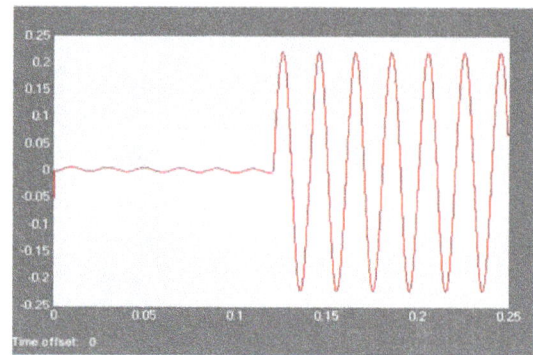

Figure 14. Reference of injected volatge by series converter of USSC to mitigate volatge sag

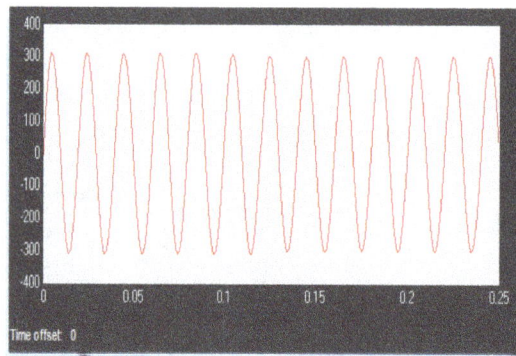

Figure 15. Voltage sag compensated by USSC using fuzzy bang-bang based controller

8. CONCLUSION

In this paper, USSC controller is derived by using Fuzzy Logic Control (FLC) based on bang-bang control to compensate the voltage sag occurred due to short circuit fault in distribution system. Of course another main reason of voltage sag is motor stating which has not been analyzed in this paper. The model is simulated in MATLAB/SIMULINK platform and USSC controller's performance is evaluated. Numerical simulation proved the effectiveness of the controller in compensating voltage sag. Simulations have been carried out to evaluate the performance of the USSC. Simulation results revealed that the USSC can mitigate effectively voltage sag. The results revealed that the USSC gives a better performance in power quality mitigation especially in voltage sag compensation and power flow control and also provide more power quality solutions as compared to the D-STATCOM and DVR.

REFERENCES

[1] PT Nguyen, TK Saha. DVR against balanced and unbalanced Voltage sags: Modeling and simulation. *IEEE-School of Information Technology and Electrical Engineering, University of Queesland*, Australia, 2004.

[2] SR Mendis, MT Bishop, JF Witte. Investigations of voltage flicker in electric arc furnace power systems. *IEEE Industry Applications Magazine*. 1996; 2(1): 28 – 34.

[3] RW Boom. Superconductive Magnetic Energy Storage for Electric Utilities--A review of the 20 year Wisconsin Program. *Proceedings of the International Power Sources Symposium*. 1991; 1: 1-4.

[4] AliZa'fari. Mitigation of Flicker using STATCOM with Three-Level 12-pulse Voltage Source Inverter, *World Academy of Science Engineering and Technology*. 2011; 73: 263-268.

[5] VJ Gosbell, D Robinson, S Perera, The Analysis of Utility Voltage Sag Data, *International Power Quality Conference*, Singapore. 2002.

[6] Yop Chung, Dong-Jun Won, Sang-Young Park, Seung-Il Moon andJong-Keun Park. The DC link energy control method in dynamic voltage restorer system. *International journal on Electric power and energy system*. 2003; 25(7): 525-531.

[7] T Larsson, C Poumarede. STATCOM, an efficient means for flicker mitigation. *IEEE Power Engineering Society Winter Meeting*. 1999; 2: 1208-1213.

[8]   Dinavahi V, Iravani R, Bonert R. Design of a real-time digital Simulator for a D-STATCOM system; *Industrial Electronics, IEEE Transactions* on. 2004; 51(5): 1001-1008.

[9]   NH Woodley, L Morgan, A Sundaram. Experience with an inverter-based dynamic voltage restorer. *IEEE Transactions on Power Delivery*. 1999; 14(3): 1181-1186.

[10] S Chandrasekhar, et al., Mitigation of Voltage flicker and reduction in THD by using STATCOM, *International Journal of Electrical and Computer Engineering (IJECE)*. 2013; 3(1): 102-108.

[11] JG Nielsen, F Blaabjerg, N Mohan. Control Strategies for Dynamic Voltage Restorer compensating Voltage Sags with Phase Jump. *Proceedings of 16th Annual IEEE Applied Power Electronics Conference and Exposition 2001*, APEC 2001; 2: 1267-1273.

[12] MA Hannan, A Mohamed. Unified Series-Shunt Compensator Modeling and Simulation, *IEEE, National Power & Energy Conference (PECon) 2004 Proceedings*, Kuala Lumpur, Malaysia.

[13] S Asha Kiranmai, M Manjula, AVRS Sarma. Mitigation of Various Power Quality Problems Using Unified Series Shunt Compensator in PSCAD/EMTDC, *16th National power systems conference*. 2010.

[14] M A Hannan, A Mohamed, A Hussain. Dynamic Phasor Modeling and EMT Simulation of USSC, *Proceedings of the World Congress on Engineering and Computer Science*, WCECS. San Francisco, USA. 2009; 1.

[15] Arnez RL, LC Zanetta. Unified power flow controller (UPFC): Its versatility in handling power flow and interaction with the network. *IIEEE/PES Asia Pacific Transmission and Distribution Conference and Exhibition*. 2002; 1338-1343.

[16] Hingorani NG, L Gyugyi. Understanding FACTS Concept and Technology of Flexible AC Transmission System. *IEEE Press*, New York. 2000.

[17] Su C, G. Joos. Series and shunt active power conditioners for compensating distribution system faults. *Procedding of the Canadian Conference on Electrical and Computer Engineering*. 2000; 1182-1186.

[18] Jin Nan, Tang Hou Jun, Yao Chen, Wu Pan. Topology and Control of Chopper Type Dynamic Voltage Regulator, *International Review of Electrical Engineering (IREE)*, February, 2011; 6(2part A): 160-168.

[19] Hendri Masdi, Norman Mariun, S. Bashi, Azah Mohamed. Voltage Sag Compensation in Distribution System due to SLG Fault Using D-STATCOM, *International Review of Electrical Engineering (IREE)*. 2010; 5(6,part B): 2836-2845.

[20] MA Hannan, A Mohamed, A Hussain, Majid al Dabbay. Development of the Unified Series-Shunt Compensator for Power Quality Mitigation, *American Journal of Applied Sciences*. 2009; 6(5): 978-986.

[21] Mridul Jha, SP Dubey. NeuroFuzzy based Controller for a Three Phase Four Wire Shunt Active Power Filter, *International Journal of Power Electronics and Drive Systems (IJPEDS)*. 2011; 1(2): 148-155.

[22] Abubakkar Siddik A, Shangeetha M. Implementation of Fuzzy Logic controller in Photovoltaic Power generation using Boost Converter and Boost Inverter. *International Journal of Power Electronics and Drive Systems (IJPEDS)*. 2012; 2(3): 249-256.

7

Operation and Control of Grid Connected Hybrid AC/DC Microgrid using Various RES

Kodanda Ram R B P U S B, M. Venu Gopala Rao

Department of Electrical and Electronics Engineering, K L University, Guntur, AP, INDIA

Keyword:	ABSTRACT

Keyword:

Fuel Cells
Hybrid Power Systems
Interfacing Converter
Micro grids
Solar Energy
Wind Energy

This paper proposes a Hybrid AC/DC Microgrid in alliance with Photo Voltaic (PV) energy, Wind Energy and Proton Exchange Membrane (PEM) Fuel cells. Microgrids are becoming increasingly attractive to the researchers because of the less greenhouse gases, low running cost, and flexibility to operate in connection with utility grid. The Hybrid AC/DC Microgrid constitutes independent AC and DC subgrids, where all the corresponding sources and loads are connected to their respective buses and these buses are interfaced using an interfacing converter. The Hybrid AC/DC Microgrid increases system efficiency by reducing the multiple reverse conversions involved in conventional RES integration to grid. A Small Hybrid AC/DC Microgrid in grid connected mode was modeled and simulated in MATLAB-SIMULINK environment. The simulation results prove the stable operation considering the uncertainty of generations and loads.

Corresponding Author:

Kodanda Ram R B P U S B,
Department of Electrical and Electronics Engineering,
K L University, Vaddeswaram
Guntur District, Andhra Pradesh, INDIA – 521456
Email: balajir1986@kluniversity.in

1.   INTRODUCTION

A Microgrid is a small grid formed by banking multiple energy resources and loads to enhance overall reliability and independent advantages. Now-a-days, it is more preferred to integrate renewable energy resources to Microgrid to lessen the CO_2 emission and fossil fuel consumption. The banked Microgrid can be operated either in connection to main grid or operated like isolated "islanded" [1]. Now-a-days, DC loads like LED's, Electric Vehicles and other Electronic Gadgets are being increasingly used due to their inherent advantages. Three Phase AC Power systems have existed for over 100 years due to their efficient transformation at different voltage levels and transmission over long distances. The inherent characteristics of rotating machines make it feasible for larger period.

To connect the conventional AC system to the renewable resources, AC Microgrids have been proposed and DC power from the various resources like PV panel, Fuel cells etc., are converted into AC in order to connect to an AC grid, which are implanted by AC/DC Converters and DC/DC Converters [2]. In an AC Grid, several converters are required for various home and office facilities to provide required DC voltages. AC/DC/AC converters are commonly used as drives in order to control the speed of AC motors in industries.

Recently DC grids are resurging due to development and deployment of renewable DC resources and their inherent advantage for DC loads in residential, commercial and industrial applications. The DC Microgrid has been proposed [3]. However, for conventional AC loads DC/AC inverters are required and AC sources are connected using AC/DC Converters.

Multiple reverse conversions required in individual AC or DC grid may add additional loss to the system operation and will make the current home and office appliances more complicated in design and operation [4]. The current research in the electric power industry is smart grid. One of the most important futures of a smart grid is the advanced structure which can facilitate the connections of different AC and DC generation systems, energy storage options and various AC and DC loads with the optimal asset utilization and operational efficiency. The power electronics converter plays a most important role to interfacing AC and DC grids, which makes future grid much smarter.

A Hybrid AC/DC Microgrid is proposed to reduce processes of multiple reverse conversions in an individual AC or DC grid and to facilitate the connection of various energy sources, storage devices and loads [5]. The advanced power electronic devices and control techniques are used to harness maximum power from renewable power sources, to minimize power transfer between AC and DC networks. PV system, PEMFC constitutes the DC Energy sources; Wind system constitutes the AC energy source, whereas Battery and Conventional Grid are used as storage devices whenever required.

2.   SYSTEM CONFIGURATION AND MODELING

Figure 1 illustrates the compact representation of proposed Hybrid Microgrid Configuration. The Hybrid Microgrid was formed by a DC sub grid and an AC sub grid. Each sub grid has its own sources elements, storage elements and loads of the same category grouped together so as to reduce the amount of power conversion required. Both sub grids are interfaced using interfacing converters. Interfacing converters are the bidirectional converters, and their major role is to provide bidirectional energy transfer between the sub grids, depending on the prevailing internal supply – demand conditions.

The formed Hybrid grid can be tied to Utility grid using an Intelligent Transfer Switch at point of common coupling as in conventional AC grids. In grid tied mode of operation, surplus energy in the internal sub grids if any can be injected to the utility grid without violating the local utility rules. Similarly, the shortfall in both the sub grids if any can be absorbed from the utility grid.

Figure 1. A Compact representation of the proposed Hybrid Microgrid

2.1. Proposed Hybrid Microgrid Configuration

PV Array(40kW) and PEM Fuel Cell (50kW) are connected to DC bus through independent DC/DC boost converter to simulate DC sources. Capacitors C_{pv} and C_{fc} are used to suppress the high frequency ripples of the PV and FC output voltage.

Also, a wind turbine generator (WTG) with DFIG (50kW) and utility grid are connected to AC bus to simulate AC Sources. In addition, a battery (65Ah) and super capacitor (0.5F) are individually connected as energy storages to DC bus through buck-boost (DC/DC) converter. The DC load was considered as pure resistive load and AC loads are considered with RLC which are dynamic in nature. Both the loads are variable between 20kW – 40kW. Rated voltages for both buses are considered as 400V. The parameters of the Hybrid Microgrid are tabulated in Table 3

2.2. Modeling of PV Panel

Figure 2 shows the equivalent circuit of a PV Panel modeled by a controlled current source. I_{pv} and V_{pv} are the terminal current and voltage of the PV panel, respectively. The current output of the panel is modeled using three Equation (1), (2), (3) [6]-[7]. The parameters that were taken into consideration for simulation are shown in Table 1.

$$I_{pv} = n_p I_{ph} - n_p I_{sat} \times \left[exp\left(\left(\frac{q}{Akt}\right)\left(\frac{V_{pv}}{n_s} + I_{pv}R_s\right)\right) - 1 \right] \tag{1}$$

$$I_{pv} = \left(I_{sso} + K_i(T - T_r)\right) . \frac{S}{100} \tag{2}$$

$$I_{sat} = I_{rr}\left(\frac{T}{T_r}\right)^3 exp\left(\left(\frac{qE_{gap}}{kA}\right) . \left(\frac{1}{T_r} - \frac{1}{T}\right)\right) \tag{3}$$

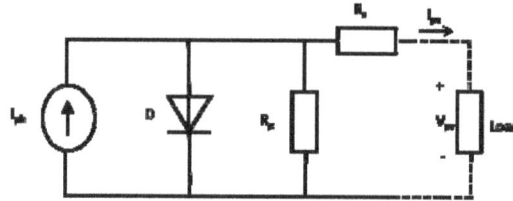

Figure 2. Equivalent circuit of a PV Panel

Table 1. Parameters of Photovoltaic Panel

Symbol	Description	Value
V_{oc}	Rated open circuit voltage	403 V
I_{ph}	Photocurrent	
I_{sat}	Module reverse saturation current	
Q	Electron charge	1.602 x 10^{-19} C
A	Ideality factor	1.50
K	Boltzmann Constant	1.38 x 10^{-23} J/K
R_s	Series resistance of a PV cell	
R_p	Parallel resistance of a PV cell	
I_{sso}	Short-circuit current	3.27 A
k_i	SC Current temperature Coefficient	1.7 e^{-3}
T_r	Reference Temperature	301.18 K
I_{rr}	Reverse Saturation current at T_r	2.0793e^{-6} A
E_{gap}	Energy of the band gap for silicon	1.1 eV
n_p	Number of cells in parallel	40
n_s	Number of cells in series	900
S	Solar Irradiation Level	0 ~ 1000 W/m
T	Surface temperature of the PV	

2.3. Modeling of Fuel Cell

Figure 3 shows the equivalent circuit of PEM Fuel cell. The ohmic, activation and concentration resistances are represented with R_{ohmic}, R_{act}, R_{conc} respectively. C is the membrane capacitance. The Membrane voltage equation is given by Equation (4).

$$V_c = \left(1 - \frac{dV_c}{dt}\right)\left(R_{act} + R_{conc}\right) \tag{4}$$

The output voltage of the PEMFC is given by (5):

$$V_{fc} = E - V_c - V_{act} - V_{ohmic} \tag{5}$$

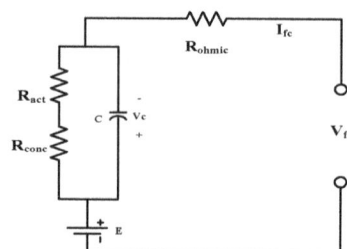

Figure 3. Equivalent circuit of PEM Fuel cell

2.4. Modeling of Battery

Battery is not very important in grid-tied mode. But, provides an energy storage in DC subgrid, which can reducethe multiple reverse conversion, whenever required. In emergency i.e., Grid Failed Condition, they plays a vital role in power balance and voltage stability. The battery was modeled using a controlled nonlinear source in series with a constant resistance. The State Of Charge (SOC) of the battery is given by Equation (6).

$$SOC\% = 100\left(1 + \frac{\int it\,dt}{Q}\right) \tag{6}$$

Where it is the extracted capacity and Q is the Maximum capacity if battery.

2.5. Modeling of Wind Turbine Generator with DFIG

In this paper, DFIG was considered as a wound rotor induction machine. The power output P_m from a WT is determined by [3]. A 50kW DFIG parameters, used in this paper are shown in Table 2.

$$P_m = 0.5\rho A C_p(\lambda, \beta)V_w^3 \tag{7}$$

Table 2. Parameters of DFIG

Symbol	Description	Value
P_{nom}	Nominal power	50 kW
V_{nom}	Nominal Voltage	400 V
R_s	Stator resistance	0.00706 pu
L_s	Stator inductance	0.171 pu
R_r	Rotor resistance	0.005 pu
L_r	Rotor inductance	0.156 pu
L_m	Mutual inductance	2.9 pu
J	Rotor inertia constant	3.1 s
n_p	Number of poles	6
$V_{dc\ nom}$	Nominal DC voltage of AC/DC/AC converter	800 V
P_m	Nominal Mechanical power	45 W

3. CONTROLLERS

The Hybrid Microgrid contains six types of converters. All the converters have to be coordinately controlled with the utility grid to supply reliable, high efficiency, high quality power for variable DC and AC loads. The controllers are presented in this section are coordinated successfully in both grid-tied. A Direct Torque Control Strategy(DTC) with feed forward voltage compensation is selected for DFIG control system [9].

3.1. Boost Converter

In grid tied mode, the control objective of the boost converter is to track the MPPT of the PV panel and Fuel Cell. The PV Panel and Fuel Cell boost converters are designed to support the DC bus voltage as shown in Figure 4. To achieve maximum power, P & O Method proposed in [6].

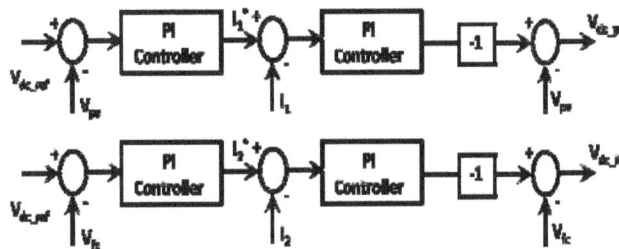

Figure 4. Control Scheme of PV Cell and PEM Fuel cell

3.2. Control of Battery

Battery has high energy density with slow charging and discharging speeds. Control scheme of Battery is shown in Figure 5.

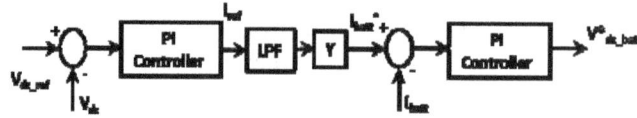

Figure 5. Control Scheme for Battery

3.3. Interfacing Converter

The objective of the interfacing converter is to interface both the sub grids i.e., AC grid and DC grid. The major role of the interfacing converter is to exchange power between the AC bus and DC bus. When operating in grid tied mode, the converter supplies given active and reactive power. The interfacing converter acts DC/AC inverter when supplying power from DC grid to AC grid and acts as AC/DC rectifier when supplying power from AC grid to DC grid whenever required. The interfacing converter works based on droop control [11]. The control scheme of interfacing converter is shown in Figure 6.

The advantages of interfacing converter cannot be realized by just relying on the droop controlled sources. The interlinking control challenges has to be carefully addressed [12].

   a) Unlike unidirectional sources, the interlinking converters has to manage bidirectional active and reactive power flows between sub grids.
   b) At any one instant, the interlinking converters have two roles to play. They appear as load to one sub grid where energy is absorbed and appear as source to other grid where energy is injected.

Figure 6. Control Scheme for Interfacing Converter

Table 3. Parameters for the Hybrid Grid

Symbol	Description	Value
C_{pv}	Capacitor across the solar panel	110 µF
L_1	Inductor for the boost converter	2.5 mH
C_d	Capacitor across the dc-link	4700 µF
L_2	Filtering inductor for the inverter	0.43 mH
R_2	Equivalent resistance of the inverter	0.3 ohm
C_2	Filtering capacitor for the inverter	60 µF
L_3	Inductor for the battery converter	3 mH
R_3	Resistance of L3	0.1 ohm
F	Frequency of AC grid	60 Hz
f_s	Switching frequency of power converters	10 kHz
V_d	Rated DC bus voltage	400 V
$V_{ll\ rms}$	Rated AC bus line voltage (rms value)	400 V
n1/n2	Ratio of the transformer	2:1
C	Capacity of Super Capacitor	0.5 F

4. SIMULATION RESULTS

The operation of Grid Connected Hybrid AC/DC Microgrid under various source and load conditions are simulated to verify the reliability.

DC RES power is supplied directly to the DC loads and AC RES power is supplied directly to AC loads. Power is balanced directly by the utility grid on AC bus and on DC bus through interfacing converter. The battery is assumed to be fully charged and operated in rest mode. DC bus voltage is controlled and maintained by utility grid through interfacing converter. AC bus voltage is directly maintained by utility grid.

The terminal voltage for change in solar irradiation is shown in Figure 7. Optimal terminal voltage of PV panel is obtained by using the standard P&O algorithm. The solar irradiance was set as 400W/m^2 from 0.0s to 0.1s, later it was linearly increased to 1000W/m^2 until 0.2s, kept constant to 0.3s, decreased to 400W/m^2 by 0.4s and keeps that value until final time 0.5s. The slow tracing speed of the standard P&O algorithm is optimized by using fuel cell in DC subgrid.

Figure 8 and Figure 9 shows the curves of the PV panel power output and solar irradiation respectively. The power output varies from 4.85kW to 13.5kW, which closely follows the solar irradiation curve assuming the fixed ambient temperature.

Figure 10 shows the voltage and current responses on AC side of interfacing converter with a fixed DC load of 20kW. It was observed that the current direction of interfacing converter was reversed before 0.3s and after 0.4s.

Figure 11 shows the voltage and current responses on AC side of interfacing converter with variable DC load from 20kW to 40kW at 0.25s with fixed solar irradiation at 750W/m^2. It can be seen that current direction was reversed at 0.25s.

Figure 12 shows the voltage response at DC bus of interfacing converter with Fuel cell shows an improved transient response when compared Figure 13 without fuel cell under same conditions

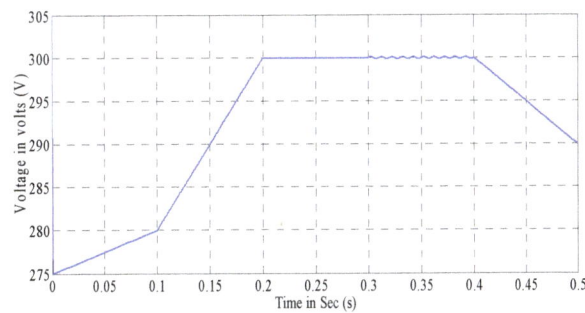

Figure 7. Terminal Voltage of PV Panel

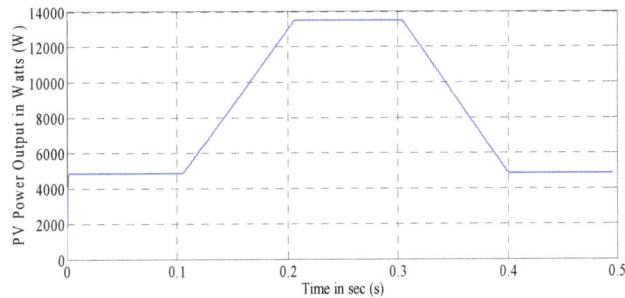

Figure 8. Power output of PV Panel

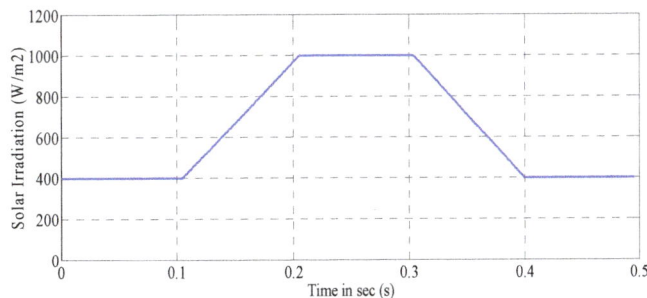

Figure 9. Solar Irradiation

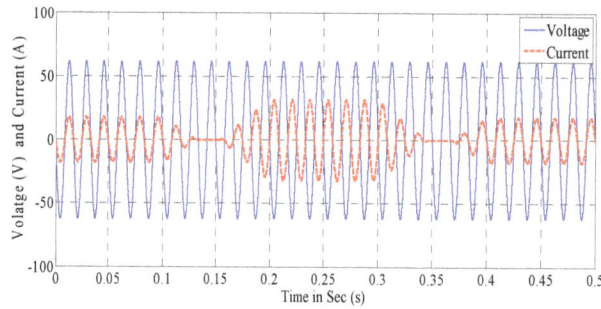

Figure 10. AC side Voltage and Current of the Interfacing Converter with Variable Solar Irradiation and Constant DC Load

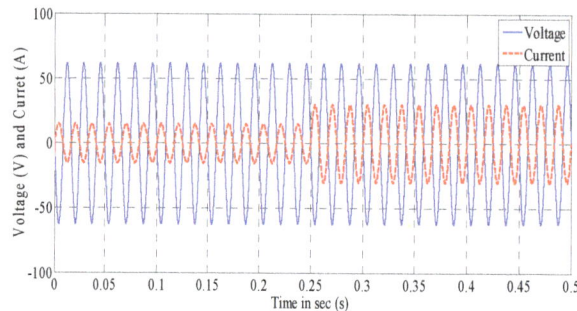

Figure 11. AC side Voltage and Current of the Interfacing Converter with Constant Solar Irradiation and Variable DC Load

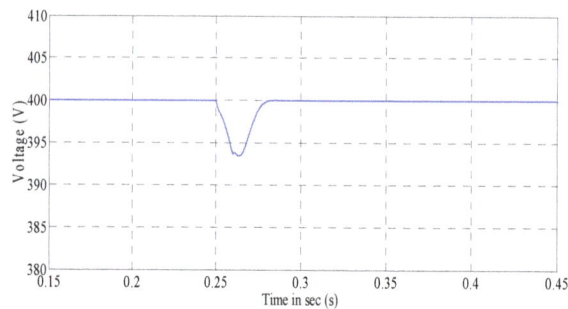

Figure 12. DC Bus Transient Response with Fuel Cell

5. CONCLUSION

A Hybrid AC/DC Microgrid is proposed and comprehensively studied in this paper. The control strategies are concisely stated to maintain stable system operation under various load and resource conditions. The control strategies are verified by using MATLAB/Simulink. Various control methods are incorporated to harness the maximum power from RES during grid connected mode and resembles stable operation.

The Interfacing Converter shows stable operation during load variations. However there will be some practical limitations, because of fast and continuous load variations. Even-though, the proposed Hybrid grid reduces the processes of DC/AC and AC/DC conversions in an individual sub grids, the theory is still challenging in the AC dominated infrastructure. The Hybrid AC/DC Microgrid has to be tested for various faults on subgrids and their effects on the other grid. The Hybrid AC/DC Microgrid is only feasible for new construction either in remote location or industries.

REFERENCES

[1] PC Loh, Ding Li, Yi Kang Chai, Frede Blaabjerg. Autonomous Control of Interlinking Converter with Energy Storage in Hybrid AC-DC Microgrid. *IEEE Trans. Industry Applications*. 2013; 49(03): 1374-1382.

[2] RH Lasseter. MicroGrids. *Proc. IEEE Power Eng. Soc. Winter meet.* 2002; 1: 305–308.

[3] ME Baran, NR Mahajan. DC Distribution for Industrial Systems: Opportunities and Challenges. *IEEE Trans. Industry Applications*. 2003; 39(06): 1596-1601.

[4] Peng Wang, X Liu, Chi Jin, PC Loh, FH Choo. A Hybrid AC/DC Micro-grid Architecture, Operation and Control. *Proc. IEEE Power and Energy Society General Meeting*. 2011: 1-7.

[5] X Liu, Peng Wang, PC Loh. A Hybrid AC/DC Microgrid and Its Coordination Control. *IEEE Trans. Smart Grid.*, 2011; 02(02): 278-286.

[6] Michael M, S Gonzalez. Development of a MATLAB/Simulink Model of a Single-Phase Grid-Connected photovoltaic System. *IEEE Trans. Energy Conversion*. 2009; 24(01): 195-202.

[7] KH Chao, CJ Li, SH Ho. Modeling and fault simulation of Photovoltaic generation systems using circuit-based model. *Proc. IEEE Int. Conf. Sustainable Energy Technol.*, 2008: 284-289.

[8] M Akbari, MA Golkar, SMM Tafreshi. Voltage Control of a Hybrid AC/DC Microgrid in stand-Alone Operation Mode. *Proc. IEEE PES innovative Smart Grid Technologies*. 363-367.

[9] X Liu, Peng Wang, PC Loh. A Hybrid AC/DC Micro-grid. *Proc. IEEE IPEC*. 2010: 746-751.

[10] M Akbari, MA Golkar, SMM Tafreshi. Voltage Control of a Hybrid AC/DC Microgrid in Grid Connected Operation Mode. *Proc. IEEE PES innovative Smart Grid Technologies*. 358-362.

[11] Chi Jin, P C Loh, Peng Wang, Yang Mi, F Blaabjerg. Autonomous Operation of Hybrid AC-DC Microgrids. *Proc. IEEE ICSET*. 2010; 1-7.

[12] PC Loh, Ding Li, YK Chai, F Blaabjerg. Autonomous Operation of Hybrid Microgrid with AC and DC Subgrids," *IEEE Trans. Power Electronics*. 2013; 28(05): 2214-2223.

8

Improvement of Power Quality using Fuzzy Logic Controller in Grid Connected Photovoltaic Cell using UPQC

K. Ramalingeswara Rao, K. S. Srikanth

Departement of Electrical and ElectronicsEngineering, K L University

ABSTRACT

Keyword:

Fuzzy logic controller
Harmonics reactive power
Active power
PI controller
Total harmonics distortion
Unified power quality
controller

In this paper, the design of combined operation of UPQC and PV-ARRAY is designed. The proposed system is composed of series and shunt inverters connected back to back by a dc-link to which pv-array is connected. This system is able to compensate voltage and current related problems both in inter-connected mode and islanding mode by injecting active power to grid. The fundamental aspect is that the power electronic devices (PE) and sensitive equipments (SE) are normally designed to work in non-polluted power system, so they would suffer from malfunctions when supply voltage is not pure sinusoidal. Thus this proposed operating strategy with flexible operation mode improves the power quality of the grid system combining photovoltaic array with a control of unified power quality conditioner. Pulse Width Modulation (PWM) is used in both three phase four leg inverters. A Proportional Integral (PI) and Fuzzy Logic Controllers are used for power quality improvement by reducing the distortions in the output power. The simulated results were compared among the two controller's strategies With pi controller and fuzzy logic controller

Corresponding Author:

K.S. Srikanth,
Departement of Electrical and Electronics Engineering,
KL University,
Greenfields, Vaddeswaram, Guntur Ditrict, Andhra Pradesh, India.
Email: srikanth.dsd@gmail.com

1. INTRODUCTION

One of the important aspects is that, power electronic devices and sensitive equipments are designed to work in non-polluted power systems. So, they would suffer from malfunctions when the supply voltage is not pure sinusoidal. As these devices are the most important cause of harmonics, inter harmonics, notches and neutral currents, the power quality should be improved. The solution to PQ problem can be achieved by adding auxiliary individual device with energy storage at its dc-link by PV-array. This auxiliary equipment has the general name of power conditioners and is mainly characterized by the amount of stored energy or stand alone supply time. That auxiliary equipment having both "shunt" and "series" inverter connected back to back by a dc-link is called the "unified power quality conditioner" (UPQC) [1]. Renewable energy resource that is Photo voltaic with UPQC is greatly studied by several researchers as a basic device to control the power quality. The work of UPQC is reducing perturbations which affect on the operation of sensitive loads [2].

UPQC is able to reduce voltage sag, swell, voltage and current harmonics using shunt and series inverters. In spite of this issue, UPQC is able to compensate voltage interruption and active power injection to grid because in its dc-link there is energy storage known as distributed generating (DG) source. The attention to distributed generating (DG) sources is increasing day by day. The important reason is that roll they will likely play in the future of power systems .Recently, several studies are accomplished in the field of connecting DGs to grid using power electronic converters. Here, grid's interface shunt inverters are considered more where the reason is low sensitiveness of DGs to grid's parameters and DG power transferring facility using this approach. Although Distributed Generating needs more controls to reduce the problems like grid power quality and reliability, PV energy is one of the distributed generation sources which provides a part of human required energy nowadays and will provide in the future scope [3]. The greatest share of applying this kind of energy in the future will be its usage in interconnected systems. Now a days, so many countries like European has caused interconnected systems development in their countries by choosing supporting policies. In this paper, UPQC and PV combined system has been presented. UPQC introduced in has the ability to compensate voltage swell and sag, harmonics and reactive power.

The UPQC is a combination of series and shunt active filters connected in cascade via a common DC link capacitor. The main purpose of a UPQC is to compensate for supply voltage power quality issues such as, sags, swells, unbalance, harmonics, and for load current power quality problems such as unbalance, harmonics, voltage dips , reactive current and neutral current

2. SYSTEM DESCRIPTION OF UPQC

UPQC has two inverters shunt (or) D-Statcom and series (or) DVR voltage source inverters which are as 3-phase 4-leg. Series inverter stands between source and coupling point by series transformer and Shunt inverter is connected to point of common coupling (PCC) by shunt transformer. Shunt inverter operates as current source and series inverter operates as voltage source.

UPQC is able to reduce current's harmonics, to compensate reactive power, voltage distortions and can compensate voltage interruption because of having PV-array as a source. Common interconnected PV systems structure is as shown in Figure 1 [4]. In this paper a new structure is proposed for UPQC, where PV is connected to DC link in UPQC as energy source [5].

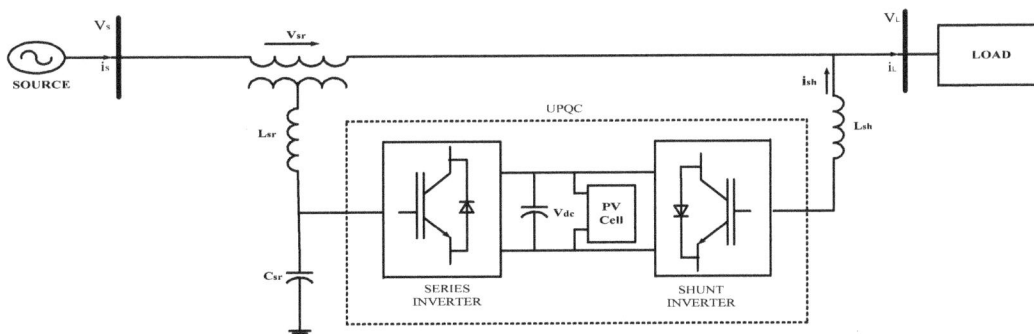

Figure 1. Configuration of proposed UPQC

3. SYSTEM DESIGN

The controlling design of proposed system is composed of two following parts:
a) Series inverter control
b) Shunt inverter control
Controlling strategy is designed and applied for two interconnected and islanding modes. In interconnected mode, source and PV provide the load power together while in islanding mode; only PV transfers the power to the load. By removing voltage interruption, system comes back to interconnected mode.

3.1. Series Inverter Controlling

The duty of the series inverter is to compensate the voltage disturbance in the source side, grid which is due to the fault in the distribution line. Series inverter control calculates the voltage reference values which are injected to grid by series controller. In order to control series controller of UPQC, load sinusoidal voltage control design and implementation strategy is proposed as shown in figure below:

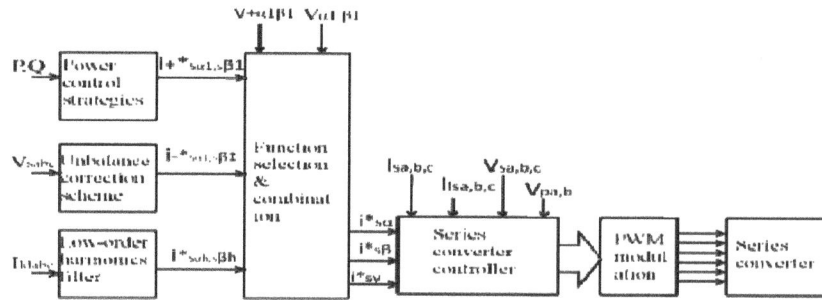

Figure 2. Block diagram of overall control structure with Series converter

The series converter is applicable for achieving multilevel control objectives [6]. Hence, the block "function selection and combination" is shown in Figure 2 is that different types of objectives can be integrated into the system by choosing appropriate reference signals i*sα,i*sβ,i*sγ. Details about the unbalance correction scheme, which is used to generate current reference for negative-sequence voltage compensation. For the power control strategy, which are used to obtain desired currents for active/reactive power transfer [7]. Due to the space limitation, they are not duplicated here. The active filter function is represented by the block "low-order harmonics filter". References is denoted by i*sαh,sβh,sγh can be obtained [8]. In order to track the desired reference signals, the rest of this section presents the main design aspects of the series and parallel converter control

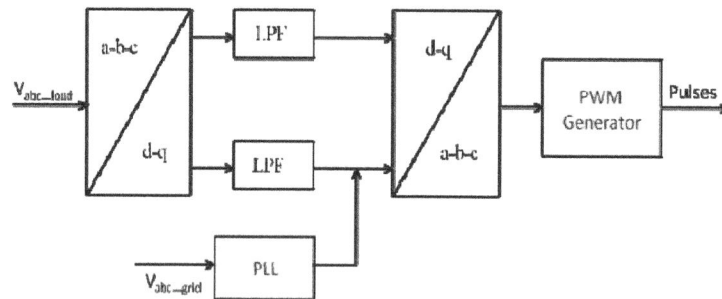

Figure 3. Control block diagram of series converter

3.2. Shunt Inverter Controlling

Shunt inverter undertakes two main operations. First is compensating both current harmonics generated by nonlinear load and reactive power, another is injecting active power generated by Photo voltaic (PV) system. The shunt inverter controlling system should be designed in a way that it would provide the ability of undertaking two above operations. Shunt inverter control calculates the compensation current for current harmonics and reactive Power when PV is out of the grid [7].

The power loss caused by inverter operation should be considered in this calculation. It has the ability of stabilizing DC-link voltage during shunt inverter operation to compensate voltage distortions [6]. The stabilization is maintained by DC-link capacitor voltage controlling loop in which fuzzy logic controller is applied. Shunt inverter control consists of the control circuit as shown in figure below.

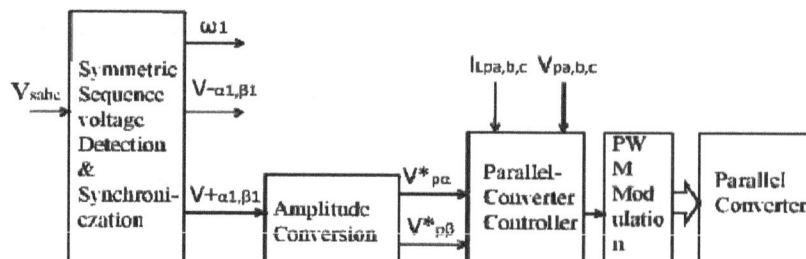

Figure 4. Block diagram of overall control structure with Parallel converter

As shown in Figure 4 , based on the fundamental positive sequence grid voltages (V+α1,V+β1) derived in the stationary frame, the amplitude conversion block first shapes the signals to per-unit quantities and then generates a set of reference signals (V*pα,V*pβ) with a specified amplitude for the parallel converter then give to PWM [4].

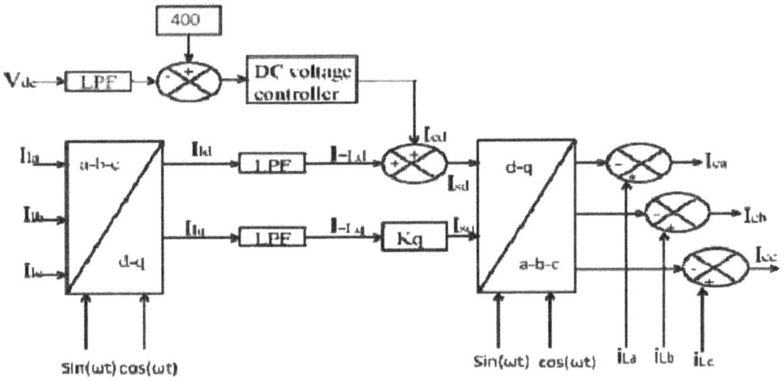

Figure 5. Control diagram of the parallel converter

4. MODELING OF PV MODULE

The PV cell is the basic unit of a photovoltaic module and it is the element in charge of transforming the sun rays or photons directly into electric power.

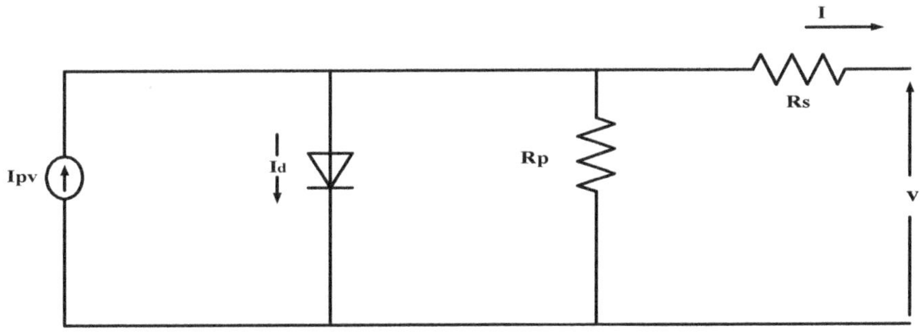

Figure 6. Equivalent circuit of a PV Cell

The equivalent circuit of a practical PV cell is shown in Figure 6. The characteristic equation of a PV cell is the output current produced by it and is expressed as:

$$I = Ipv - Io\left(e^{\left(\dfrac{V + RsI}{Vta}\right)} - 1 \right) - \dfrac{V + RsI}{Rp} \qquad (1)$$

Where, IPV=Current generated by the incident solar radiation
I0=Reverse saturation or leakage current of the diode
Vt =Thermal voltage of PV module with Ns PV cell connected in series
= NsKT/Q
K=Boltzmann constant=1.3806503 x 10-23J/K
Q=Electron Charge=1.60217646 x 10-19 C
T=Temperature in Kelvin
a=Diode ideality constant (1<a<1.5)

PV cells connected in parallel increases the total output current of the PV module where as cells connected in series increases the total output voltage of the cell. The open circuit voltage/temperature coefficient (KV), the short circuit current/temperature coefficient (KI), and the maximum experimental peak output power (Pmax, e).These information are always given at standard test condition i.e. at 1000W/m2 irradiation and 250C temperature. The other information like the light generated diode, saturation current, diode ideality constant, parallel and series resistance which are not noticed in manufacturer datasheet but necessary for the simulation purpose can be evaluated as follows [7].

The current produce by the event solar radiation is depends linearly on the solar irradiation and is also influenced by the temperature according to the following Equation [6].

$$Ipv = \left(Ipv, n + K1\Delta T\right)\frac{G}{Gn} \tag{2}$$

Where,
IPV, n is the light generated current at the nominal condition i.e. at 25oC and 1000W/m2
$T \Delta$ =Actual temperature-Nominal temperature in Kelvin
G=Irradiation on the device surface
Gn=Irradiation at nominal irradiation

The diode saturation current I0 and its addition on the temperature may be expressed as [4]:

$$Io = Io, n\left[\frac{Tn}{T}\right]^3 \exp\left[\frac{qEg}{aK}\left(\frac{1}{Tn} - \frac{1}{T}\right)\right] \tag{3}$$

Where Eg is the band gap energy of the semiconductor and Io, n is the nominal saturation current and is expressed as [3]:

$$Io, n = \frac{Isc, n}{\exp\left(\frac{Voc, n}{aVt, n}\right) - 1} \tag{4}$$

Where Voc, n=Nominal open circuit voltage of the PV module

Lastly the series and parallel resistance of the PV cell can be calculated by any iteration method.

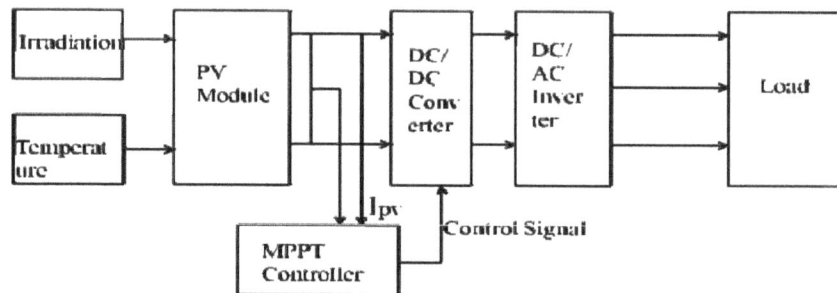

Figure 7. Complete block diagram of PV Module with MPPT Controller

Figure 7 shows the complete block diagram of a PV module with a MPPT controller and feed power to the load through a dc/dc converter. MPPT controller takes the output current and voltage of the PV module as its input and based on the control algorithm it gives appropriate command to the converter to interface the load with the PV module.

5.   MAXIMUM POWER POINT TRACKING

Maximum Power Point tracking controller is basically used to operates the Photovoltaic modules in a manner that allows the load connected with the PV module to extract the maximum power which the PV module capable to produce at a given atmospheric conditions. PV module has a single operating point where the values of the current and voltage of the cell result in a maximum output power. It is a big task to operate a PV module consistently on the maximum power point and for which many MPPT algorithms have been developed [5]. The most popular among the available MPPT techniques is Perturb and Observe (P&O) method. This method is having its own advantages and disadvantages. The aim of the present work is to improve the (P&O). MPPT controller and then the fuzzy control has introduced on it to improve its overall performance.

6. PERTURB & OBSERVE TECHNIQUE (P&O) FOR MAXIMUM POWER POINT TRACKING

Currently the most popular MPPT algorithm is perturb and observe (P&O), where the current/voltage is repeatedly perturbed by a fixed amount in a given direction, and the direction is alternated only the algorithm detects a drop in power. Here if there is an improve in power, the subsequent perturbation should be kept in the same direction to reach the MPP and if there is a decrease in power then the perturbation should be reversed. In the proposed work each perturbation of the controller gives a reference voltage which is compared with the instantaneous PV module output voltage and the error is fed to a PI controller which in turns decides the duty cycle of the DC/DC converter as shown in Figure 8. The process of perturbation is repeated periodically until the MPP is reached. Hence at every instant of PV-array we are determining MPP and correspondingly capacitor-DC links voltage is charged.

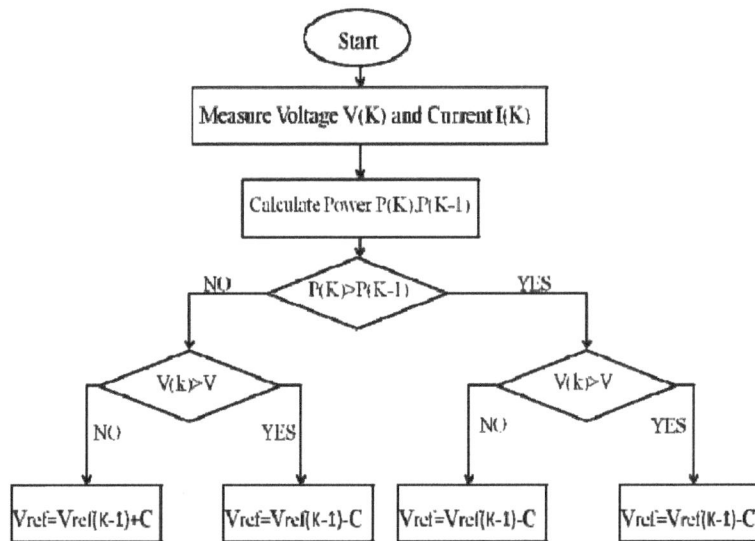

Figure 8. Algorithm for Maximum Power Point tracking by Perturb and Observe method

7. FUZZY LOGIC CONTROLLER

Fuzzy logic control mostly consists of three stages:
a) Fuzzification
b) Rule base
c) Defuzzification

During fuzzification, numerical input variables are converted into linguistic variable based on a membership functions. For these MPP techniques the inputs to fuzzy logic controller are taken as a change in power w.r.t change in current E and change in voltage error C. Once E and C are calculated and converted to the linguistic variables, the fuzzy controller output, which is the duty cycle ratio D of the power converter, can be search for rule base table. The variables assigned to D for the different combinations of E and C is based on the intelligence of the user. Here the rule base is prepared based on P&O algorithm.

In the defuzzification stage, the fuzzy logic controller output is converted from a linguistic variable to a numerical variable still using a membership function. MPPT fuzzy controllers have been shown to perform well under varying atmospheric conditions. However, their influence depends a lot on the intelligence of the user or control engineer in choosing the right error computation and coming up with the rule base table. The comparison for error E and change in code C are given as follows:

$$E = \frac{P(K) - P(K-1)}{I(K) - I(K-1)}$$

(5)

$$C = V(K) - V(K-1)$$

(6)

8. FUZZY CONTROLLER

The general structure of a complete fuzzy control system is given in Figure 9. The plant control 'u' is inferred from the two state variables, error (e) and change in error (Äe) The actual crisp input are approximates to the closer values of the respective universes of its course. Hence, the fuzzyfied inputs are described by singleton fuzzy sets. The elaboration of this controller is based on the phase plan. The control rules base are designed to assign a fuzzy set of the control input u for each combination of fuzzy sets of e and de. The Table 1 is as shown in below:

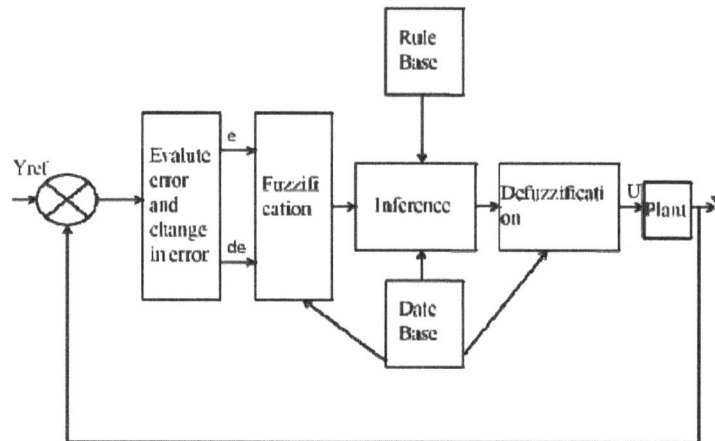

Figure 9. Basic structure of fuzzy control system

Table 1. Fuzzy Rules

code \ error	NL	NM	NS	Z	PS	PM	PL
NL	PL	PL	PL	PL	NM	Z	Z
NM	PL	PL	PM	PL	PS	Z	Z
NS	PL	PM	PS	PS	PS	Z	Z
Z	PL	PM	PS	Z	NS	NM	NL
PS	Z	Z	NM	NS	NS	NM	NL
PM	Z	Z	NS	NM	NL	NL	NL
PL	Z	Z	NM	NL	NL	NL	NL

Here,   NL=Negative Large
NM=Negative Medium
NS=Negative Small
Z=Zero
PS=Positive Small
PM= Positive Medium
PL= Positive Large

9. EMPLOYED CONfiGURATION OF THE GRID-INTERFACING CONVERTER SYSTEM

In this case, UPQC finds the ability of injecting power using PV to sensitive load during source voltage interruption. Figure 1 shows the configuration of proposed system. In this designed system, two Operational modes are studied as Interconnected mode: where PV transfers power to load and source and Islanding mode: where the source voltage is interrupted and PV provides a part of load power separately.

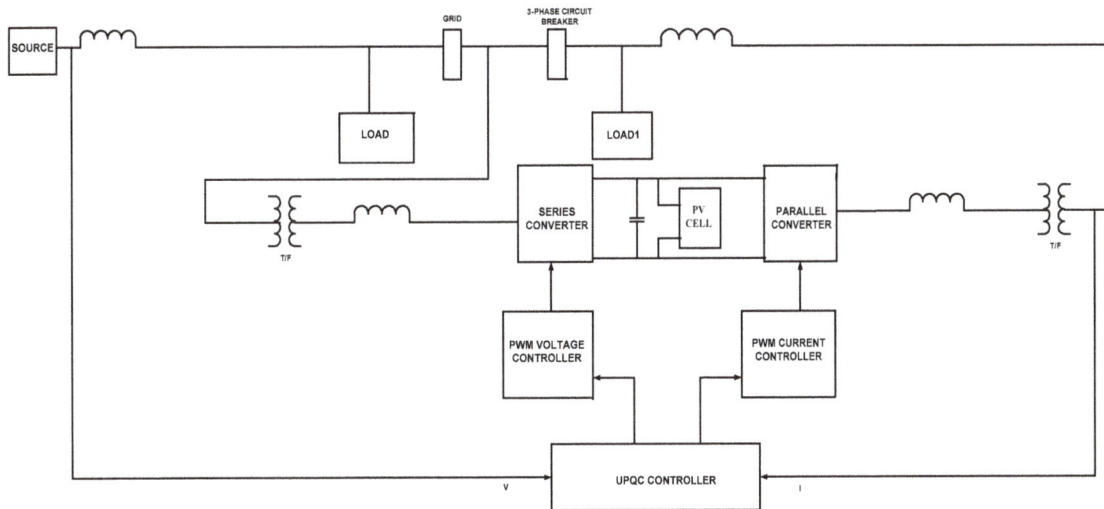

Figure 10. Grid interfacing converter ystem

10. RESULTS
10.1. Experimental Results of the Series-parallel System under Unbalanced Voltage Dips

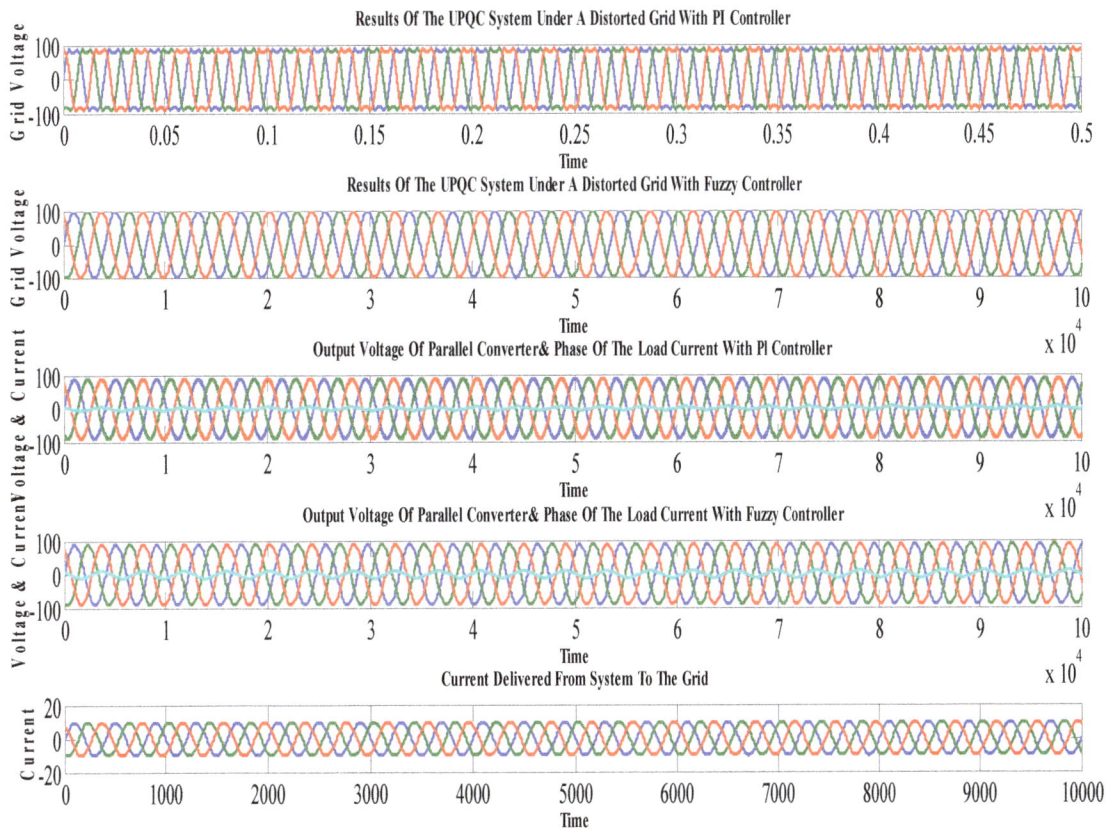

Figure 11. UPQC system under distorted condition

From the above Figure 11 results under distorted grid voltage side a harmonics is obtained with PI controller. These harmonics are eliminated in grid voltage side by applying fuzzy controller so in turn pure sinusoidal wave is obtained. The magnitude of output voltage of parallel converter is reduced by using fuzzy controller compared to PI controller.At 0.00045sec when PV out-ages, source current returns to sinusoidal mode after passing he transient state. It can understood that, before PV outages, voltage has $180°$ phase difference with its current and PV injects current to source in addition to providing load that is islanding mode. After PV outages, it is seen that, current and voltage are in same phase and UPQC compensates current harmonics and power factor. The THD factor in grid voltage side the difference is 7.03% by comparing both controllers.

10.2. Experimental Results of the Series-parallel System under Unbalanced Voltage Dips

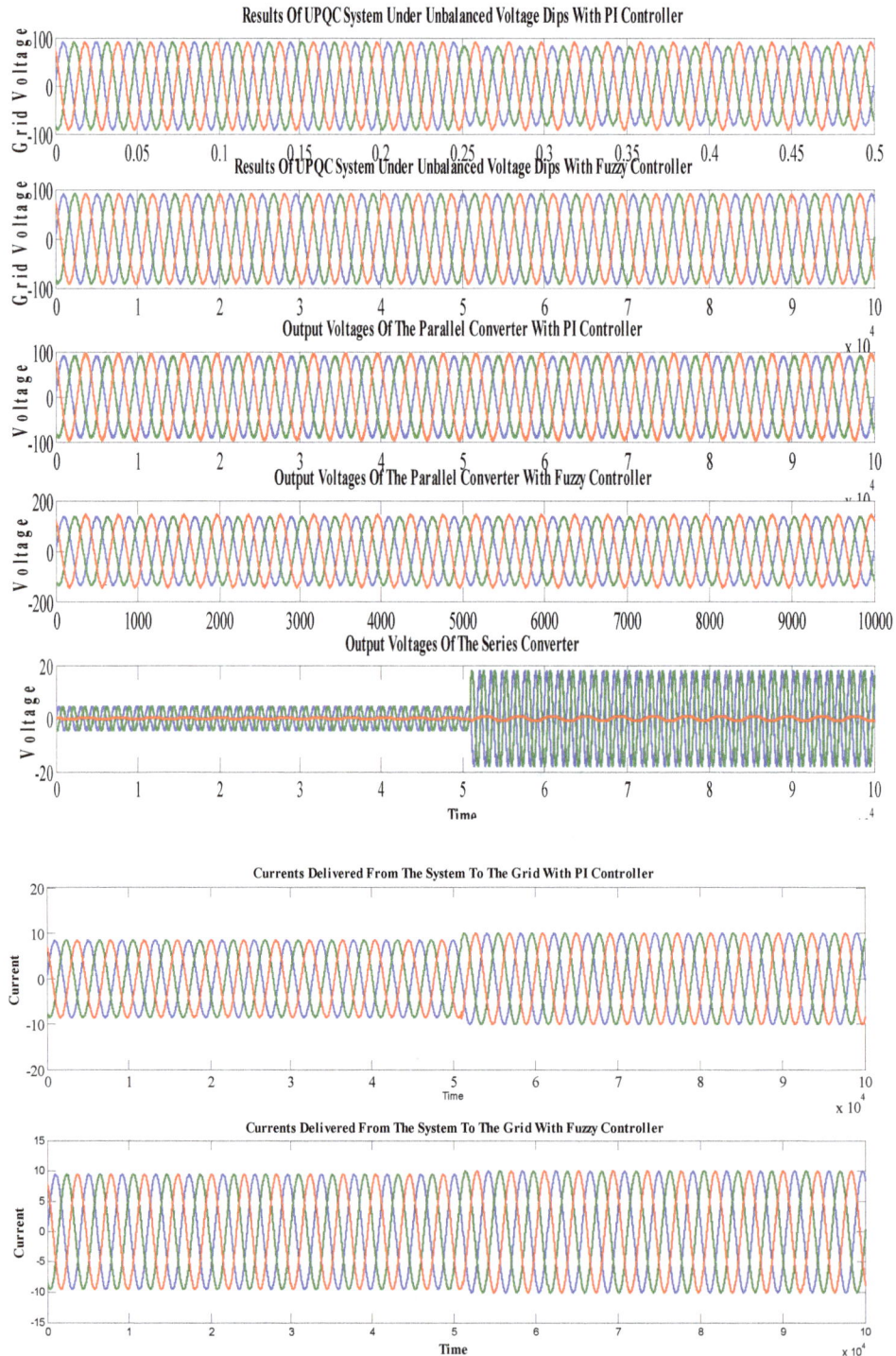

Figure 12. UPQC system under unbalanced voltage dips

From the above Figure 12 results under unbalanced voltage dips in grid voltage side, voltage dips are obtained with PI controller that is eliminated by using fuzzy controller and a pure sinusoidal wave is obtained. The current delivered from the system to the grid at 0.0005 sec there is change in magnitude value in PI controller that will be reduced by applying fuzzy controller. THD factor in grid voltage side difference is 7.06% by comparing both controllers and also in series converter THD is 1.36% .

10.3. Output Voltages of the Parallel Converter Tested under a Single Phase Nonlinear Load

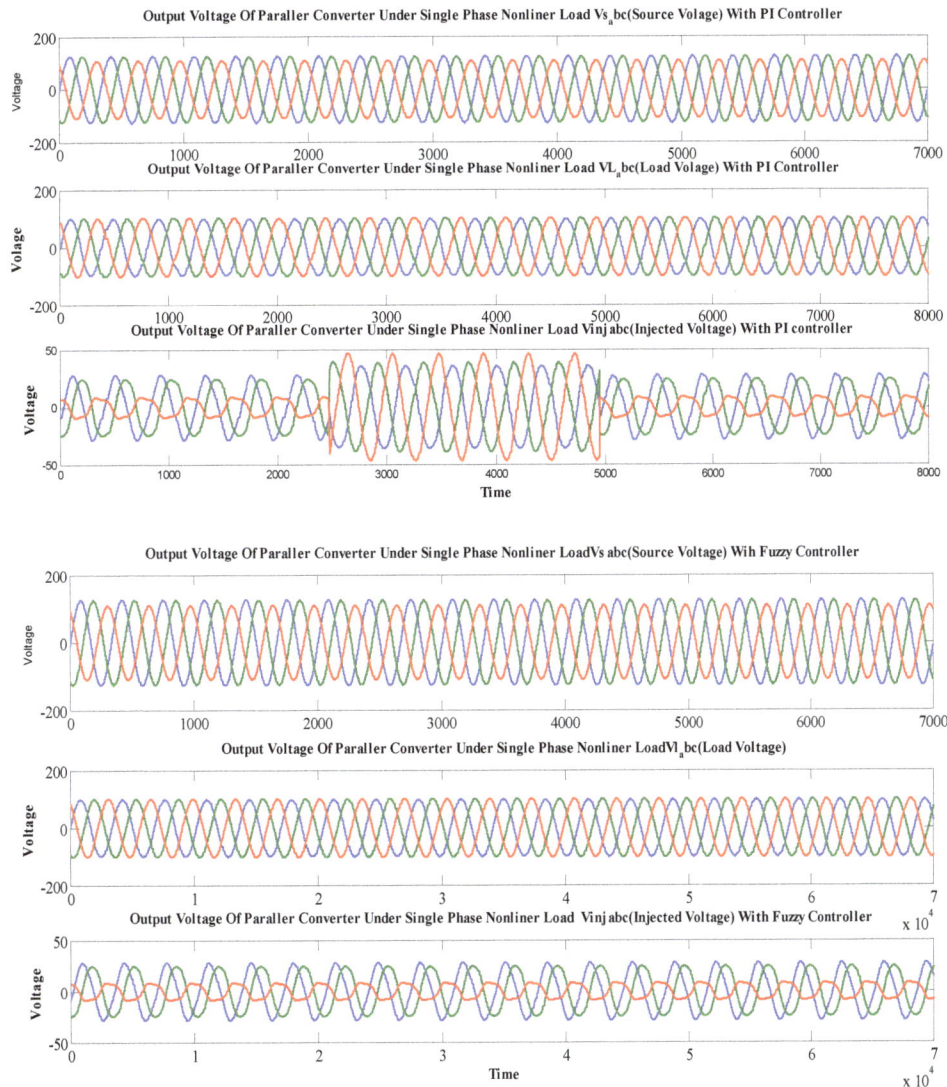

Figure 13. Output voltage of parallel converter under single phase non-linear load

From the above Figure 13 result single phase nonlinear load at source voltage V s with PI controller at the time period during 2000sec to 4000sec there is voltage sag. This sag is reduced by applying fuzzy controller as shown in the above figure. When voltage (Vinjecj) is injected with PI controller at single phase nonlinear load the obtained voltage swell is eliminated by using fuzzy controller.

11. CONCLUSION

In this paper, the results of analyzing combined operation of UPQC and PV is explained. The designed system is composed of series and shunt inverters, PV module and DC/DC converter which can compensate the swell, voltage sag, interruption and reactive power and harmonics in both islanding and interconnected modes. The advantages of proposed system is reducing the expense of PV interface inverter connection to grid because of applying UPQC shunt inverter and also is the ability of compensating the voltage interruption using UPQC because of connecting PV array to DC link. In this proposed system, P&O method is used to achieve the maximum power point of PV array. Along with Advanced compensation of faulted voltage from source, Fuzzy is more advantageous than PI controller because of its faster response. The operation of fuzzy logic is much simpler when the fault occurs at the source due to its rule during the type of fault obtained in the source voltage, need less space to establish and finally most important thing we have to concern it is very less in cost compared to PI controller. The simulation results obtained for the Grid interfacing using series and parallel converter system with conventional PI controller and Fuzzy logic controller are shown above.

REFERENCES

[1] KR Padiyar. Facts controllers in power transmission and distribution. New age international (P) Limited, Publishers. 2008.

[2] F Wang, JL Duarte, MAM Hendrix. *Grid-Interfacing Converter Systems with Enhanced Voltag Quality for Microgrid Application Concept and Implementation*. IEEE. 2013.

[3] F Wang, JL Duarte, M Hendrix. High performance stationary frame filters for symmetrical sequences or harmonics separation under avariety of grid conditions. Proc IEEE APEC. 2009: 1570-1576.

[4] Akagi H, H Fujita. A new power line conditional for harmonic compensation in Power systems"IEEE Transaction on Power Delivery. 1995; 10(3): 1570-1575.

[5] H Fujita, H Akagi. The unified power quality conditioner: the integration of series- and shunt-active filters. *IEEE Trans. Power Electron.*, 1999; 13(2): 315-322.

[6] Ulapane NNB, Dhanapala CH, Wickramasinghe SM, Abeyratne SG, Rathnayake N, Binduhewa PJ. *Extraction of parameters for simulating photovoltaic panels*. 6th IEEE International Conference on Industrial and Information Systems (ICIIS). 2011; 539-544.

[7] Villalva MG, Gazoli JR, Filho ER. Comprehensive Approach to Modeling and Simulation of Photovoltaic Arrays. *IEEE Transactions on Power Electronics*. 2006; 24(5):1198-1208.

[8] Esram T, Chapman PL. Comparison of Photovoltaic Array Maximum Power Point Tracking Techniques. *IEEE Transactions on Energy Conversion*. 2008; 22(2): 439-449.

9

Active and Reactive Power Control of a Doubly Fed Induction Generator

Zerzouri Nora, Labar Hocine

Department of Electrical Engineering, Badji Mokthar University Annaba, Algeria

ABSTRACT

Keyword:

Doubly Fed Induction
Generator (DFIG)
Wind Turbine
Active and Reactive Power
Control

Wind energy has many advantages, it does not pollute and it is an inexhaustible source. However, the cost of this energy is still too high to compete with traditional fossil fuels, especially on sites less windy. The performance of a wind turbine depends on three parameters: the power of wind, the power curve of the turbine and the generator's ability to respond to wind fluctuations. This paper presents a control chain conversion based on a double-fed asynchronous machine (D.F.I.G). To improve the transient and steady state performance and the power factor of generation, a stator flux oriented vector control scheme is used in this work. The vector control structure employs conventional PI controllers for the decoupled control of the stator side active and reactive power. The whole system is modeled and simulated using Matlab/Simulink and the results are analyzed.

Corresponding Author:

Zerzouri Nora
Departement of Electrical Engineering,
Badji Mokthar University Annaba,
Université Badji Mokhtar -Annaba- B.P.12, Annaba, 23000 Algeria.
Email: Zerzouri_karima@yahoo.fr

1. INTRODUCTION

Wind energy is one of the most important and promising source of renewable energy all over the world, mainly because it reduces the environmental pollution caused by traditional power plants as well as the dependence on fossil fuel, which have limited reserves. Electric energy, generated by wind power plants is the fastest developing and most promising renewable energy source [1]. Off-shore wind power plants provide higher yields because of better conditions. With increased penetration of wind power into electrical grids, wind turbines are largely deployed due to their variable speed feature and hence influencing system dynamics. But unbalances in wind energy are highly impacting the energy conversion and this problem can be overcome by using a Doubly Fed Induction Generator (DFIG) [2]. Doubly fed wound rotor induction machine with vector control is very attractive to the high performance variable speed drive and generating applications. In variable speed drive application, the so called slip power recovery scheme is a common practice here the power due to the rotor slip below or above synchronous speed is recovered to or supplied from the power source resulting in a highly efficient variable speed system. Slip power control can be obtained by using popular Static Scherbius drive for bi directional power flow. Advantage of the DFIG is that the power electronic equipment used a back to back converter that handles a fraction of (20-30%) total system power. The back to back converter consists of two converters. Grid Side Converter (GSC) and Rotor Side Converter (RSC) connected back to back through a dc link capacitor for energy storage purpose [2].

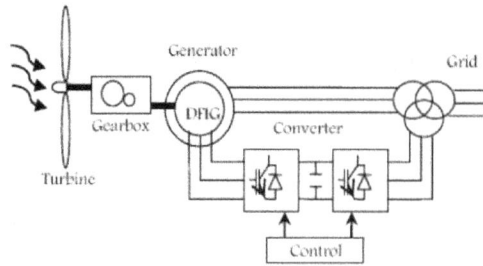

Figure 1. Wind energy conversion chain

2. WIND TURBINE MODEL RESEARCH

Wind turbines produce electricity by using the power of the wind to drive an electrical generator. Wind passes over the blades, generating lift and exerting a turning force. The rotating blades turn a shaft inside the nacelle, which goes into a gearbox. The gearbox increases the rotational speed to that which is appropriate for the generator, which uses magnetic fields to convert the rotational energy into electrical energy.

The power contained in the wind is given by the kinetic energy of the flowing air mass per unit time [3], [4].

$$P_{air} = \frac{1}{2}\rho S v^3 \tag{1}$$

Where P_{air} the power contained in wind (in watts) , ρ is the air density (1.225 kg/m3 at 15°C and normal pressure), S is the swept area in (square meter), and v is the wind velocity without rotor interference, ideally at infinite distance from the rotor (in meter per second). Although (1) gives the power available in the wind, the power transferred to the wind turbine rotor is reduced by the power coefficient C_p

$$C_p = \frac{P_{wind\,turbine}}{P_{air}} \tag{2}$$

A maximum value of C_p is defined by the Betz limit, which states that a turbine can never extract more than 59.3% of the power from an air stream. In reality, wind turbine rotors have maximum C_p values in the range 25-45%. It is also conventional to define a tip speed ratio λ as [5], [6]:

$$\lambda = \frac{\omega R}{v} \tag{3}$$

Where ω is rotational speed of rotor (in rpm), R is the radius of the swept area (in meter).The tip speed ratio λ and the power coefficient C_p are the dimensionless and so can be used to describe the performance of any size of wind turbine rotor.

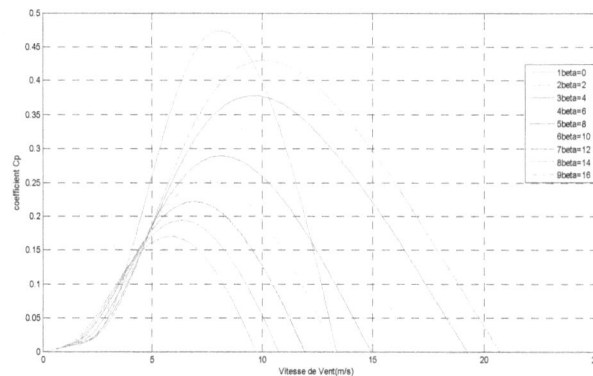

Figure 2. The typical curves of Cp versus λ for various values of the pitch angle β

3. DFIG MODELING AND POWER CONTROL

3.1. Principe of Operation

The machine stator winding is directly connected to the grid and the rotor winding is connected to the rotor-side VSC by slip rings and brushes. A wide range of variable speed operating mode can be achieved by applying a controllable voltage across the rotor terminals. This is done through the rotor-side VSC. The applied rotor voltage can be varied in both magnitude and phase by the converter controller, which controls the rotor currents. The rotor side VSC changes the magnitude and angle of the applied voltages and hence decoupled control of real and reactive power can be achieved.

3.2. Mathematical Model of DFIG

For a doubly fed induction machine, the Concordia and Park transformation's application to the traditional a,b,c model allows to write a dynamic model in a d-q reference frame as follows [7]:

$$
\begin{cases}
V_{ds} = R_s I_{ds} + \frac{d\phi_{ds}}{dt} - \phi_{qs}\omega_s \\
V_{qs} = R_s I_{qs} + \frac{d\phi_{qs}}{dt} + \phi_{ds}\omega_s \\
V_{dr} = R_r I_{dr} + \frac{d\phi_{dr}}{dt} - \phi_{qr}(\omega_s - \omega_r) \\
V_{qr} = R_r I_{qr} + \frac{d\phi_{qr}}{dt} + \phi_{dr}(\omega_s - \omega_r)
\end{cases}
\tag{4}
$$

The flux équations are:

$$
\begin{cases}
\phi_{ds} = L_s I_{ds} + M I_{dr} \\
\phi_{qs} = L_s I_{qs} + M I_{qr} \\
\phi_{dr} = L_r I_{dr} + M I_{ds} \\
\phi_{qr} = L_r I_{qr} + M I_{qs}
\end{cases}
\tag{5}
$$

Where

ω_s: synchronous angular frequency

ω_r: rotor angular frequency

Rs, Rr: equivalent resistances of stator and rotor windings, respectively

Ls, Lr, M: self and mutual inductances of stator and rotor windings, respectively

The motion equations are given as follows:

$$
\frac{d\omega_r}{dt} = \frac{C_m - C_e}{J}
\tag{6}
$$

$$
C_e = \frac{3}{2} PM(I_{qs} I_{dr} - I_{ds} I_{qr})
\tag{7}
$$

$$
\omega_g = s\omega_s = \omega_s - \omega_r
\tag{8}
$$

Where

ω_g: slip angular frequency

s: slip

C_m: mechanical torque provided to the wind turbine

C_e: electromagnetic torque

J: moment of inertia

3.3. Establishment of the Control Strategy

Neglecting the resistance of the generator stator winding, the phase difference between stator flux and stator voltage vector is just 90°. Therefore, utilizing the stator flux-oriented to align the stator flux vector position with d-axis, the flux equation is:

$$
\begin{cases}
\phi_{ds} = \phi_s \\
\phi_{qs} = 0
\end{cases}
\tag{9}
$$

To keep the stator flux ɸs constant, the voltage equations can be expressed as:

$$\begin{cases} V_{ds} \approx \frac{d\phi_{ds}}{dt} = 0 \\ V_{ds} \approx \frac{d\phi_{ds}}{dt} = V_s \end{cases} \tag{10}$$

Where Vs is the space vector amplitude of stator voltage. The active and reactive powers of stator can be derived as:

$$\begin{cases} P_s = \frac{3}{2}(V_{ds}I_{ds} + V_{qs}I_{qs}) \approx \frac{3}{2}V_sI_{qs} \\ Q_s = \frac{3}{2}(V_{qs}I_{ds} - V_{ds}I_{qs}) \approx \frac{3}{2}V_sI_{ds} \end{cases} \tag{11}$$

According to (10), while DFIG is connected to an infinite grid, the stator voltage is considered a constant. The stator current is the only controlled quantity. Therefore, the DFIG output power to grid can be controlled by the stator current, which achieves the goal of independent control for the DFIG active and reactive power output. Due to the stator windings are directly connected to the power systems and the effect of the stator resistance is very small.

Substituting (9) into (5), d-q axis stator current can be calculated as:

$$\begin{cases} I_{ds} = \frac{MI_{dr} - \phi_{ds}}{L_s} \\ I_{qs} = \frac{M}{L_s}I_{qr} \end{cases} \tag{12}$$

Substituting "(12)" into "(4)", the rotor voltage can be express as:

$$\begin{cases} V_{dr} = R_rI_{dr} + \delta L_r\frac{dI_{dr}}{dt} - \omega_g\delta L_rI_{qr} \\ V_{qr} = R_rI_{qr} + \delta L_r\frac{dI_{qr}}{dt} + \omega_g\delta L_rI_{dr} + \frac{M}{L_s}\phi_{ds} \end{cases} \tag{13}$$

Where $\delta = 1 - \frac{M^2}{L_rL_s}$ is the leakage factor.

The control variables Vdr and Vqr of the rotor voltage can be obtained from "(13)". The influence of the cross-coupling between the d-q axis components of rotor current on system performance is small, which can be eliminated by adopting some control law. The model of the vector control of the rotor-side converter obtained from the above analysis is shown in Figure 3.

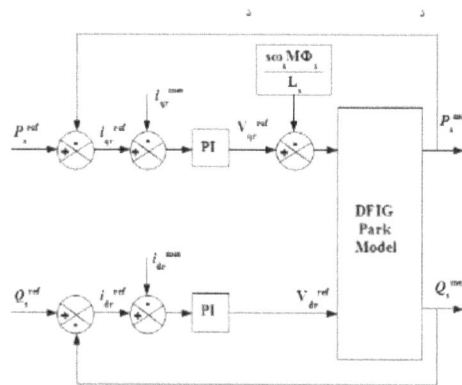

Figure 3. Power control of the DFIG

4. SIMULATION RESULTS

The structure of the DFIG wind energy system is illustrated in Figure 1. The DFIG connected directly to the grid through the stator, and its speed is controlled via a back-to-back PWM converter. The parameters of the DFIG are given in Table 1. A speed wind profile is applied to the system Figure 4.

Table 1. 3MW WTG Induction Machine Parameters

Parameter	Value
Rotor resistor per phase	2,97mΩ
Rotor resistor per phase	3,82mΩ
Inductance of the stator winding	121 mH
Inductance of the retor winding	57,3 mH
Mutual Inductance	12,12 mH
Number of pole pairs	2
inertia	114 kg.m^2
Rated power	3MW
Rated voltage	690V

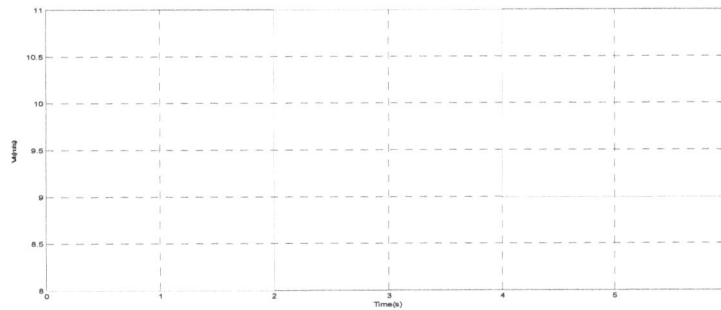

Figure 4. Wind speed profile

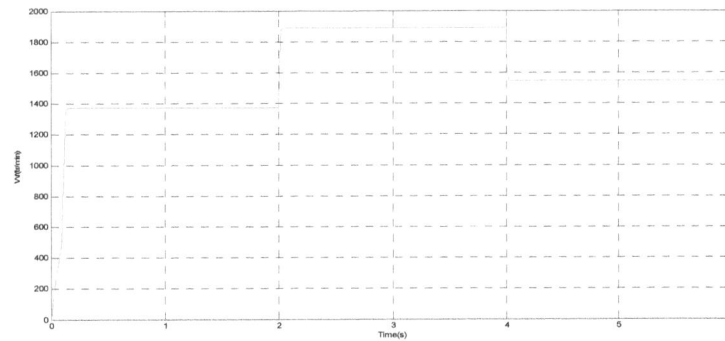

Figure 5. Mechanical speed of the DFIG

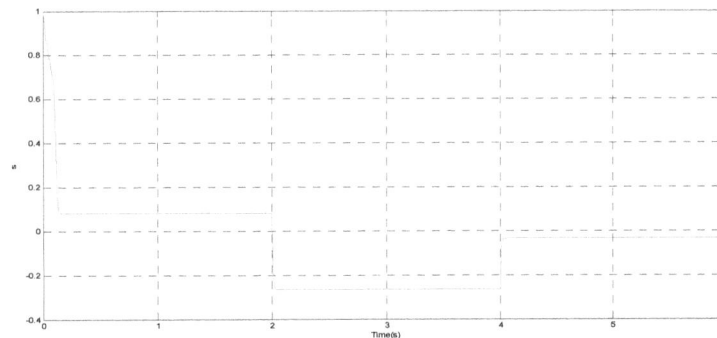

Figure 6. Rotor slip

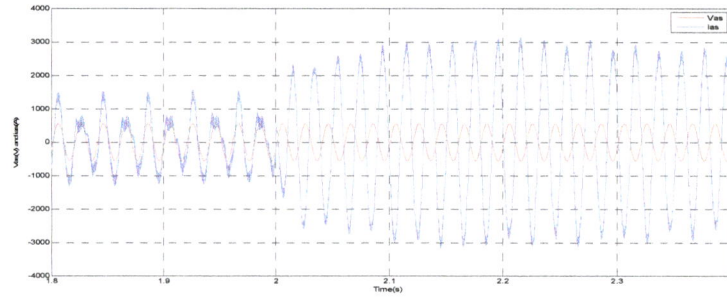

Figure 7. Stator current and voltage

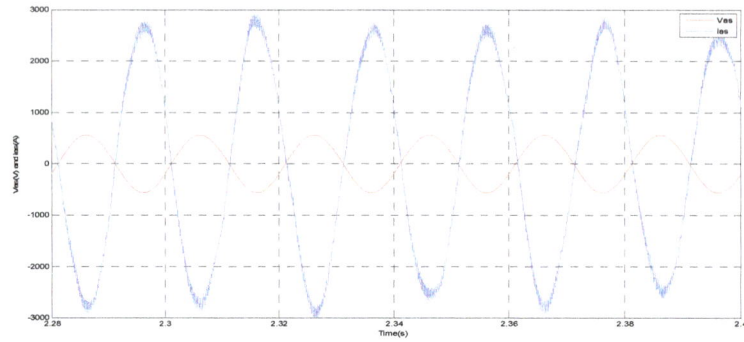

Figure 8. Zoom stator current and voltage

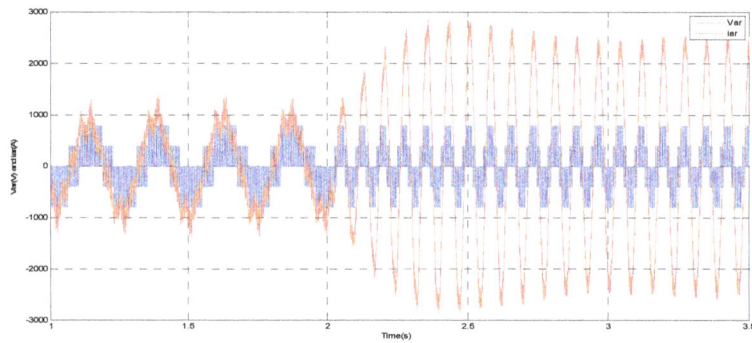

Figure 9. Rotor current and voltage

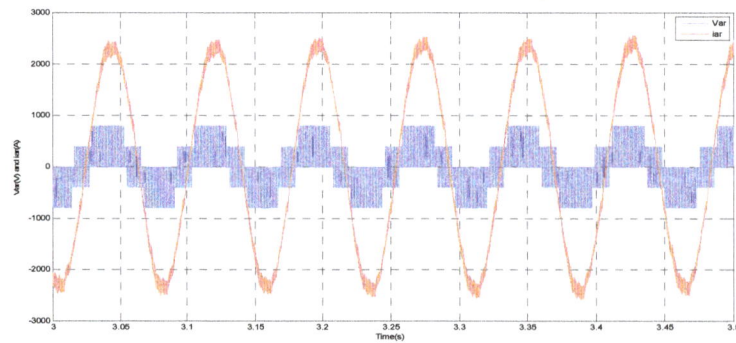

Figure 10. Zoom rotor current and voltage

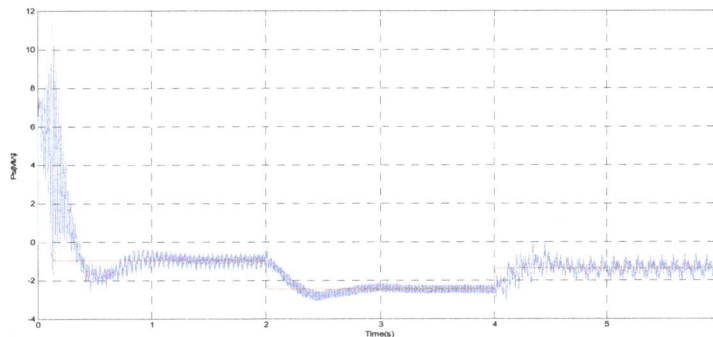

Figure 11. Stator active power

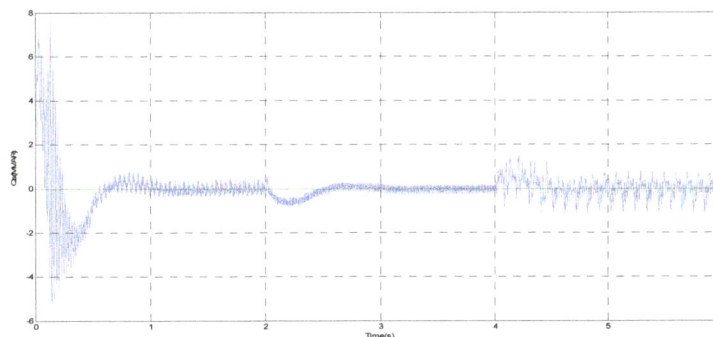

Figure 12. Stator reactive power

Figure 7 shows the zoom of the waveform of the stator voltage and current are in phase opposition. This confirms that the DFIG is sending active power to the grid. We can see that the current and voltage are in phase when the machine acts the motor. Figure 6 shows the generator slip, below synchronous speed the slip is positive and the machine acts as motor, above synchronous speed the slip is negative and machine acts as generator. Figures 11 and 12 illustrate respectively the stator active power and reactive power. We can see the robustness of the power control of the DFIG. Figures 9 and 10 show the rotor voltage and current waveforms. The frequency of these voltage and current, vary according to the slip s.

The active power of DFIG increase from 1MW to the power 2.5MW and the reactive power remains 0Mvar, which signified the reactive power output is not affected. The simulation result indicates that the active and reactive power decoupled control is achieved and the performance is good.

5. CONCLUSION

This paper presents the doubly fed induction generator used in variable-speed wind power generation. And a control structure using standard proportional integral PI controller and a field-oriented control strategy based on a reference frame rotating synchronously with the rotor flux for variable speed wind turbines using doubly fed induction generator and for obtaining injected rotor voltages is described and simulated. Hence results are determined sub-synchronous and super synchronous speeds and the active and reactive power control is achieved by the RSC and GSC. For the purpose of future extension instead of standard PI controllers fuzzy controllers etc. can be used.

REFERENCES

[1] A Babaie Lajimi, S Asghar Gholamian, M Shahabi. Modeling and Control of a DFIG-Based Wind Turbine During a Grid Voltage Drop. *ETASR - Engineering, Technology & Applied Science Research.* 2011; 1(5): 121-125.
[2] MA Mossa. Field Orientation Control of a Wind Driven DFIG Connected to the Grid. *Wseas Transactions On Power Systems.* 2012; 4(7).
[3] Hachemi Glaoui, Harrouz Abdelkader, Ismail Messaoudi, Hamid Saab. Modeling of Wind Energy on Isolated Area" *International Journal of Power Electronics and Drive System (IJPEDS).* 2014; 4(2): 274~280.

[4] Yu Ling, Xu Cai. Rotor current dynamics of doubly fed induction generators during grid voltage dip and rise. *Electrical Power and Energy Systems.* 2013; 44: 17–24.

[5] Srinath Vanukuru, Sateesh Sukhavasi. Active & Reactive Power Control Of A Doubly Fed Induction Generator Driven By A Wind Turbine. *International Journal of Power System Operation and Energy Management.* ISSN *(PRINT):* 2011; 1(2): 2231–4407.

[6] Sai Sindhura K, G Srinivas Rao. Control And Modeling Of Doubly Fed Induction Machine For Wind Turbines.*Int. Journal of Engineering Research and Applications.* 2013; 3(6): 532-538.

[7] Belabbas Belkacem, Tayeb Allaoui, Mohamed Tadjine, Ahmed Safa. Hybrid Fuzzy Sliding Mode Control of a DFIG Integrated into the Network. *International Journal of Power Electronics and Drive System (IJPEDS).* 2013; 3(4): 351~364.

10

An Adaptive Neuro-Fuzzy Inference Distributed Power Flow Controller (DPFC) in Multi-Machine Power Systems

G. Madhusudhana Rao*, V. Anwesha Kumar**, B.V. Sanker Ram*

* Professor, Departement of Electrical and Electronics Engineering, TKRCET

** Research Scholar, Departement of Electrical and Electronics Engineering, JNTUH

ABSTRACT

Keyword:

DPFC
Genetic Algoritm (GA)
Optimal location
Optimal Settings
UPFC

A well-prepared abstract enables the reader to identify the basic content of a document quickly and accurately, to determine its relevance to their interests, and thus to decide whether to read the document in its entirety. The Abstract should be informative and completely self-explanatory, provide a clear statement of the problem, the proposed approach or solution, and point out major findings and conclusions. The Abstract should be 100 to 200 words in length. The abstract should be written in the past tense. Standard nomenclature should be used and abbreviations should be avoided. No literature should be cited. The keyword list provides the opportunity to add keywords, used by the indexing and abstracting services, in addition to those already present in the title. Judicious use of keywords may increase the ease with which interested parties can locate our article.

Corresponding Author:

G. Madhusudhana Rao,
Departement of Electrical and Electronics Engineering,
TKR College of Engineering and Technology,
Medbowli, Meerpet, Hyderabad-500097.
Email: gurralamadhu@gmail.com

1. INTRODUCTION

FACTS Technology is concerned with the management of active and reactive power to improve the performance of electrical networks. The concept of FACTS technology [4], [8]-[20] embraces a wide variety of tasks related to both networks and consumers problems, especially related to power quality issues, where a lot of power quality issues can be improved or enhanced with an adequate control of the power flow.

By FACTS, operator governs the phase angle, the voltage profile at certain buses and line impedance. Power flow is controlled and it flows by the control actions using FACTS [4], [8]-[20] devices, which include:

a) Static VAR Compensators (SVC)
b) Thyristor Controlled Series Capacitors (TCSC)
c) Static Compensators (STATCOM)
d) Static Series Synchronous Compensators (SSSC)
e) Unified Power Flow Controllers (UPFC)

2. UNIFIED POWER FLOW CONTROLLERS (UPFC)

The UPFC may be considered to be constructed of two VSCs sharing a common capacitor on their DC side and a unified control system. A simplified schematic representation of the UPFC is given in Figure 1.

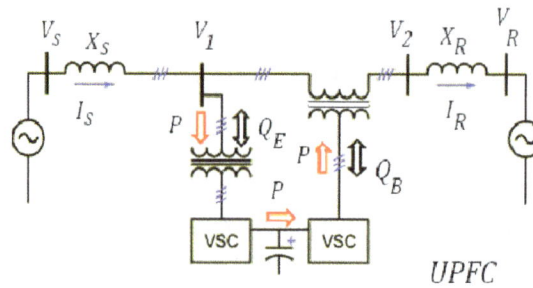

Figure 1. Schematic diagram for the UPFC

The UPFC gives simultaneous control of real and reactive power flow and voltage amplitude at the UPFC terminals. Additionally, the controller may be adjusted to govern one or more of these criteria in any combination or to control none of them. This technique permits with the combined application of controlling the phase angle with controlled series reactive compensations and voltage regulation, but also the real-time change from one mode of compensation into another one to handle the actual system contingencies more effectively. For instance, series reactive compensation may be altered by phase-angle control [2] or vice versa. This can become essentially important at relatively big numbers of FACTS devices will be applied in interconnected power grids, and compatibility and coordination control can own to be save in the face of devices failures and system changes.

2.1. DPFC Modeling

To enable the control of the DPFC [22], controllers for individual DPFC converters are needed. This chapter addresses the basic control system of the DPFC [21], which is composed of shunt control and series control that are highlighted in Figure 2.

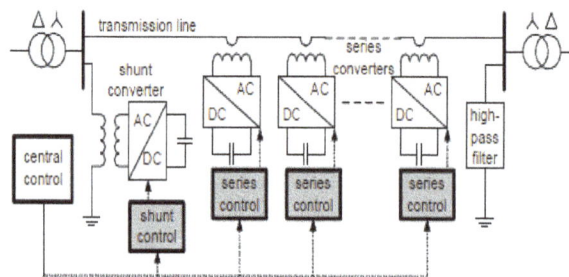

Figure 2 DPFC Basic Control

The functions of the series control can be summarized as:

a)   Maintain the capacitor DC voltage of its own converter by using the 3rd harmonic frequency components.

b)   Generate the series voltage at the fundamental frequency that is prescribed by the central control.

The functions of the shunt control are:

a)   Inject a constant 3rd harmonic current into the line to supply active power for series converters.

b)   Maintain the capacitor DC voltage of the shunt converter by absorbing active power from the grid at the fundamental frequency.

c)   Inject reactive voltage at the fundamental frequency [1] to the grid as prescribed by the central control.

To design a DPFC control scheme, the DPFC must first be modeled. This section presents such modeling of the DPFC. As the DPFC serves the power system, the model should describe the behavior of the DPFC at the system level, which is at the fundamental and the 3rd harmonic frequency. The modeling of the switching behavior of converters is not required. The modeling of the DPFC consists of the converter modeling and the network modeling. Due to the use of single-phase series converters, they are modeled as a single-phase system. To ensure that the single-phase series converter model is compatible with the three-phase network model, the network is modeled as three single-phase networks with 120° phase shift. Figure 3 gives the flow chart of the DPFC modeling process, which leads to six separated models.

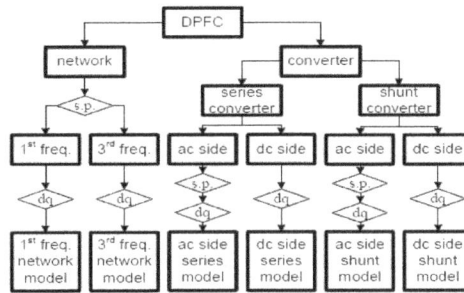

Figure 3. DPFC Modeling process flow chat

Two tools are employed for the DPFC modeling: the superposition theorem and Park's transformation [1]. As is well known, the transmission network is a linear system and the superposition theorem can therefore be applied. However, for the converter, certain approximations are needed for the application of the superposition theorem. Within the flow chart, the diamond shapes with 's.p.' indicate the process of applying the superposition theorem, and the shapes with 'dq' represent the process of Park's transformation. Because Park's transformation is designed for analysis of signals at a single frequency and the DPFC signal consists of two frequency components, the superposition theorem is first used to separate the components. Then, the component at different frequencies are subjected to Park's transformation and analyzed separately. Park's transformation, which is widely used in electrical machinery analysis, transforms AC components into DC.

The principle of Park's transformation is to project the AC signal in vector representation on to a rotating reference frame, referred to as the 'd-q frame'. The frequency of the rotation is chosen to be the same as the frequency of the AC signal. As a result, the voltages and the current in the d-q reference are constant in steady-state. The components at different frequencies are transformed into two independent rotating reference frames at different frequencies. The components at the fundamental frequency are 3-phase components, so Park's transformation can be applied directly. However, as Park's transformation is designed for a 3-phase system, a variation is required before its application to a single-phase system. The reason for this is that the 3rd harmonic component of a three phase system can be considered a single-phase component, as its components are all in phase ('zero-sequence').

2.2. Adaptive Neuro-Fuzzy Inference Systems (ANFIS)

Jang and Sun introduced the adaptive Neuro-Fuzzy inference system. This system makes use of a hybrid-learning rule to optimize the fuzzy system parameters of a first order Sugeno system. The Sugeno fuzzy model (also known as TSK fuzzy model) was presented to save a systematic method to produce fuzzy rules of a certain input-output data set. Figure 4 shows the architecture of two inputs, two-rule first-order ANFIS Sugeno system, the system has only one output.

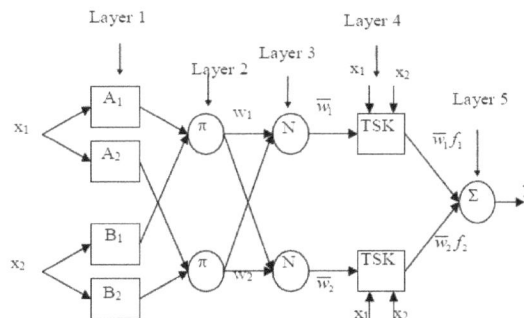

Figure 4. Two-input, two-rule first order Sugeno ANFIS System

The first layer of the ANFIS has adaptive nodes with each node has its function:

$O_{1,i} = \mu A(x_1)$, for i-1,2 or $O_{1,i} = \mu B2(x_2)$, for I = 3,4

Where x1 and x2 are the inputs; and Ai and Bi are linguistic labels for the node. And $O_{1,I}$ is the membership grade of a fuzzy set A (= A_1, A_2, B_1 or B_2) to define the degree of applying the input to the set A.

The second layer has fixed nodes, where its output is the product of the present signals to act as the firing power of a rule.

$O_{2,i} = w_i = \mu A(x_1) \mu B2(x_2)$, for i = 3,4

The third layer also has fixed nodes; the ith node computes the ratio of the ith rules firing strength to the rules' firing strengths sum:

$O_{3,i} = \overline{W_i} = \dfrac{w_i}{w_1 + w_2}$,   i=1,2.

The nodes of the forth layers are adaptive nodes, each with a node function:

$O_{4,i} = \overline{W_i} f_i = \overline{W_i}(P_i x_1 + q_i x_2 + r_i)$

Where w_i is a normalized firing strength produced by layer 3; {p_i, q_i, r_i) is the parameter set of the node, and pointed to consequent parameters.

There is a single node in the fifth layer, which is a fixed node, which calculates the resultant output as the summation of all signals.

Overall output = $O_{5,1} = \sum_i \overline{W_i} f_i = \dfrac{\sum_i w_i f_i}{\sum_i w_i}$

The contributions of the paper start with formatting, deriving, coding and programming the network equations required to link DPFC steady-state and dynamic models to the power systems [3]. One of the other contributions of the paper is deriving GA applications on DPFC to achieve real criteria on a real world sub-transmission network. An enhanced GA technique is proposed by enhancing and updating the working phases of the GA including the objective function formulation and computing the fitness using the diversity in the population and selection probability. The simulations and results show the advantages of using the proposed technique. Integrating the results by linking the case studies of the steady-state and the dynamic analysis [5] is achieved. In the dynamic analysis section, a new idea for integrating the GA with ANFIS to be applied on the control action procedure is presented. In addition to, packages of Software for genetic algorithm and adaptive Neuro-fuzzy system are developed. In other related work, GA only was used to enhance the system dynamic performance considering all working range of power system at a time that gave a difficulty and inability in some cases to reach the solution criteria. In this paper, for every operating point GA is used to search for controllers' parameters, parameters found at certain operating point are different from those found at others. ANFISs are required in this case to recognize the appropriate parameters for each operating point.

2.3. Proposed Adaptive Neuro-Fuzzy Inference Distributed Power Flow Controller (DPFC)

The DPFC modeling and control are simulated in the Mat lab Simulink. The schematic of the DPFC system in the simulation is shown in Figure 5. To simplify the calculation, one set of series converters is used to represent the distributed converters.

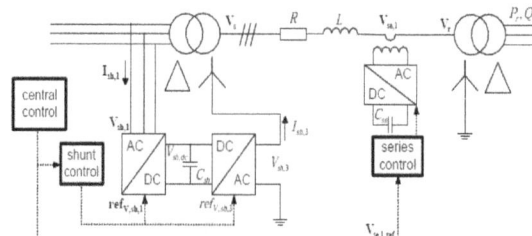

Figure 5. DPFC system in the simulation

The capability of injecting a controllable 360° series voltage is signified by the independent control of the active and reactive power flows at the receiving end. As shown, the active and reactive power can be independently controlled, which indicates that the DPFC is capable of injecting the 360° controllable voltage at the fundamental frequency. The transients are caused by the variation in the DC voltages of the series converters [6]. The DC voltages of both the series and the shunt converters are well maintained during operation. The proposed structure of a DPFC Shunt Converter Control DC voltage regulator is shown in Figure 6.

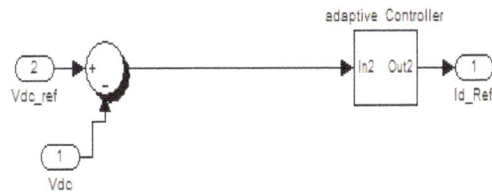

Figure 6. DPFC Shunt Converter Control DC voltage regulator

The proposed structure of Adaptive Neuro-Fuzzy Inference Distributed Power Flow Controller (DPFC) is shown in Figure 7.

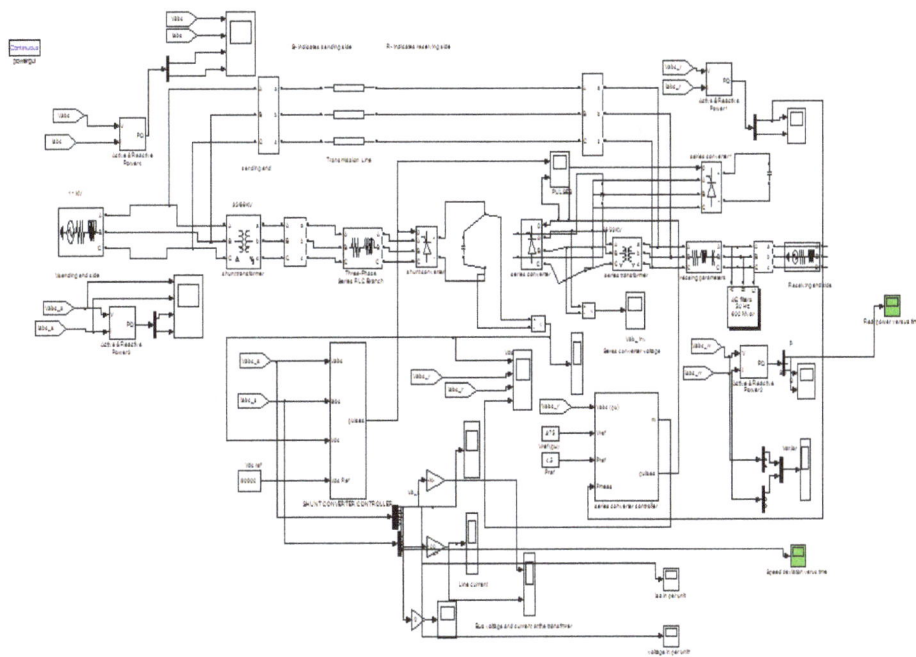

Figure 7. Adaptive Neuro-Fuzzy Inference Distributed Power Flow Controller (DPFC)

3. SIMULATION RESULTS AND DISCUSSIONS

In this section, the DPFC model is created and simulated on Mat lab/Simulink. All the simulations are based on single-phase per-unit system. One shunt converter and two single phase series converters are built and tested.

The system under consideration is simulated under different operating conditions to investigate its transient stability performance and to demonstrate the effectiveness of the proposed controller. The contingency under consideration is a three phase fault at the sending end of one of the transmission lines when the generator is operating at different power levels. The fault is considered to occur between t=0.2s and t=0.3s. The fault is cleared with the operation of transmission line reclosure.

The following case studies were undertaken to make the assessments and shown in Figure 8 to Figure 9.

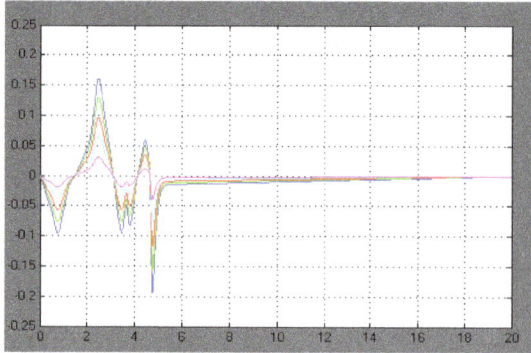

Figure 6-1: Speed deviation versus time

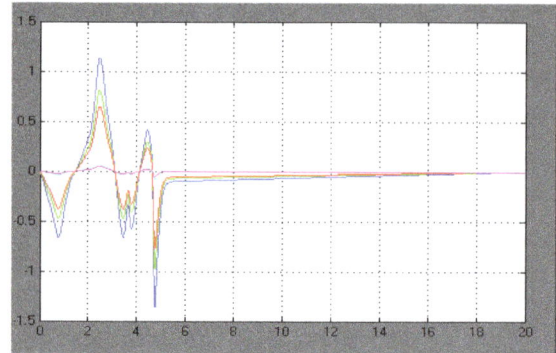

Figure 6-2: Power angle versus time

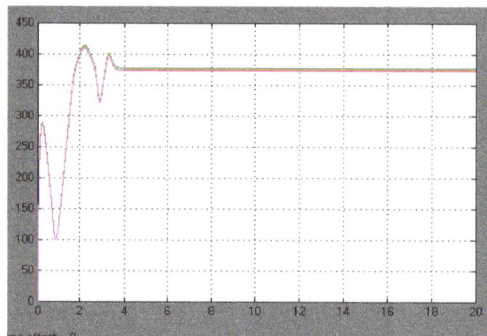

Figure 6-3. Real power versus time

4.   CONCLUSION

The DPFC is modeled in the d-q frame by using Park's transformation. The components of the DPFC in AC quantity are transformed into DC quantity. The components in different frequencies are then separately modeled. This model is a good representation of the behavior of the DPFC at the system level and can be used to design the parameters of the DPFC control. Based on the DPFC model, the shunt control and the series control are developed. The functions of these controls are to maintain the DC capacitor voltages of the converters and to ensure the required voltages and currents are injected from the central control. The DPFC basic control and model are simulated in Mat lab Simulink. The simulation results show that the DPFC is able to control the active and reactive power flows independently and that during operation, the DC voltages of the converters are well maintained. Communication between the central control and the series converters is also considered. To increase the reliability of the DPFC during communication failure, the reference signals in DC quantities are used instead of in AC quantities. The line current is selected as the rotation reference frame because it can be easily measured by the series converters without extra cost. During communication failure, the series converter can use the last received setting to continue operation, thereby increasing the system's reliability. This communication method is also tested in Mat lab Simulink. It shows that in steady-state, communication in DC quantities [6]-[7] has the same result as in AC quantities. During communication failure, the series converter of the DPFC can maintain synchronization with the system.

REFERENCES

[1]   Elgerd. *Electric Energy System Theory: An Induction,* McGRAWHILL BOOK COMPANY, USA. 1971.
[2]   Prabha Kundur. *Power System Stability and Control,* McGRAW-HILL BOOK COMPANY, USA. 1993.
[3]   JP Barret, P Bornard, B Meyer. *Power System Simulation,* CHAPMAN & HALL, UK. 1997.
[4]   M Chamia. *Market Driven vs. Technology Driven Approach to Flexible AC Transmission System.* Proceedings: FACTS Conference I--The Future in High-Voltage Transmission, TR100504, Research Project 3022, EPRI, 1992; 5.1.

[5] K Takahshi, RM Maliszewski, F Meslier, P Wallace, L Salvaderi, T Watanabe. *Impacts of Demand-Side Power Electronics Technologies on Power System Planning.* Proceedings of CIGRE Power Electronics in Electric Power Systems, Tokyo. 1995: 310-01.

[6] J Arrillaga. *High Voltage Direct Current Transmission,* PETER PEREGRINUS LTD., UK. 1983.

[7] HG Hingorani *High-Voltage DC Transmission.* IEEE Spectrum. 1996: 63-72.

[8] HG Hingorani. *Flexible AC transmission.* IEEE Spectrum. 1993: 40-45.

[9] L Gyugyi. *Unified Power Flow Control Concept for Flexible ac Transmission Systems.* Proc TEE, Pt. C. 1992; 139(4): 323-331.

[10] Y Sekine, K Takahashi, T Hayashi. *Application of Power Electronics Technologies to the 21st Century's Bulk Power Transmission in Japan.* Flexible AC Transmission Systems SpecialIssue, International Journal of Electrical Power & Energy Systems. 1995; 17(3): 181- 194.

[11] B Avramovic, L H Fink. *Energy Management Systems and Control of FACTS,* Flexible AC Transmission Systems Special Issue. International Journal of Electrical Power & Energy Systems. 1995; 17(3): 195-198.

[12] W.A.Mittlestadt. *Considering in Planning Use of FACTS Devices on a Utility System,* Proceedings: FACTS Conference I--The Future in High-Voltage Transmission, TR-100504, Research Project 3022, EPRI, 1992: 4.2.

[13] J Maughn. *Evaluation of the Flexible AC Transmission System (FACTS) Technologies on the Southern Electric System,* Proceedings: FACTS Conference I--The Future in High-Voltage Transmission, TR-100504, Research Project 3022, EPRI. 1992.

[14] TJ Hammons. *Flexible AC Transmission Systems (FACTS),* Electric Machines and Power Systems. 1997; 25: 73-85.

[15] A Davriu, G Douard, P Mallet, PG Therond. *Taking Account of FACTS Investment Selection Studies.* Proceedings of CIGRE Power Electronics in Electric Power Systems, Tokyo. 1995: 220-05.

[16] Ross Baldick, Edward Kahn, *Contract Paths, Phase-shifters, and Efficient Electricity Trade.* IEEE Transactions on Power Systems. 1997; 12(2): 749-755.

[17] J Zaboszky. *On the Road Towards FACTS,* Flexible AC Transmission Systems Special Issue. *International Journal of Electrical Power & Energy Systems.* 1995; 17(3): 165-172.

[18] EPRI Report: *Flexible ac Transmission Systems (FACTS): Scoping Study,* Vol. 1. Part 1: Analytical Studies, EL-6943. 1990.

[19] ARM Tenorio, JB Ekanayake, N Jenkins. *Modeling of FACTS Devices.* The six international conference on AC and DC transmission, IEE. UK. 1996: 340-345.

[20] E Larsen. *Control Aspects of FACTS Applications,* Proceedings: FACTS Conference I-- The Future in High-Voltage Transmission, TR-100504, Research Project 3022, EPRI, 1992: 2.1.

[21] P Ramesh, M Damodara Reddy. *Power Transfer Capability & Reliability Improvement in a Transmission Line using Distributed Power- Flow Controller.* International Journal of Electrical and Computer Engineering (IJECE). 2012; 2(4): 553-562.

[22] D Divan, H Johal. *Distributed facts A new concept for realizing grid power flow control.* Proc. IEEE 36th Power Electron. Spec. Conf. (PESC), 2005: 8–14.

11

Proposed Method for Shoot-Through in Three Phase ZSI and Comparison of Different Control Techniques

Byamakesh Nayak *, Saswati Swapna Dash**, Subrat Kumar***
* School of Electrical Engineering, KIIT University, Bhubaneswar
** Department of Electrical Engineering, YMCA University of science and technology, Faridabad
***Department of operation and control (Electrical), Bharat Petroleum Corporation limited, MMBPL, Mathura

Keyword:

FFT analysis
Maximum boost
Maximum constant boost
Shoot-through
Simple Boost
Switching losses
THD
Z-Source inverter

ABSTRACT

This paper presented the new methodology for different control techniques applied to three phase Z-source inverter for minimisation of switching losses. The procedure for proposed control techniques and its effects on the performance of operation of three phase Z-source inverter are analyzed. The graphs for voltage gain and voltage stress are drawn for different control methods. The flow-chart for the symmetrical and unsymmetrical control techniques for creating pulse signals for switches of three phase inverter are shown. All the methods are studied and compared with each other. The Total harmonic distortion (THD) of output voltage of both the control methods has been analyzed using FFT analysis. The experiments done and the results shown for capacitor voltage, load current and load line voltage for simple boost and constant boost control techniques are presented using MATLAB/ Simulink.

Corresponding Author:

Saswati Swapna Dash
Department of Electrical Engineering,
YMCA University of Science and Technology, Faridabad
Email: reachtoswapna@gmail.com

1. INTRODUCTION

The conventional voltage-source inverter (VSI) is used in industries to control the speed of AC motor drive, which consist of diode rectifier at front end, DC link capacitor and inverter bridge[1]. Similarly, in order to transfer energy from PV array into utility grids, the voltage-source inverter is used to convert the DC voltage into AC voltage. VSI is a buck converter that can only produce an AC voltage limited by DC link voltage, which is roughly equal to 1.35 times the input line voltage, if three phase diode bridge rectifier is used at front end [2], [3]. Inrush and harmonic current from the diode bridge rectifier can decreases the efficiency and produces pulsating torque which creates the noise and vibration of ASD system. Low power factor is another issue of the traditional ASD system. Performance and reliability can be achieved by overcoming the three important factors like miss-gating from EMI can cause shoot-through that leads to destruction of the inverter, the dead time that is needed to avoid shoot-through, which increases the complexity of control technique, an output LC filter is needed for providing a sinusoidal voltage compared with the current source inverter, which causes additional power loss and control complexity [4]. There are eight states in one cycle of operation of voltage source inverter. Out of which six states are called active states in which the DC link voltage is impressed across the load and two zero states where the load terminals are shorted through either the lower and upper three device respectively. The voltage across the load is zero in two zero state conditions. Amplitude modulation control the width of zero states and thus the voltage across the load is regulated but remain well below the DC-link voltage. Therefore, VSI has only one control variable i.e.modulation index which is used for buck the voltage across the load. The Z-source inverter employs an X-shape network before the traditional voltage-source inverter bridge. During shoot-through period the capacitor voltage is boosted up by receiving the energy from inductor, while producing no voltage to load. It should be emphasized that both the shoot-through zero state and the two traditional zero states short the load terminals and produce zero voltage across the load, thus preserving the same PWM properties and voltage waveform to the load. The only difference is that the shoot-through zero state boost the capacitor voltage, where as the traditional zero states do not have boosting capability. For the same output voltage the total harmonic distortion is less in Z-source inverter compared to traditional voltage-source inverter because of active-state voltage across the load is the capacitor output voltage [5], [6]. There are four ways of introducing shoot through in a Z-source inverter, out of which one method is very common in traditional method, where all the switches of all three legs are made ON at a time.but in this method the main disadvantage is the swiching losses occurs during the switching action of switches. This paper presents a new technique of shoot through which will no doubt minimize the switching losses fulfilling the shoot through purposes in a Z-source inverter during null state period. The Z-source inverter circuit analysis, criteria for choosing the value of passive parameters, proposed control techniques for providing shoot-through. Simulation results are included to prove the concept. Figure 1 shows the main circuit configuration of the Z-source inverter with 3-phase load. Similar to that of the traditional voltage source inverter.

2.   CIRCUIT CONFIGURATION, OPERATING PRINCIPLES AND MODES OF OPERATION

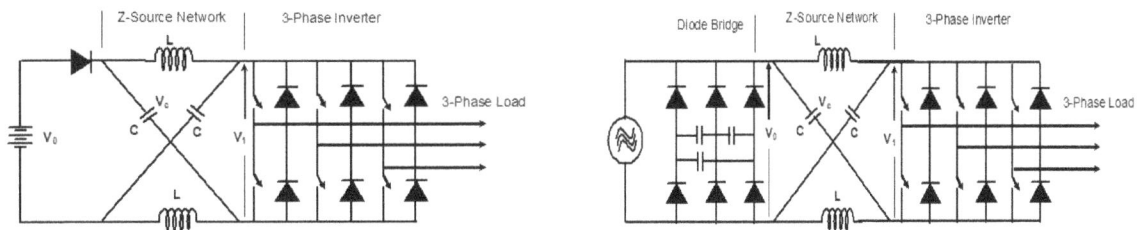

Figure 1. (a, b) Z-source inverter with 3- phase load.

The Z-source inverter circuit consists of three parts: a three-phase, single-phase diode bridge rectifier or battery depending upon the availability of input system, DC-link circuit , and an inverter bridge. Small input capacitors are connected to the diode bridge rectifier if diode bridge rectifier is used instead of battery. For battery as input, a diode is connected before dc-link to oppose the flow of energy towards source during shoot-through period. The dc link circuit of Z-source inverter is different from traditional voltage-source inverter and it consist of symmetrical X shape network consist of series inductance and shunt capacitance ($L_1=L_2=L$, $C1=C_2=C$). For a diode bridge rectifier as input Fig1(b), at any instant of time, only two phases that have the largest potential difference may conduct, therefore as viewed from Z-source network, the diode bridge can be modeled as a dc source in series with two diode, acts just like as battery with diode. The operating control technique of switching of the inverter is such a way that the inverter operates in three modes: active mode, traditional zero-state mode and shoot-through zero state mode. The modes of operation are explained assuming the operation of Z-source inverter is only three dynamic states.

2.1. Active-state -1

The inverter bridge is operating in one of the six traditional active vectors (100,010,110,001,101,011), thus acting the output as a current source viewed from the Z-source circuit. In this mode the power is taken from the source and feed to the load. The continuous flow of input current reduces the harmonic current. This mode of operation is shown in Figure 2(a).

2.2. Zero-state-1

The inverter bridge is operating in one of the two traditional zero vectors (111,000) and shorting through either the upper or lower three devices, thus acting as a open circuit viewed from the Z-source circuit. In this mode, the input is connected to the impedance network and input current is the inductor current, which contribute to the line current's harmonic reduction. This state may be completely or partially compensated by shoot-through state depending on control technique applied to inverter switches. Completely compensated mode is named as 3-mode operation whereas partially compensated mode is called as 2-mode operation in this paper. This mode of operation is shown in Figure 2(b).

2.3. Shoot-through- state-1

The inverter bridge is operating in one of the seven shoot-through states. During this mode the input is disconnected and load is shorted through Z-source network. Seven shoot-through are achieved by turning all switches or five switches or 4 switches at a time. The control technique is such that the shoot-through state is inserted during the period of zero-state without affecting the period of active state. It can be seen that the shoot-through interval is only a fraction of switching cycle; therefore it needs a small capacitor to suppress voltage. During this period the energy stored in the inductor is transferred to the capacitor and hence the capacitor voltage is boosted up. Depending on how much a boost voltage is needed, the shoot-through interval is determined. This mode of operation is shown in Figure 2(c).

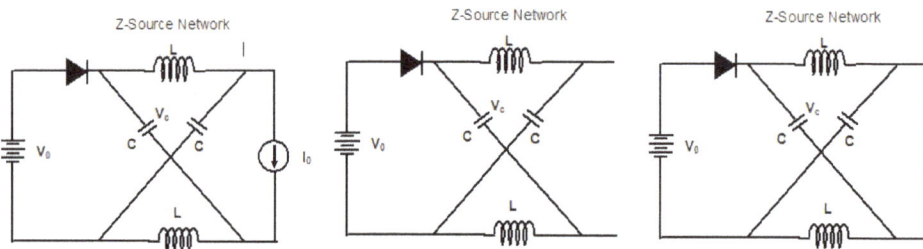

Figure 2. (a, b, c) Different modes of operation of Z-source inverter.

3. DIFFERENT CONTROL TECHNIQUES

Figure 3(a, b) shows the traditional PWM switching sequence based on the triangular carrier signals compared with the 3 sinusoidal signals with a phase difference of 120^0 for $M_a = 0.5$. In every switching cycle, the zero states (000 or 111) are created along with active states. In Z-source inverter, without affecting the active states, the shoot-through states is allocated in zero-state intervals evenly to boost the voltage. If the shoot-through state is completely allocated in zero-state interval then the operation is called two -mode operation, otherwise the operation is called three-mode operation.

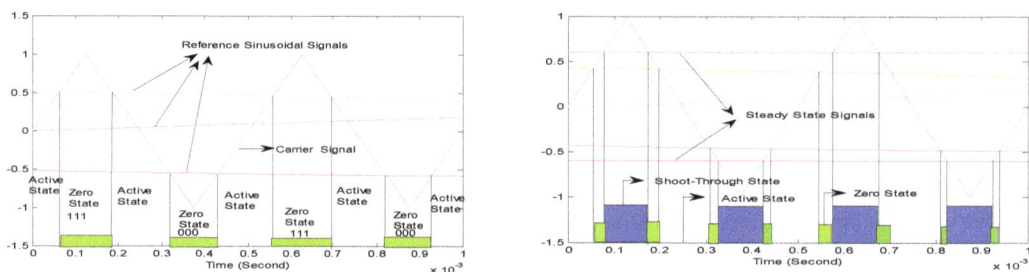

Figure 3. (a, b) Switching Techniques of traditional VSI and ZSI using simple boost control.

3.1. Simple boost control

In this method, the shoot-through time per switching cycle is kept constant, thus having a constant boost factor [6]. From the figure of VSI switching cycle it is confirmed that the zero state is produced when all the sinusoidal waveforms are less or greater than carrier signal. Therefore, to provide shoot –through in zero state the other two steady signals, whose amplitudes are equal to amplitude of sinusoidal waveform and one is negative magnitude of other are compared with triangular signal. The proposed procedures can be used to produce different shoot-through combinations. Different shoot-through combinations can be obtained by including the comparative output of steady-state signals and carrier signal. Symmetrical switching occurs when the above comparative outputs are used for all the legs of the inverter. In this case, the shoot –through period is produced by turning on all the switches. By doing this, the switching frequencies of all the switches are doubled as compared to VSI inverter which increases the switching losses. Unsymmetrical switching occurs when the comparative outputs are used in any one leg or any two legs. In this case the switching frequencies of switches have different values. For example, if the shoot-through is created in a-phase then the switching frequency of a -phase switch is twice the other switches of b and c phases. As a result, the switching losses can be minimized by providing the shoot-through in one phase only. However , at the same time the current stress in each switches during shoot-through is three times(4 switches on) or two times (5 switches on) when compared with symmetrical switching. Another advantage of this control method is that the dc inductor current and capacitor voltage have no ripples that are associated with the output frequency, because of shoot through period is constant over one switching cycle. Figure 3(b) shows how the shoot-through state is included in zero-state interval without affecting the active state intervals.[7]-[11]. The average DC-link voltage across the inverter bridge is same as the capacitor voltage can be represented as :

$$V_1 = V_0 = \frac{T_1}{T_1 - T_0} V_0 \tag{1}$$

The peak DC-link voltage across the inverter bridge is represented as

$$\hat{V}_1 = \frac{T}{T_1 - T_0} V_0 \tag{2}$$

Where,

$$B = \frac{T}{T_1 - T_0} = \frac{1}{1 - \frac{2T_0}{T}} \geq 1 \tag{3}$$

Since $T_1 + T_0 = T$ is the boost factor resulting from the shoot- through zero state. T_0 is the shoot-through zero state interval, T is the switching period and T_1 is the combination of active-state and zero-state intervals. $\hat{V}_1$ is the input voltage appeared before Z-source network. V_c is the capacitor voltage which is same as the average dc-link voltage V_0 appeared after Z-source network. On the other side, the output peak phase voltage from the inverter can be expressed as :

$$\hat{V}_{ao} = M_a \frac{\hat{V}_1}{2} = M_a B \frac{V_0}{2} \tag{4}$$

Where, M_a is the modulation index.The peak DC-link voltage across the inverter bridge is represented as voltage stress V_{stress} of the inverter.

$$V_{stress} = BV_0 \tag{5}$$

Let $\frac{T_0}{T} = D_0$ (shoot-through duty ratio) and $M_a = \frac{T_1}{T}$ (assuming no zero-state interval),then $M_a = 1 - D_0$ (When the magnitude of steady state signal is same as the amplitude of sinu-soid signal)

$$B = \frac{1}{1 - 2D_0} \tag{6}$$

$$G = M_a B = \frac{M_a}{2M_a - 1} \tag{7}$$

The ratio of the voltage stress to the equivalent to the equivalent dc voltage denoted as voltage stress for same output voltage ($V_{so} = BV_0 / GV_0$) for the simple boost control is as:

$$V_{so} = 2 - \frac{1}{G} \tag{8}$$

It can be concluded from the above equations that :

- The modulation index and shoot-through duty ratio are interdependence with each other if magnitude of steady-state signal is equal to the amplitude of sinusoidal signal and the ranges of M_a and B are lying in between 0.5 to 1 and 0.5 to 0 respectively.
- This method is used to boost the output voltage, theoretically to infinity but practically it is limited to 3 to 4 times due to parasitic elements of impedance network and switches.

To make the modulation index and shoot-through duty ratio independent with each other and to control by two degrees of freedoms M_a and B for boost and buck the output voltage, the steady-state signal is to be controlled and it should be greater than the peak of sinusoidal signal.

3.2.   Maximum constant boost control

In order to reduce the voltage stress and increase the modulation index from 1 to $2/\sqrt{3}$, the Maximum constant boost control technique is used. A sketch map of Maximum constant boost control method is shown in Fig.4(b). The flow charts of symmetrical and unsymmetrical control technique for maximum constant boost control are same as the simple boost control, except the reference signals.

$$V = V_m \sin wt + 1/6 \sin 3wt \tag{9}$$
$$V = V_m \sin(wt - 120°) + 1/6 \sin 3wt \tag{10}$$
$$V = V_m \sin(wt + 120°) + 1/6 \sin 3wt \tag{11}$$

For phase A, B and C respectively.

$$\frac{T_0}{T} = \frac{2 - \sqrt{3}M_a}{2} \tag{12}$$

$$B = \frac{1}{1 - 2D_0} = \frac{1}{\sqrt{3}M_a - 1} \tag{13}$$

$$G = M_a B = \frac{M_a}{\sqrt{3}M_n - 1} \tag{14}$$

Theoretically, the gain is infinite when $M_a = 0.577$. Therefore the ranges of M_a and B are lying in between 0.577 to $2/\sqrt{3}$ and 0.423 to 0 respectively. The ratio of the voltage stress to the equivalent to the equivalent dc voltage denoted as voltage stress for same output voltage ($V_{so} = BV_0/GV_0$) for the maximum constant boost control is as:

$$V_{so} = \sqrt{3} - \frac{1}{G} \tag{15}$$

3.3. Maximum boost control method

In order to completely eliminate the zero state and thus maximize the voltage boost and minimize the voltage stress for the same output voltage the maximum boost control method is used. The shoot through state is achieved when the triangular carrier signal is either greater than the maximum curve of three sinusoidal references or smaller than the minimum of the references. Figure 4 (a, b, c) show the modulation technique to provide shoot-through and driver signals for the six switches of simple boost control, maximum constant boost control and maximum boost control. With taking average of varying shoot through times from Figure 4(b) the boost factor, voltage gain and voltage stress are given by following equations.

$$B = \frac{1}{1 - 2D_0} = \frac{\pi}{3\sqrt{3}M_a - \pi} \tag{16}$$

$$G = M_a B = \frac{M_a \pi}{3\sqrt{3}M_a - \pi} \tag{17}$$

In this control, the ranges of M_a and B are lying in between 0.604 to 1 and 0.4 to 0 respectively. The ratio of the voltage stress to the equivalent to the equivalent dc voltage denoted as voltage stress for same output voltage ($V_{so} = BV_0/GV_0$) for the maximum constant boost control is as:

$$V_{so} = \frac{3\sqrt{3}}{\pi} - \frac{1}{G} \tag{18}$$

The range of M_a and the output frequency are also deciding factors for selection of control technique.[12], [13]. The waveform of different control techniques and the graphs for Voltage gain and Voltage stress comparison of different control methods are shown in figure 4(a, b, c) and figure 5(a, b).

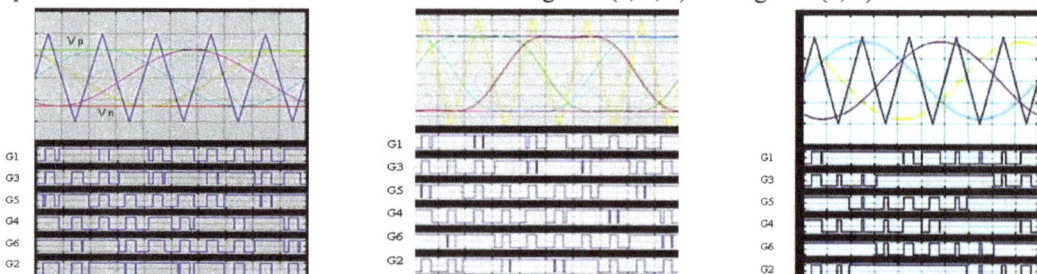

Figure 4. (a, b, c) Waveform of simple boost maximum constant boost and maximum boost respectively.

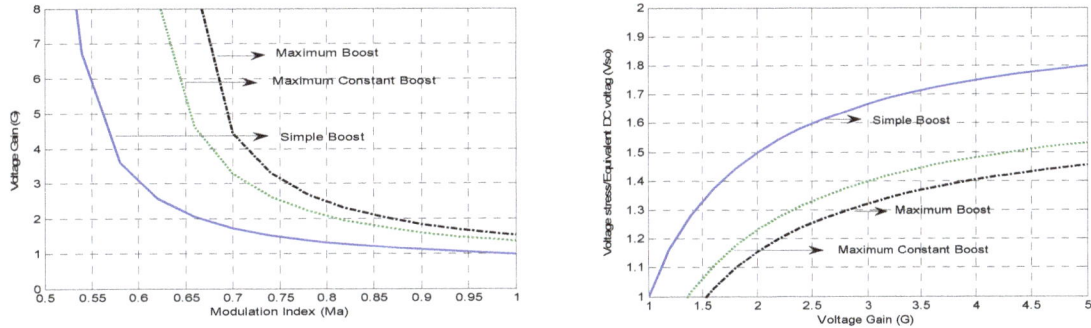

Figure 5 (a, b) Voltage gain and Voltage stress comparison of different control methods respectively.

The flow charts of symmetrical and unsymmetrical control technique for creating pulse signals of switches of inverter of simple boost control are shown in Figure 6 (a,b,c) and Figure 7 (a,b,c) respectively. The flow charts are designed for generating pulses for switches of ZSI. Similar switching phenomena can be implemented to other control techniques.

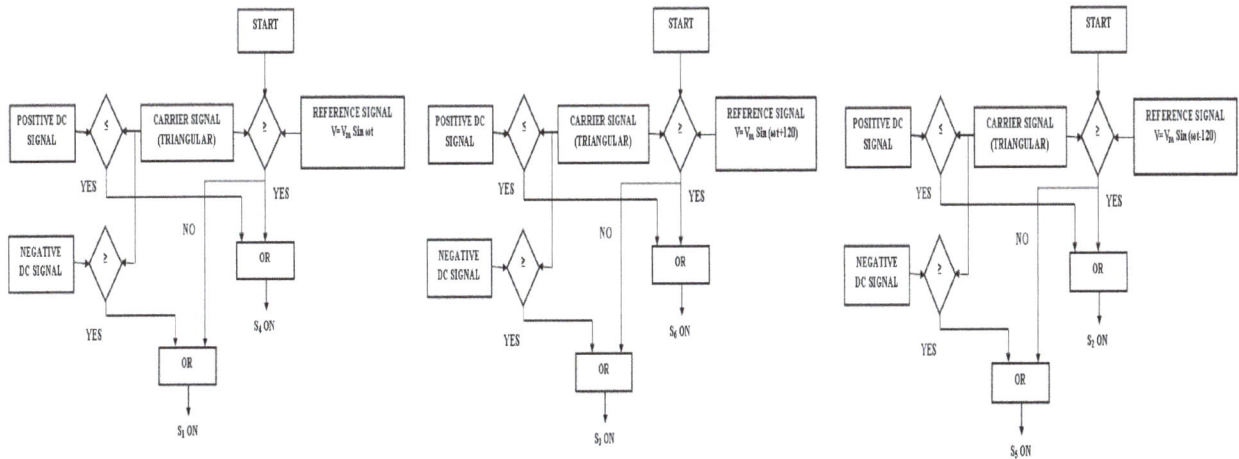

(a) (b)   (c)

Figure 6. (a, b,c) Flow charts of simple boost symmetrical control technique.

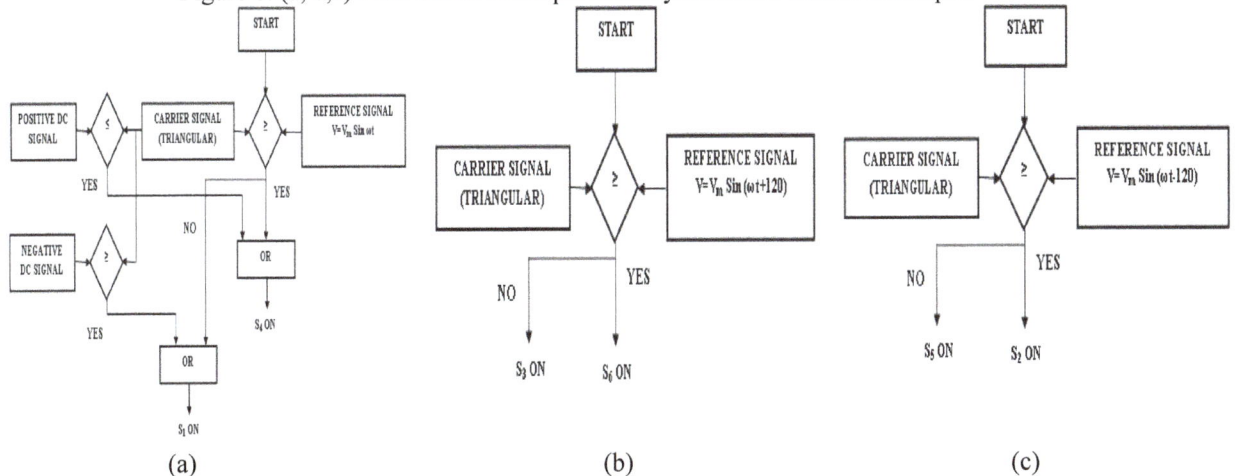

(a)   (b)   (c)

Figure 7. (a, b,c) Flow charts of simple boost unsymmetrical control technique.

4. OPERATION OF INVERTER IN 3 MODES KEEPING AMPLITUDE MODULATION CONSTANT

Different modes operation is achieved by simple boost control and maximum constant boost control. In the above control methods, shoot through is provided by using pair of steady state values (straight lines) unequal to the peak maximum and minimum of the sinusoidal reference signals and non-sinusoidal signals (sinusoidal signals with injected third harmonic signal) for simple boost and maximum constant boost control respectively. The advantage of the above methods is the output voltage is controlled by independent control of amplitude modulation and shoot-through duty ratio (two degrees of control). The unsymmetrical control technique and without affecting the active state is the advantageous control technique is the proposed control technique and can be carried out by using pair of steady state values (straight lines) greater than peak maximum and minimum of the sinusoidal reference signals and non-sinusoidal signals for simple boost and maximum constant boost control respectively and used for comparing it to one leg signal for providing different width of shoot-through by turning on one switch of that leg. Based on highest voltage boost requirement, the amplitude modulation is chosen and makes to be constant. The voltage boost is controlled by changing the magnitude of straight line and it must be greater or equal to amplitude modulation.

5. SIMULATION RESULTS

To verify and compare different control methods, Matlab simulations were performed having following parameters. $L_1 = L_2 = 0.5$mH, $C_1 = C_2 = 2$mF, $V_0 = 100$V (dc), Switching frequency = 2 KHz, 3-phase load: R/phase = 50 ohm, L/phase = 2 mH. The simulation results of simple boost and maximum constant boost are shown in Figure 7(a,b,c) and Figure 8(a,b,c) respectively. The amplitude modulation is kept constant, 0.5 for simple boost and 0.5774 for maximum constant boost. The shoot-through duty ratio of both control techniques keeps at 0.6,so that the amplitude modulation and shoot-through duty ratio are independent with each other. We observe from figures, that the capacitor voltage and peak value of dc link inverter voltage of simple boost control method are same as produced by maximum constant boost control method. Further, it can be observed that there is no overshoot of capacitor voltage for both proposed techniques (shoot through duty ratio is greater or equal than amplitude modulation. It can be seen that the fundamental output line voltage (rms) of inverter is about 228V of simple boost and 260V of maximum constant boost through FFT analysis of output voltage for the same dc link inverter voltage. The above analysis indicates that the voltage stress across the inverter switches is higher of simple boost control technique than the maximum boost control technique. However, in maximum constant boost control method, the total harmonic distortion (THD) is 3.31% of fundamental, which is higher than simple boost control whose THD is about 1.36%of fundamental. In maximum constant boost control, the 3rd and 5th harmonic components of output voltage are about 3.22% and 0.27%, whereas in simple boost control method the above values are .18% and 0.9% of fundamental .This is because of the third harmonic component is injected in reference signal in maximum constant boost control method for generation of driver signals of switches.

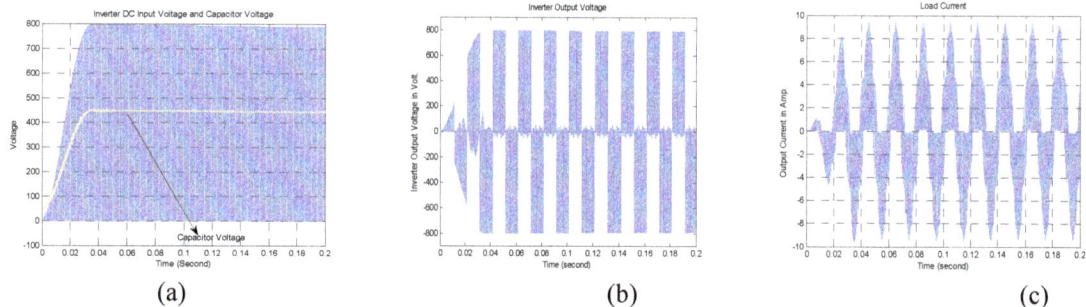

| (a) | (b) | (c) |

Fig.7(a, b, c) Simulated results of capacitor voltage load current and load line voltage of simple boost control technique.

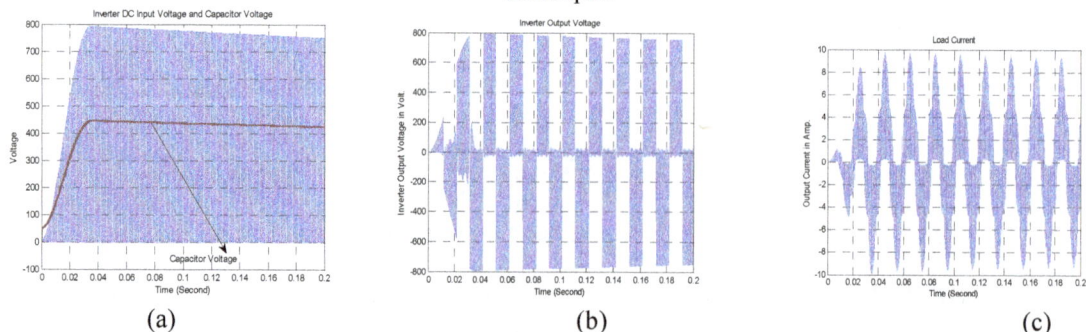

| (a) | (b) | (c) |

Fig.8(a, b, c) Simulated results of capacitor voltage load current and load line voltage of maximum constant boost control technique.

6. CONCLUSION

Three control methods for providing shoot-through in 3-phase Z-source inverter has been analyzed and compared in this paper. The boost factor, Voltage gain, Overshoot of capacitor voltage, Voltage stress across the switches and THD of output voltage have been analyzed. Simulation of Z-source 3-phase inverter under simple boost and maximum constant using straight lines for providing shoot-through) have been presented, showing overshoot in capacitor voltage would not be occurred if the value of straight lines is greater than the peak value of sinusoidal for simple boost and peak value of non-sinusoidal (combination of fundamental and third harmonic sinusoid) signal.

REFERENCES

[1]   F.Z.Peng, *et al.*, "Z-Source Inverter for Motor Drives," *IEEE Transaction on Power Electronics*, vol. 20, pp. 857-863, July 2005.

[2]   F.Z.Peng, *et al.*, "Z source inverter for adjustable speed drives," *IEEE Power Electronics Letter*, vol. 1, pp. 33-35, June 2003.

[3]   Yi Huang, *et al.*, "Z-Source Inverter for Residential Photovoltaic Systems," *IEEE Transaction on Power Electronics*, vol. 21, pp. 1776-1782, Nov. 2006.

[4]   F.Z.Peng, "Z-source inverter," *IEEE Transactions on Industry Applications*, vol. 39, pp. 504-510, March/April 2003.

[5]   F.Z.Peng, *et al.*, "Maximum boost control of Z-source inverter," *IEEE Transactions on Power Electronics*, vol. 20, no. 4, pp 883-838, July/Aug. 2005.

[6]   M.Shen, *et al.*, "Constant Boost Control of the Z-Source Inverter to Minimize Current Ripple and Voltage Stress," *IEEE Transactions on Industry Applications*, vol. 42, pp. 770-778, May/June 2006.

[7]   S.rajkaruna,L.Jaywickrama, "Steady-State Analysis and Designing Impedance Network of Z-Source Inverters," *IEEE Transactions on Industrial Electronics*, vol. 57, pp. 2483-2491, July 2010.

[8]   B.Y. Husodo, *et al.*, "Analysis and Simulations of Z-Source Inverter Control Methods, " *IEEE IPEC 2010.*

[9]   Poh.Chiang Loh, *et al.*, "Transient Modeling and Analysis of Pulse-width_modulated Z-Source Inverter," *IEEE Transactions on Power Electonics*, vol. 22, pp. 498-507, March 2007.

[10] B.K. Nayak and S. S. Dash, "Transient modeling of Z-source chopper used for adjustable speed control of DC motor drive," *IEEE Fifth Power India Conference(PICONF-2012)*, pp. 1-6, Dec.2012.

[11] B.K.Nayak, Saswati Swapna Dash, "Battery Operated Closed Loop Speed Control of DC Separately Excited Motor by Boost-Buck Converter," *IEEE International conference on power electronics (IICPE-2012)*, Dec.2012.

[12] Gokhan Sen and Malik Elbuluk, "Voltage and Current Programmed Modes in control of the Z-Source Converter," *IEEE Transactions on Industry application*, vol. 46, pp. 680-686, March/April 2010.

[13] B.K.Nayak, Saswati Swapna Dash, "Performance Analysis of Different Control Strategies in Z-source Inverter," *ETASR-Engineering, Technology & Applied Science Research*, vol. 3, pp. 391-395, Feb.2013.

Fuzzy-PI Torque and Flux Controllers for DTC with Multilevel Inverter of Induction Machines

N. M. Nordin[1], N. R. N. Idris[2], N. A. Azli[3], M. Z. Puteh[4], T. Sutikno[5]

[1,2,3]Faculty of Electrical Engineering, Universiti Teknologi Malaysia, Johor Bahru, Malaysia
[4]MIMOS Berhad, Technology Park Malaysia, Kuala Lumpur, Malaysia
[5]Department of Electrical Engineering, Universitas Ahmad Dahlan, Yogyakarta, Indonesia

ABSTRACT

Keyword:

CMLI
Direct torque control
Fuzzy logic control
Induction machines
Multilevel inverter

In this paper the performance of flux and torque controller for a Direct Torque Control of Cascaded H-bridge Multilevel Inverter (DTC-CMLI) fed induction machines are investigated. A Fuzzy-PI with fixed switching frequency is proposed for both torque and flux controller to enhance the DTC-CMLI performance. The operational concepts of the Fuzzy-PI with the fixed switching frequency controller of a DTC-MLI system followed by the simulation results and analysis are presented. The performance of the proposed system is verified via MATLAB/Simulink©. The proposed system significantly improves the DTC drive in terms of dynamic performance, smaller torque and flux ripple, and lower total harmonic distortion (THD).

Corresponding Author:

Nik Rumzi Nik Idris,
Departement of Electrical Power Engineering,
Faculty of Electrical Engineering, Universiti Teknologi Malaysia,
81310 Skudai, Johor, Malaysia.
Email: nikrumzi@ieee.org / nikrumzi@fke.utm.my

1. INTRODUCTION

The superior performance of DTC in dynamic response and simple control configuration which is originally introduced in [1], has made it one of the most popular research topics in electrical drive systems. Since the application of high-power medium voltage in AC drives has shown rapid development, the use of multilevel inverters in DTC scheme has become an important structure for further development and improvement. Various technical papers have shown better performance of DTC scheme using multilevel inverters [2-31].

By employing the multilevel inverter, the choices of voltage vectors that can be used to control the torque and flux are increased. Different approaches have been proposed for DTC scheme using multilevel inverter; hysteresis-based controller and non-hysteresis-based controller such as space vector modulation (SVM)[8, 13, 16, 18, 22, 27-29, 31], predictive control strategy [10, 12, 30] and fuzzy logic controller (FLC)[7, 9, 11, 22].

The implementation of the hysteresis-based control strategy has lead to a high torque ripple especially in discrete implementation even with small hysteresis band. This is due to the delay in the sampling time. On top of that, the variable switching frequency of the switching devices which leads to unpredictable harmonics current is also produced by implementing the hysteresis-based control strategy.

As a result, some researchers have chosen to use non-hysteresis-based control strategies to overcome these drawbacks. Significant improvements in terms of flux and torque ripple and switching frequency are accomplished by using these control strategies; however the use of complex mathematical equations and algorithms has led to the computational burden and complexity of the DTC-MLI scheme especially when the level of voltage is increased.

In [6] the multilevel inverter in the DTC scheme employs a multilevel hysteresis controller. Although the results have shown some significant improvements, as mentioned earlier, by using the hysteresis-based controller the switching frequency of the power devices varies while the torque and flux ripple can still be considered as high. A fuzzy-PI based controller for DTC was initially introduced in[32]. The controller has been used to replace the hysteresis controller while maintaining the use of a look-up table. However the proposed controller has been applied to a 3-phase conventional inverter.

In this work a fuzzy-PI with the fixed switching frequency controller is utilized in the look-up table based DTC drive. The proposed controller consists of a fuzzy logic controller and a triangular carrier waveform. The fixed switching frequency is obtained by comparing the fuzzy logic output with the triangular waveforms. A 5-level cascaded H-bridge multilevel inverter (CMLI) is employed in this scheme. The multiple isolated input DC sources of the CMLI are particularly suitable for electric vehicle (EV) applications since the power source for an EV can be obtained from the battery modules. In the proposed strategy, the fuzzy-PI with the fixed switching frequency controller will replace the multilevel hysteresis controller for torque and flux control. Based on the proposed controller output together with the flux position, an appropriate voltage vector can be selected from the look-up table. Figure 1 shows a proposed system block diagram.

In this paper, the operational concepts of fuzzy-PI with the fixed switching frequency controller followed by the simulation results and analysis on the performance of the proposed system are presented. The results have shown that better dynamic performance, smaller torque and flux ripples, constant device switching frequency and lower THD in the phase current are achieved.

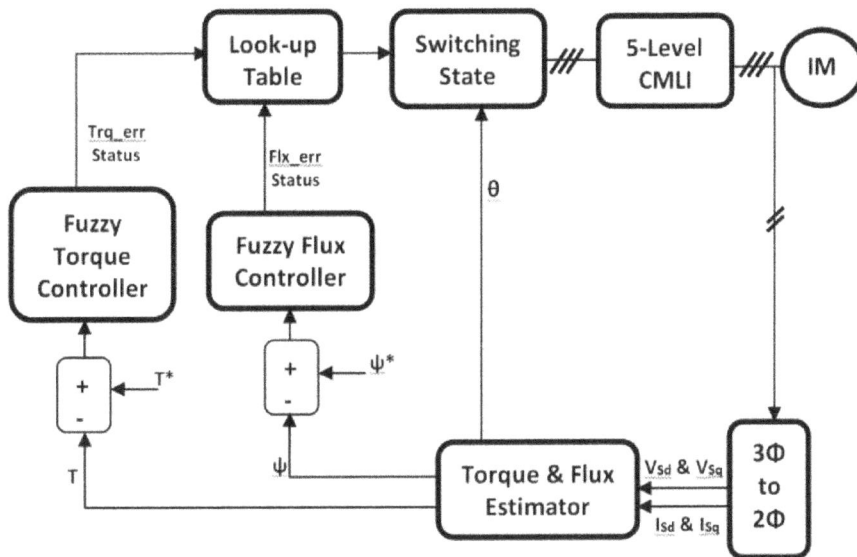

Figure 1. Proposed system block diagram

2. BASIC PRINCIPLE OF DIRECT TORQUE CONTROL (DTC)

In Figure 2 the block diagram of DTC basic control which originally introduce by [1] is shown. By having an instantaneous value of torque and flux which are calculated from the measured terminal variables of induction machine (IM), both torque and flux errors can be determined. Based on the errors, an optimum required voltage vectors are selected from the look-up tables to drive the IM.

The flux estimation is based on stator voltage vector equation in stationary reference frame which can be written as

$$\overline{v}_s = R_s \overline{i}_s + \frac{d\overline{\psi}_s}{dt} \tag{1}$$

Over the small period of time, it is assumed that the voltage drop across the stator resistance can be neglected. Therefore the equation can be rewrite as

$$\overline{v}_s = \frac{\Delta \overline{\psi}_s}{\Delta t} \tag{2}$$

It's clearly shows that the change in stator flux linkage vector, $\Delta \overline{\psi}_s$ is directly affected by the selection of stator voltage vector. Hence the stator flux locus can be controlled by selecting a suitable voltage vectors.

As for the torque estimation, it based on the stator-rotor flux angle movement. The relationship between stator-rotor flux angle and electromagnetic torque in stationary reference frame can be written as

$$T_e = \frac{3}{2} P \overline{\psi}_s \times \overline{i}_s = \frac{3}{2} P |\psi_s| |\psi_r| \sin \theta_{sr} \tag{3}$$

Where $|\psi_s|$ and $|\psi_r|$ are the magnitudes of stator flux and rotor flux linkages respectively and θ_{sr} is the stator-rotor flux angle. When a voltage vector is applied, the stator flux linkage will move faster than rotor flux linkage, where the rotor flux motion is lag behind the stator flux rotation. This is due to the to the rotor and stator leakage inductances. Therefore the stator-rotor flux angle, θ_{sr} (hence torque) is affected by the selection of appropriate voltage vector.

The abovementioned principle of both stator flux and stator-rotor angle movement are affected by the variation of stator voltage is used in DTC scheme to achieve a desired flux trajectory and torque response.

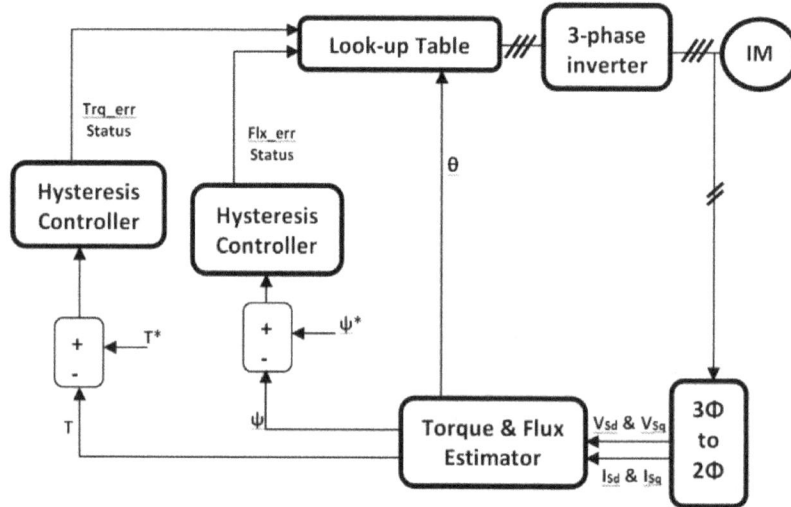

Figure 2. Conventional DTC block diagram

3.   5 LEVEL CASCDED H-BRIDGE MULTILEVEL INVERTER

5-level CMLI consist of 2 H-bridge inverter (cell) connected in cascaded form with separated DC source as shown in Figure 3. As for three-phase IM, each phase is fed by one 5-level CMLI. The configuration of three-phase IM fed by 5-level CMLI is shown in Figure 4.

The number of voltage level, L for CMLI can be obtained by

$$L = 2n + 1 \tag{4}$$

Where n is a number of cell per phase. Each cell is connected in series. Therefore the total output voltage of each phase can be determined by

$$V_{aN} = \sum_{n=1}^{2} V_{an} \tag{5}$$

$$V_{bN} = \sum_{n=1}^{2} V_{bn} \tag{6}$$

$$V_{cN} = \sum_{n=1}^{2} V_{cn} \tag{7}$$

V_{aN}, V_{bN} and V_{cN} is the phase output voltage with respect to the neutral, N. By considering each cell produce {$-V_{DC}$, 0, V_{DC}}, based on (5), (6) and (7), a 5-level output voltage for each phase is

$$V_{aN} = V_{bN} = V_{cN} = \{-2V_{DC}, -V_{DC}, 0, V_{DC}, 2V_{DC}\} \tag{8}$$

The output voltages, $\mathbf{V_S}$ generated by the inverter can be expressed in space phasor form as given in (9).

$$V_S(t) = 2/3 \ (V_{aN}(t) + aV_{bN}(t) + a^2 V_{cN}(t)) \tag{9}$$

Where $a = e^{j2\pi/3}$ and $a^2 = e^{j4\pi/3}$. In d-q form, the output voltages can be defined as

$$V_{sd} = \frac{1}{3}(2V_{aN} - V_{bN} - V_{cN}) \tag{10}$$

$$V_{sq} = \frac{1}{\sqrt{3}}(V_{bN} - V_{cN}) \tag{11}$$

Based on 5-level multilevel inverter, $5^3 = 125$ combinations of phase voltage with $3L(L-1) + 1 = 61$ voltage vectors can be generated. This can give more degrees of freedom in choosing voltage vectors for control purposes compared to the conventional 3-phase inverter. The higher the level of multilevel inverter, the more the voltage vectors generated. Figure 5 illustrates the voltage vectors generated by 5-level CMLI on a d-q plane.

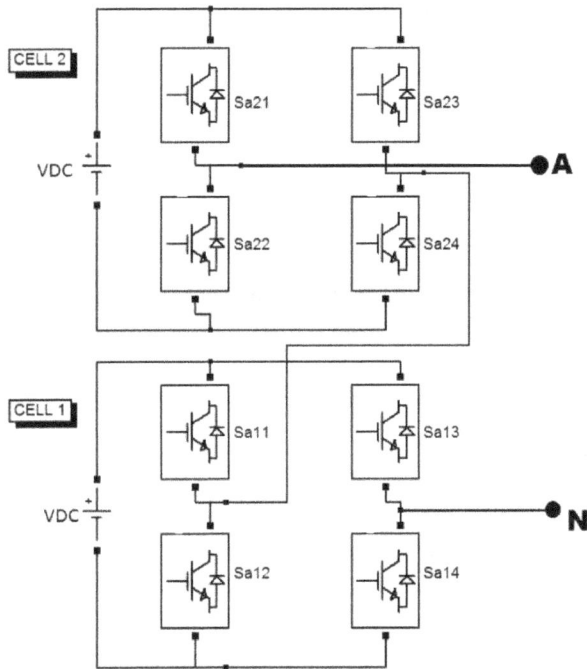

Figure 3. 5-level cascaded H-bridge multilevel inverter (CMLI)

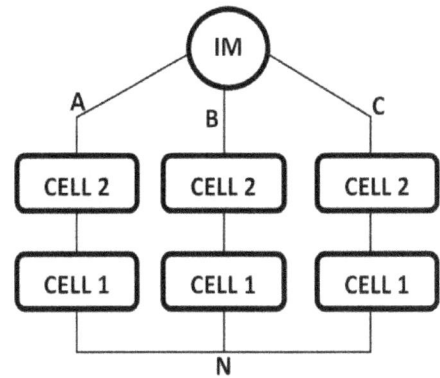

Figure 4. Three-phase IM fed by 5-level CMLI

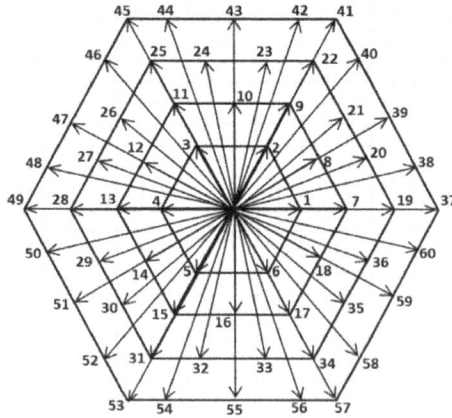

Figure 5. Voltage vector generated by 5-level CMLI

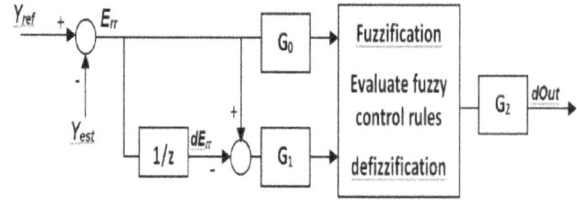

Figure 6. General structure of the Fuzzy-PI controller

4.   THE PROPOSED CONTROLLER

The proposed controller is employed as an alternative to the hysteresis-based controller with the benefit of operating at a fixed switching frequency with low torque and flux ripple. Prior knowledge of rotor parameters is not required in designing a fuzzy-PI based controller. However, the fuzzy-PI performance is highly affected by the normalization gain selection of a typical fuzzy logic controller.

The fuzzy-PI controller structure is based on the traditional PI controller [33]. The general structure of the Fuzzy-PI controller is as shown in Figure 6. It has two inputs; system error, E_{rr} and the change of system error, dE_{rr}, where $E_{rr} = Y_{ref} - Y_{est}$ and $dE_{rr} = E_{rr}(t) - E_{rr}(t-\Delta t)$. Before these input values are fed into the fuzzy controller, both values need to be normalized by using the *input normalization gain* (G_0 and G_1). As for the fuzzy controller output, *dOut*, the value need to be denormalized using the *output denormalization gain*, G_2 before it can be used as a comparator input.

4.1   Fuzzy-PI torque controller

The proposed fuzzy-PI torque controller (FTC) consists of 6 triangular waveform generators, 6 comparators and a fuzzy-PI controller. The six triangular waveform generators generate 3 pairs of triangular waveforms with the same magnitude but with different DC offset. Each pair (C_{Upper} and C_{Lower}) is 180° out of phase. In principle, the proposed controller will produce the same output as an 8-level hysteresis controller in [6], which can be either one of the following torque error status; 3, 2, 1, +0.5, -0.5, -1, -2, -3. The number of levels, however, must be realistic enough for implementation purposes. In other words, the higher the number of levels, the faster in terms of processor requirement is needed for implementation. By comparing the triangular waveforms with the fuzzy-PI controller output, a fixed switching frequency can be achieved.

Based on the general structure of fuzzy-PI controller, the FTC has two inputs; torque error, T_{err} and the change of torque error, dT_{err}, where $T_{err} = T_{ref} - T_{est}$ and $dT_{err} = T_{err}(t) - T_{err}(t-\Delta t)$ and one output, dT_O. Figure 7 shows the block diagram of the FTC.

4.2   Fuzzy-PI flux controller

The proposed fuzzy-PI flux controller (FFC) consists of 4 triangular waveform generators, 4 comparators and a fuzzy-PI controller. The four triangular waveform generators generate 2 pairs of triangular waveforms with the same magnitude but with different DC offset and similar to the FTC, each pair (C_{Upper} and C_{Lower}) is also 180° out of phase. In principle, the proposed controller will produce the same output as a 5-level hysteresis controller in [6], which can be either one of the following torque error status; 2, 1, 0, -1, -2. Again, the number of levels chosen must be practical for implementation purposes.

Based on the general structure of fuzzy-PI controller, the FFC has two inputs; flux error, Flx_{err} and the change of flux error, $dFlx_{err}$, where $Flx_{err} = Flx_{ref} - Flx_{est}$ and $dFlx_{err} = Flx_{err}(t) - Flx_{err}(t-\Delta t)$ and one output, $dFlx_O$. Figure 8 shows a block diagram of FFC.

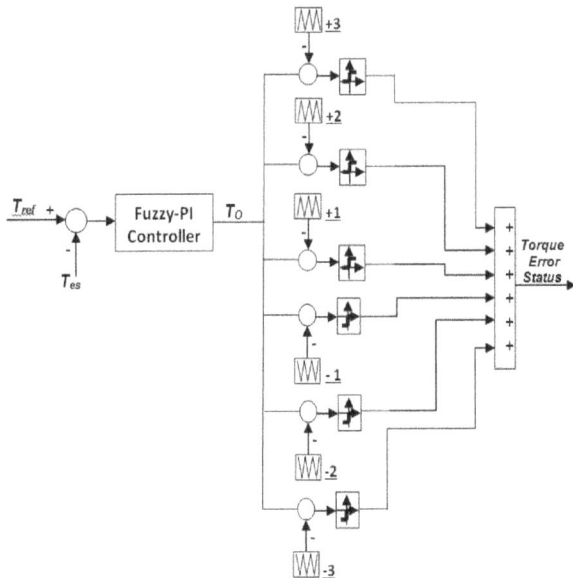

Figure 7. Fuzzy torque controller (FTC) block diagram

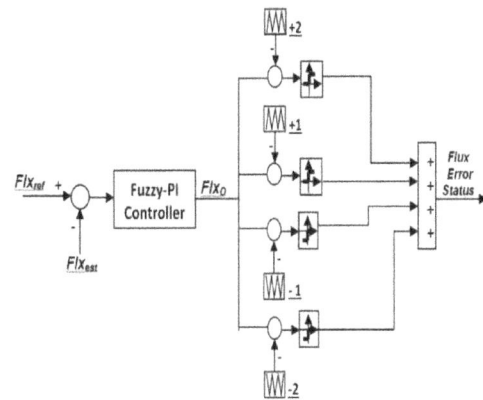

Figure 8. Fuzzy flux controller (FFC) block diagram

4.3 Fuzzification of inputs and outputs

Both FTC and FFC are using the same fuzzy control properties. The universe of discourse (UOD) for the inputs and outputs are divided into 7 overlapping subsets, +3 to -3. Table 1 shows the definitions of the fuzzy subsets in UOD.

Figure 9 shows the inputs and output membership functions of the fuzzy controller. Both inputs and outputs are connected through the rules that emulate an ideal second order system response which is expected to have good performance (response time, overshoot, damping factor and etc.) and system stability. By using the IF-THEN rules, the inputs and outputs are connected using 49 rules since 7 overlapping subsets are used. These rules are summarized and presented in the decision table as shown in Table 2.

Table 1. Fuzzy subsets definitions

Symbol	Linguistic Terms	Fuzzy Subsets
PL	Positive Large	+3
PM	Positive Medium	+2
PS	Positive Small	+1
Z	Zero	0
NS	Negative Small	-1
NM	Negative Medium	-2
NL	Negative Large	-3

Table 2. Decision table of fuzzy control rules

dT_{err}/Flx_{err} \ T_{err}/Flx_{err}	PL	PM	PS	Z	NS	NM	NL
PL	PL	PL	PM	PM	PS	PS	Z
PM	PL	PM	PM	PS	PS	Z	NS
PS	PM	PM	PS	PS	Z	NS	NS
Z	PM	PS	PS	Z	NS	NS	NM
NS	PS	PS	Z	NS	NS	NM	NM
NM	PS	Z	NS	NS	NM	NM	NL
NL	Z	NS	NS	NM	NM	NL	NL

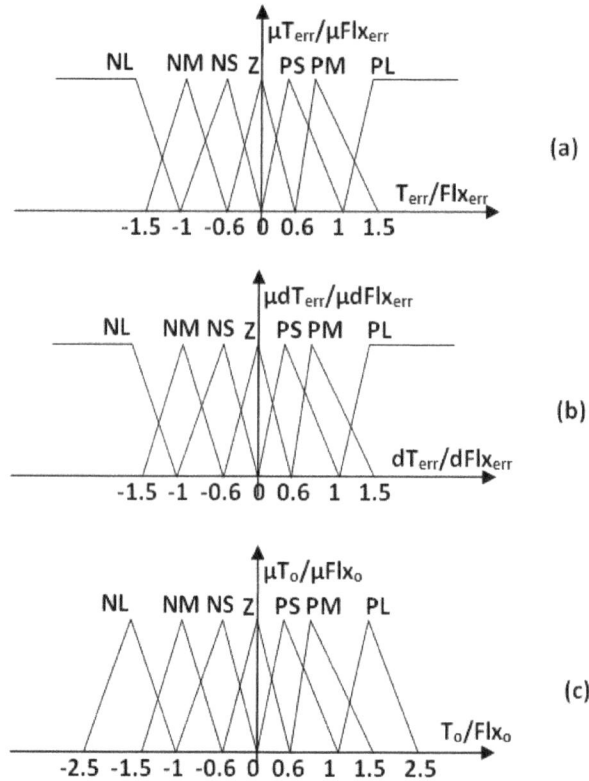

Figure 9. Membership functions of the fuzzy controller
(a) First input variable (b) Second input variable (c) Output variables

4.4   Fuzzy controller gain selection

In order to get a smaller torque and flux ripple and fixed switching frequency, the fuzzy controller gain must be properly selected to fulfill two main criteria of torque and flux controller:

i)   Produce a good dynamics of flux and torque response.

ii) Absolute slope of the torque and flux controller output (T_O and Flx_O) must not exceed the absolute slope of the triangular carrier to avoid multiple intersection and hence extremely high switching frequency.

By using trial and error method, several combinations of gain (G_0, G_1, G_2) have been found to achieve the desired response. In general, lower values of G_0 and G_1 must be accompanied with a higher value of G_2 and vice versa.

The characteristic of each gain towards the system's response can be clearly understood by referring to Table 3. In this table, the changes of G_0 and G_1 will affect the torque and flux response time and the controller output slope. Therefore the G_2 value needs an opposite adjustment to compensate the incremental or decremental effect produced by G_0 and G_1 in order to achieve the desired response. Fig. 10 shows an example of fuzzy controller output (T_O or Flx_O) with proper gain value.

Table 3. General response of fuzzy controller gain with the increasing and decreasing values of G0, G1 and G2.

↑↑ and ↓↓ Denotes highly effected   ↑ and ↓ Denotes moderately effected

	Gain	Torque & Flux Response		Slope of T_O and Flx_O
		Response time	Ripple	
G_0	Increasing	↑↑	↑↑	↑↑
	Decreasing	↓↓	↓↓	↓↓
G_1	Increasing	↓	↓	↑↑
	Decreasing	↑	↑	↓↓
G_2	Increasing	↑↑	↑↑	↑↑
	Decreasing	↓↓	↓↓	↓↓

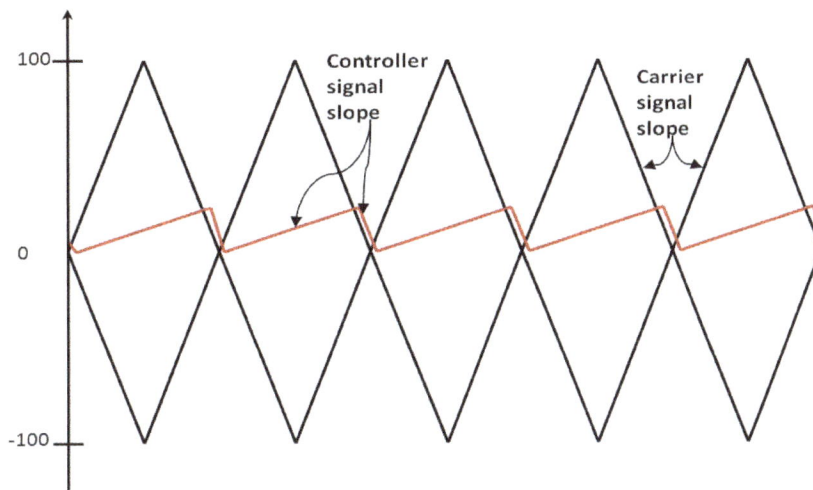

Figure 10. Fuzzy controller output with proper gain

5.   SIMULATION RESULTS

The proposed system and a hysteresis-based system have been simulated using MATLAB/Simulink for validation purposes. Both systems use the same induction machines parameters as shown in Table 4.

For the hysteresis-based system, the flux and torque hysteresis band is set to 10% of the rated values. As for the Fuzzy-PI-based controller, the values of G_0, G_1 and G_2 are set to 0.15, 0.01, 400 and 4, 0.008, 3000, for fuzzy torque and fuzzy flux controller respectively. The triangular carrier waveforms frequencies for flux and torque fuzzy-PI based controller are 5 kHz and 10 kHz respectively.

5.1   Torque and flux step response

A step reference torque and flux at $t = 0.01$s has been applied to the DTC-CMLI system. Figure 11 and Figure 12 show a step response of flux and torque for both controllers respectively. It's clearly indicated that the proposed controller responds faster than the hysteresis controller where both torque and flux with proposed controller achieved steady state at 0.0171s and 0.145s respectively while both torque and flux with hysteresis controller achieved steady state at 0.0176s and 0.0151s respectively.

In the proposed controller (DTC-CMLI with fuzzy-PI controller), the designated fuzzy rules first filtered and processed the error (torque and flux) to ensure the large error will appropriately trigger the correct triangular. Consequently, the voltage vector with highest torque and flux increment (decrement) is selected for large errors hence produce faster torque and flux response.

As for DTC-CMLI with hysteresis controller, the errors are directly controlled by hysteresis controller which produces a slower response compared to proposed controller.

Table 4. Parameter used in simulation

Induction Motor parameters		Cascaded H-Bridge Multilevel Inverter parameters	
Nominal Power	1.5 kW	V_{DC} for each cell	120 V
$V_{line-line}$ (rms)	400Vrms 50 Hz	Power electronic device	IGBT
Stator Resistance	3 Ω		
Rotor Resistance	3.793 Ω	Reference value	
Stator Inductance	0.322188 H	Torque reference	9 N.m
Rotor Inductance	0.330832 H	Flux reference	0.9 Wb
Mutual Inductance	0.3049 H		
Inertia	0.02799 kg.m^2		
Friction Factors	0.01025 N.m.s		
Pole	2 pole		

5.2 Torque and flux ripple

Figure 13 and Figure 14 show the flux and torque peak-to-peak ripples respectively for both proposed controller and hysteresis controller. These figures clearly indicate the reduction of torque and flux ripple as much as 10% in the proposed controller compared to the hysteresis controller.

Using the proposed controller in the system, the designated fuzzy rules for torque and flux controller are properly monitored and corrected the level of errors to ensure that the output fuzzy controller signals are within the appropriate triangle level hence appropriate voltage vector either to increase or decrease torque or flux is selected and been applied consecutively within a triangle waveform period.

However by using the hysteresis controller in the system, the voltage vector is chosen based on the comparison of raw error signal with the hysteresis band which the voltage vector for torque and flux increasing or decreasing are applied for entire switching period.

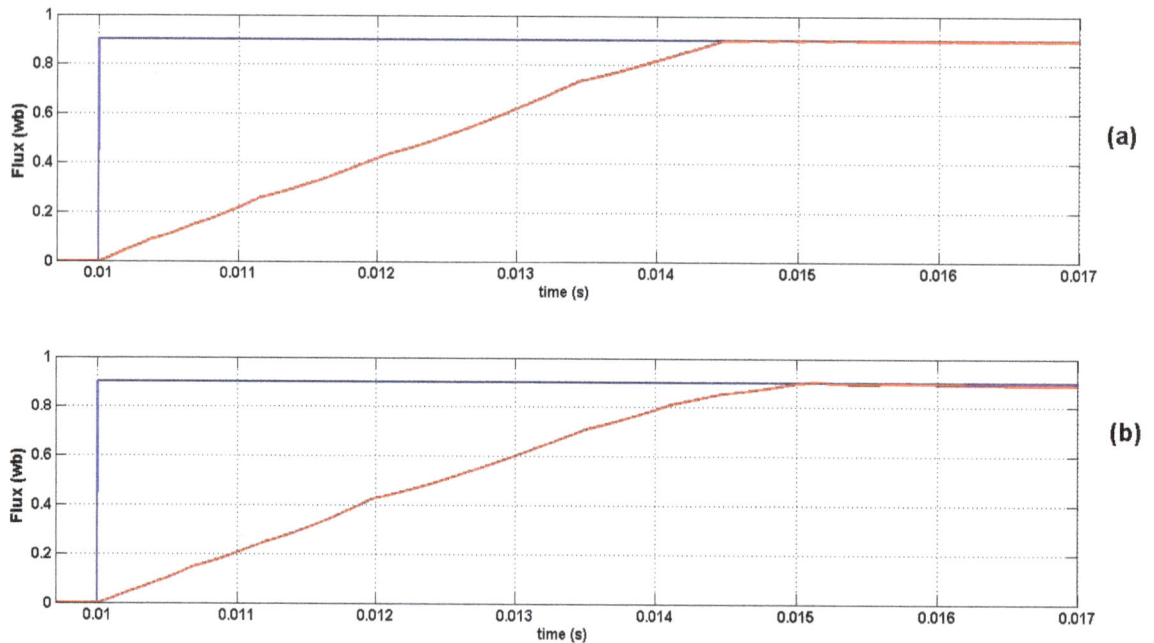

Figure 11. Stator flux step response (a) Proposed controller (b) Hysteresis controller

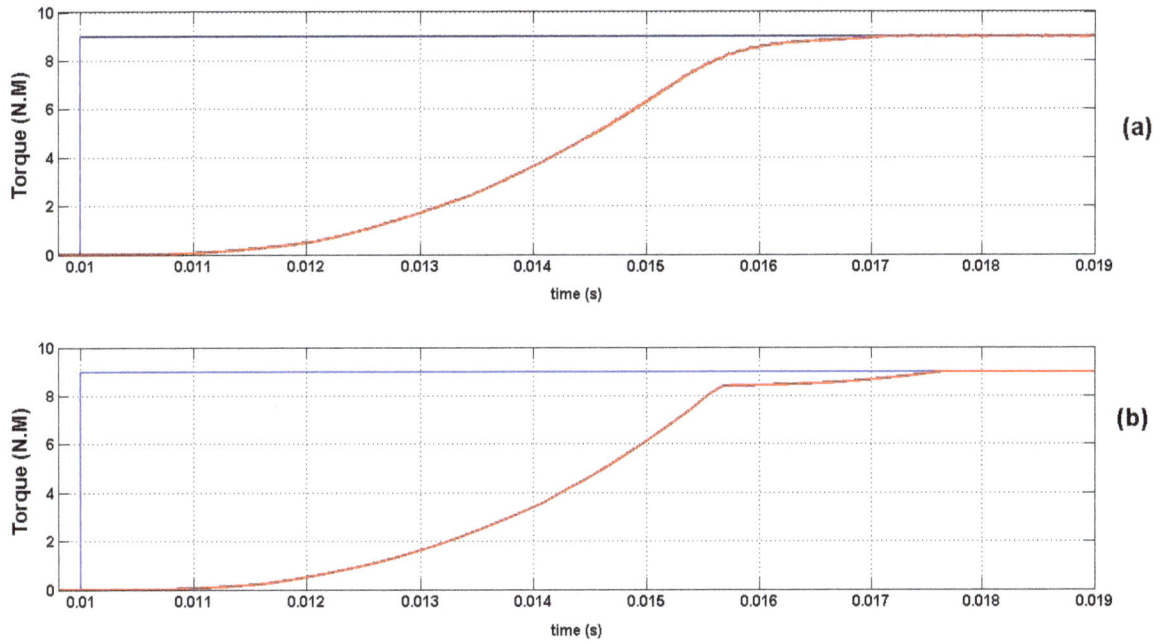

Figure 12. Torque step response (a) Proposed controller (b) Hysteresis controller

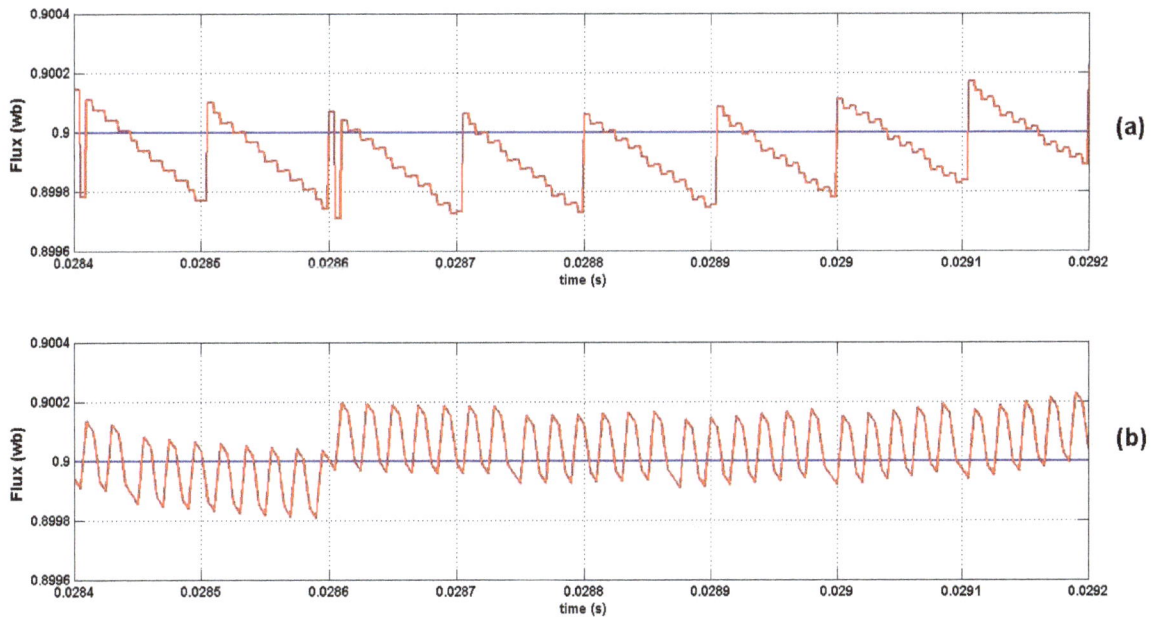

Figure 13. Stator flux ripple (a) Proposed Controller (b) Hysteresis Controller

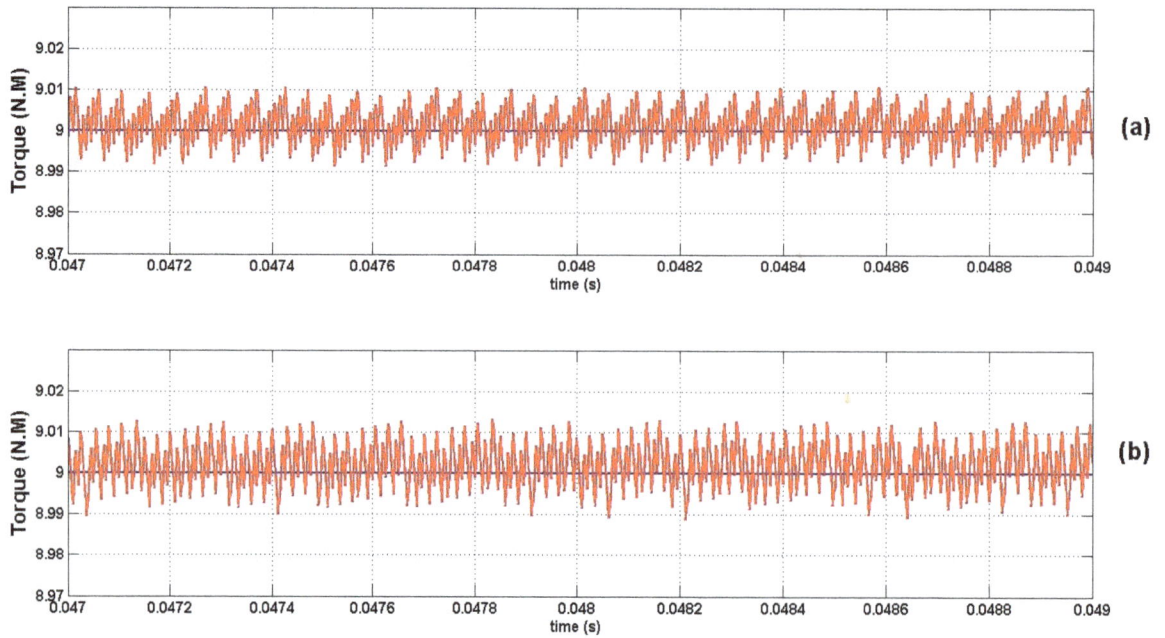

Figure 14. Torque ripple (a) Proposed Controller (b) Hysteresis Controller

5.3   Switching frequency and THD in stator phase current

Figure 15 (a) and (b) show the frequency spectrum of the switching pattern, S_{b23} for both proposed controller and hysteresis controller respectively. DTC-CMLI with hysteresis controller suffers from the varying device switching frequency. Variable switching frequency is undesirable since the switching capability of the inverter is not fully utilized. Furthermore create an unpredictable harmonics in current flow. Based on the spectrum in Figure 15 (b), the harmonics are widely distributed.

On the contrary, DTC-CMLI with proposed controller offers a constant switching frequency. Based on the spectrum in Fig. 15(a), the harmonics are concentrated around the carrier frequency and it's multiple. The first harmonic for the proposed controller appear around 10 kHz which is the triangular frequency of the torque loop controller. This is due to the torque switching is more dominant than flux switching in DTC control scheme [32].

Figure 16 and Figure 17 show the stator phase current and stator phase current harmonic spectrum respectively for both DTC-CMLI with proposed controller and DTC-CMLI with hysteresis controller. Based on these figures, DTC-CMLI with proposed system with proposed controller produced a smoother sinusoidal stator phase current with a lower current THD compared to the DTC-CMLI with hysteresis controller.

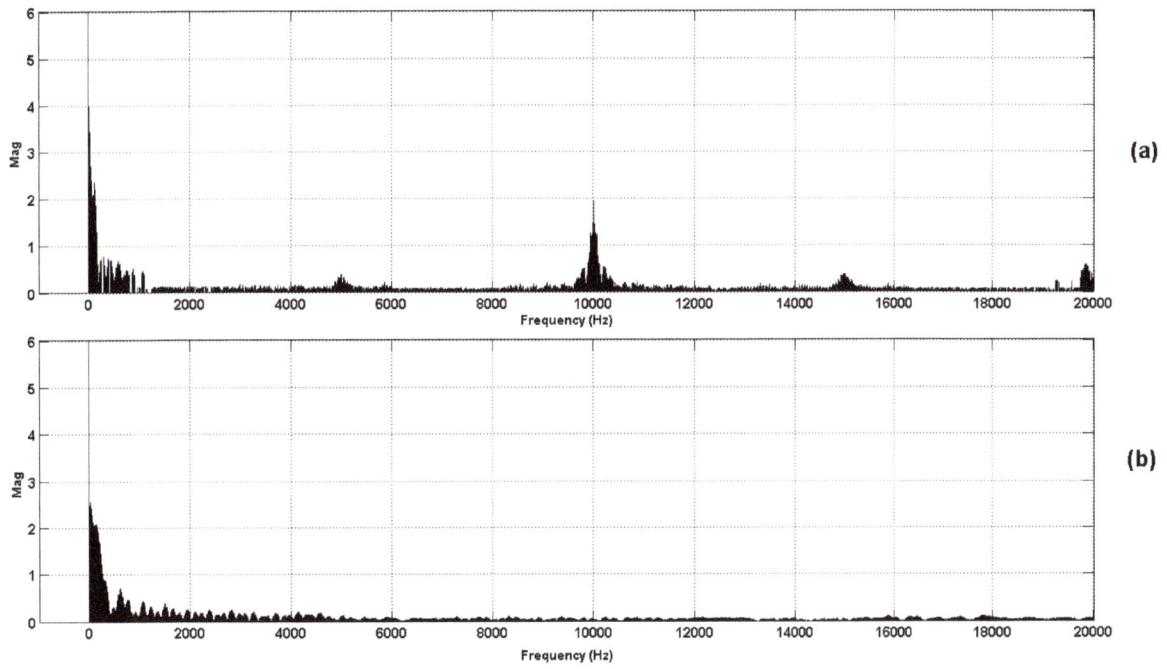

Figure 15. Frequency spectrum of the switching pattern, S_{b23} (a) Proposed controller (b) Hysteresis controller

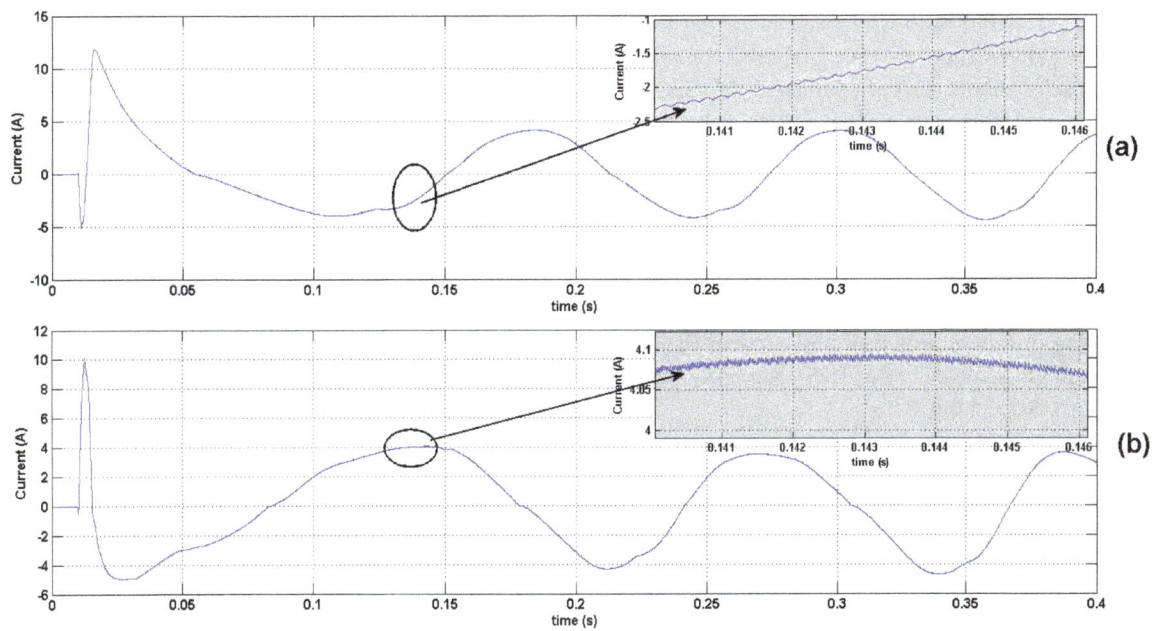

Figure 16. Stator phase current (a) Proposed controller (b) Hysteresis controller

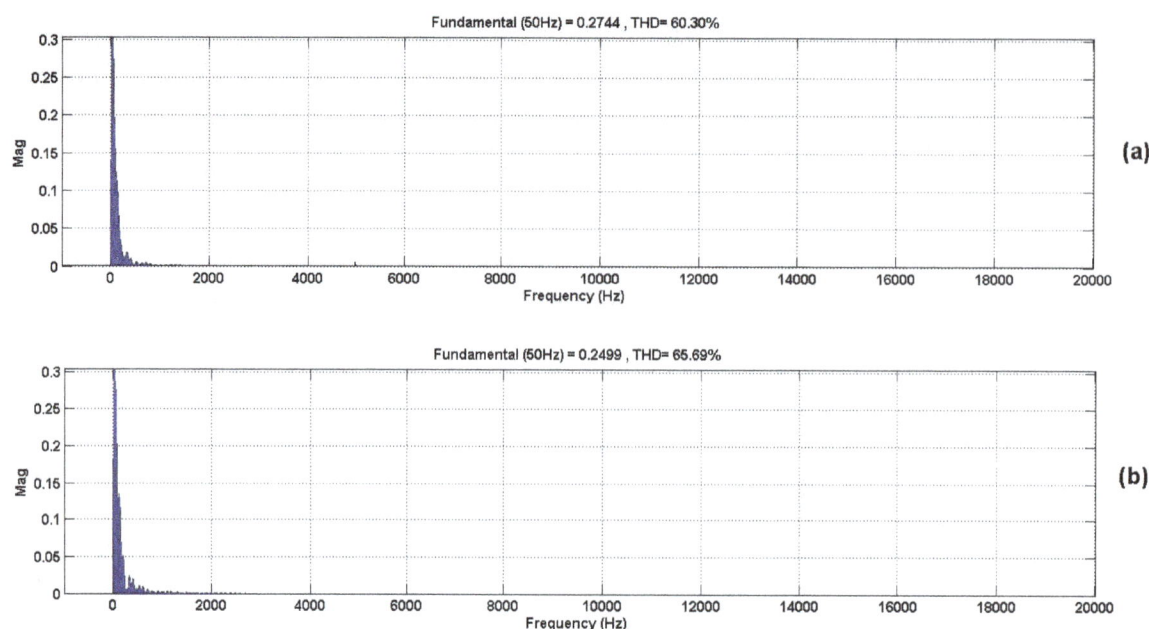

Figure 17. Stator phase current THD (a) Proposed controller (b) Hysteresis controller

6.   CONCLUSION

A Fuzzy-PI with fixed switching frequency based torque and flux controller to enhance the DTC-CMLI performance has been proposed. Its operational concept has been presented. The performance of the proposed systems has been validated through simulations. From the results, with the proposed controller, the torque and flux ripples have been reduced and is capable of producing good dynamic performance. Regardless of the operating conditions, the devices' switching frequency is able to be maintained at the triangular frequency and producing low THD in stator phase current.

ACKNOWLEDGMENT

The authors would like to thank the Ministry of Higher Education of Malaysia for the financial funding of this project with vote number 78590 and the Research Management Center (RMC) of Universiti Teknologi Malaysia for their supports on research management and funding with vote number Q.J1300002525.00H87.

REFERENCES

[1]   I. Takahashi and T. Noguchi, "A new quick-response and high-efficiency control strategy of an induction motor," *IEEE Transactions on Industry Applications*, pp. 820-827, 1986.

[2]   A. Alqudah, *et al.*, "Control of variable speed drive (VSD) based on diode clamped multilevel inverter using direct torque control and fuzzy logic," in *Applied Electrical Engineering and Computing Technologies (AEECT), 2013 IEEE Jordan Conference on*, 2013, pp. 1-6.

[3]   N. F. Alias, *et al.*, "Improved performance of Direct Torque Control of Induction Machine with 3-level Neutral Point Clamped multilevel inverter," in *Electrical Machines and Systems (ICEMS), 2013 International Conference on*, 2013, pp. 2110-2114.

[4]   Y. Zhang, *et al.*, "An improved direct torque control for three-level inverter-fed induction motor sensorless drive," *Power Electronics, IEEE Transactions on*, vol. 27, pp. 1502-1513, 2012.

[5]   N. A. Azli, *et al.*, "Direct Torque Control Of Multilevel Inverter-Fed Induction Machines – A Survey," *Journal of Theoretical and Applied Information Technology*, vol. 41, pp. 060-070, 2012.

[6] N. M. Nordin, *et al.*, "Direct Torque Control with 5-level cascaded H-bridge multilevel inverter for induction machines," in *IECON 2011 - 37th Annual Conference on IEEE Industrial Electronics Society*, Melbourne, VIC Australia, 2011, pp. 4691-4697.

[7] A. Mortezaei, *et al.*, "Direct torque control of induction machines utilizing 3-level cascaded H-Bridge Multilevel Inverter and fuzzy logic," in *Applied Power Electronics Colloquium (IAPEC), 2011 IEEE* Johor Bahru, Malaysia, 2011, pp. 116-121.

[8] Y. Zhang, *et al.*, "An Improved Direct Torque Control for Three-Level Inverter-Fed Induction Motor Sensorless Drive," *Power Electronics, IEEE Transactions on*, p. 1, 2010.

[9] L. Youb, *et al.*, "A new fuzzy logic direct torque control scheme of induction motor for electrical vehicles application," in *XIX International Conference on Electrical Machines (ICEM), 2010* Rome, 2010, pp. 1-6.

[10] P. Urrejola, *et al.*, "Direct torque control of an 3L-NPC inverter-fed induction machine: A model predictive approach," in *IECON 2010 - 36th Annual Conference on IEEE Industrial Electronics Society* Glendale, AZ, 2010, pp. 2947-2952.

[11] S.-X. Liu, *et al.*, "A novel fuzzy direct torque control system for three-level inverter-fed induction machine," *International Journal of Automation and Computing*, vol. 7, pp. 78-85, 2010.

[12] F. Khoucha, *et al.*, "Hybrid cascaded H-bridge multilevel-inverter induction-motor-drive direct torque control for automotive applications," *Industrial Electronics, IEEE Transactions on*, vol. 57, pp. 892-899, 2010.

[13] H. Alloui, *et al.*, "A three level NPC inverter with neutral point voltage balancing for induction motors Direct Torque Control," in *XIX International Conference on Electrical Machines (ICEM), 2010*, Rome, 2010, pp. 1-6.

[14] H. F. A. Wahab and H. Sanusi, "Simulink Model of Direct Torque Control of Induction Machine," *American Journal of Applied Sciences*, vol. 5, pp. 1083-1090, 2008.

[15] R. Zaimeddine, *et al.*, "An Improved Direct Torque Control Strategy for Induction Motor Drive," *International Journal of Electrical and Power Engineering, Medwell Journals*, vol. 1, pp. 21 - 27, 2007.

[16] Y. Wang, *et al.*, "Direct Torque Control with Space Vector Modulation for Induction Motors Fed by Cascaded Multilevel Inverters," 2006, pp. 1575-1579.

[17] S. Kouro, *et al.*, "Direct Torque Control With Reduced Switching Losses for Asymmetric Multilevel Inverter Fed Induction Motor Drives," 2006.

[18] X. del Toro Garcia, *et al.*, "New DTC Control Scheme for Induction Motors fed with a Three-level Inverter," *AUTOMATIKA-ZAGREB-*, vol. 46, p. 73, 2005.

[19] M. Bendyk, *et al.*, "Investigation of direct torque control system fed by modified cascade of multilevel voltage source inverter," *IEEE Compatibility in Power Electronics, 2005*, pp. 265-272, 2005.

[20] J. Rodríguez, *et al.*, "Direct torque control with imposed switching frequency in an 11-level cascaded inverter," *IEEE Transactions on Industrial Electronics*, vol. 51, pp. 827-833, 2004.

[21] J. Rodr guez, *et al.*, "Simple direct torque control of induction machine using space vector modulation," *Electronics Letters*, vol. 40, 2004.

[22] X. del Toro, *et al.*, "Direct torque control of an induction motor using a three-level inverter and fuzzy logic," in *Industrial Electronics, 2004 IEEE International Symposium on*, 2004, pp. 923-927 vol. 2.

[23] M. A. M. Prats, *et al.*, "A switching control strategy based on output regulation subspaces for the control of induction motors using a three-level inverter," *IEEE Power Electronics Letters*, vol. 1, pp. 29-32, 2003.

[24] C. A. Martins, *et al.*, "Switching frequency imposition and ripple reduction in DTC drives by using a multilevel converter," *IEEE Transactions on Power Electronics*, vol. 12, pp. 286 -297, 2002.

[25] M. F. Escalante, *et al.*, "Flying capacitor multilevel inverters and DTC motor drive applications," *IEEE Transactions on Industrial Electronics*, vol. 49, pp. 809-815, 2002.

[26] Z. Tan, *et al.*, "A direct torque control of induction motor based on three-level NPCinverter," 2001.

[27] U. Patil, *et al.*, "Torque Ripple Minimization in Direct Torque Control Induction Motor Drive Using Space Vector Controlled Diode-clamped Multi-level Inverter," *Electric Power Components and Systems*, vol. 40, pp. 792-806, 2012.

[28] F. Khoucha, *et al.*, "A comparison of symmetrical and asymmetrical three-phase H-bridge multilevel inverter for DTC induction motor drives," *Energy Conversion, IEEE Transactions on*, vol. 26, pp. 64-72, 2011.

[29] G. Sheng-wei and C. Yan, "Research on torque ripple minimization strategy for direct torque control of induction motors," in *International Conference on Computer Application and System Modeling (ICCASM), 2010* Taiyuan 2010, p. V1.

[30] T. R. Obermann, *et al.*, "Deadbeat-direct torque & flux control motor drive over a wide speed, torque and flux operating space using a single control law," in *Energy Conversion Congress and Exposition (ECCE), 2010 IEEE* Atlanta, GA 2010, pp. 215-222.

[31] S. Kouro, *et al.*, "High-performance torque and flux control for multilevel inverter fed induction motors," *IEEE Transactions on Power Electronics*, vol. 22, pp. 2116-2123, 2007.

[32] F. Patkar, *et al.*, "A New Fuzzy-PT-based Torque Controller for DTC of Induction Motor Drives," 2006, pp. 62-66.

[33] A. G. Perry, *et al.*, "A design method for PI-like fuzzy logic controllers for DC–DC converter," *Industrial Electronics, IEEE Transactions on*, vol. 54, pp. 2688-2696, 2007.

13

Sliding Mode Backstepping Control of Induction Motor

Othmane Boughazi, Abdelmadjid Boumedienne, Hachemi Glaoui

Faculty of Sciences and technology, BECHAR University B.P. 417 BECHAR, 08000

ABSTRACT

Keyword:

Backstepping
Induction motor
Proportional-integral (PI)
Sliding mode control
Sliding mode control

This work treats the modeling and simulation of non-linear system behavior of an induction motor using backstepping sliding mode control (BACK-SMC). First, the direct field oriented control IM is derived. Then, a sliding for direct field oriented control is proposed to compensate the uncertainties, which occur in the control. Finally, the study of Backstepping sliding controls strategy of the induction motor drive. Our non linear system is simulated in MATLAB SIMULINK environment, the results obtained illustrate the efficiency of the proposed control with no overshoot, and the rising time is improved with good disturbances rejections comparing with the classical control law.

Corresponding Author:

Hachemi Glaoui
Faculty of Sciences and technology,
BECHAR University,
B.P. 417 BECHAR, 08000
Email: mekka60@gmail.com

1. INTRODUCTION

The development of induction motor drives hasconsiderably accelerated in order to satisfy the increasing need of various industrial applications in low and medium power range. Indeed, induction motors have simple structure, high efficiency and increased torque/inertia ratio. However, their dynamical model is nonlinear, multivariable, coupled, and is subject to parameter uncertainties since the physical parameters are time-variant. The design of robust controllers becomes then a relevant challenge [1]-[2].

Induction motor drives control has been an activeresearch domain over the last years. Different control techniques such as Field-Oriented control (FOC), feedback linearization control, sliding mode control passivity approach, and adaptive control have been reported in the literature [3]. The FOC ensures partial decoupling of the plant model using a suitable transformation and then PI controllers are used for tracking regulation errors. The high performance of suchstrategy may be deteriorated in practice due to plantuncertainties [4]-[5]. Exact input-output feedback linearization of induction motors model can be obtained using tools from differential geometry. This method cancels the nonlinear terms in the plant model which fails when the physical parameters varies [6]-[7]. By contrast, passivity-based control does not cancel all the nonlinearities but enforce them to be passive, i.e. dissipating energy and hence ensuring tracking regime [8]-[10]. Sliding Mode Control (SMC) is widely applied because of its easiness and attractive robustness properties [11]-[12].

Otherwise, the conventional PI controllers are the most common algorithms used in industry today. Theirattractiveness is due to their structure simplicity and the industrial operators acquaintance with them. Several PI controllers have been proposed in the literature for linear and nonlinear processes [5], [15]. Nevertheless, PI controllers fundamental deficiency is the lack of asymptotic stability and robustness proofs for a given nonlinear system.

Therefore, this paper proposes to deal with this deficiency by proposing a robust nonlinear PI controller for an induction motor drive with unknown load torque. The controller is derived by combining a backstepping procedure with a sliding mode. More precisely, the controllers are determined by imposing the current-speed tracking recursively in two steps and by using appropriate gains that are nonlinear functions of the system state.The advantage ofBackstepping sliding mode control is its robustness and ability to handle the non-linear behaviour of the system.

The model of the induction motor, and shows the direct field-oriented control (FOC) of induction motor in Section (2). Section (3) showsthe development of sliding technique control design. Section (4) shows the development of Backstepping technique controldesign. The Speed Control ofinduction machine by Backstepping sliding mode controllers design is given in section (5). Simulation results using MATLAB SIMULINK of different studied cases is defined in Section (6). Finally, the conclusions are drawn in Section (7).

2. MATHEMATICAL MODE OF IM

The used motor is a three phase induction motor type (IM) supplied by an inverter voltage controlled with Pulse Modulation Width (PWM) techniques. A model based on circuit equivalent equations is generally sufficient in order to make control synthesis. The dynamic model of three-phase, Y-connected induction motor can be expressed in the d-q synchronously rotating frame as [13]:

$$
\begin{cases}
\dfrac{di_{ds}}{dt} = a_1.i_{ds} + w_s.i_{qs} + a_2.\phi_{dr} + b.V_{ds} \\[2mm]
\dfrac{di_{qs}}{dt} = a_1.i_{qs} - w_s.i_{ds} + a_3.\phi_{dr}.w_m + b.V_{qs} \\[2mm]
\dfrac{d\phi_{dr}}{dt} = a_4 i_{ds} + a_5 \phi_{dr} \\[2mm]
\dfrac{dw_m}{dt} = a_6 \left(i_{qs}.\phi_{dr}\right) + a_7.w_m + a_8 C_r
\end{cases}
\tag{1}
$$

Where σ is the coefficient of dispersion and is given by:

$$
\sigma = 1 - \frac{L_m^2}{L_s L_r}, \; b = \frac{1}{sig.L_s}, \; a_1 = -b\left(R_s + \left(\frac{L_m}{L_r}\right)^2.R_r\right), \; a_2 = \frac{L_m.R_r}{\left(sigL_s.L_r^2\right)},
$$

$$
a_3 = -b\frac{L_m}{L_r} \; a_4 = \frac{L_m.R_r}{L_r}, a_5 = \frac{-R_r}{L_r}, a_6 = \frac{P^2.L_m}{J.L_r}, a_7 = \frac{f_c}{J}, a_8 = \frac{p}{J}
\tag{2}
$$

3. SLIDING MODE CONTROL

A Sliding Mode Controller (SMC) is a Variable Structure Controller (VSC) [10]. Basically, a VSC includes several different continuous functions that can map plant state to a control surface, whereas switching among different functions is determined by plant state represented by a switching function [7]. Without lost of generality, consider the design of a sliding mode controller for the following first-order System.

$$
x^{\bullet} = A(x,t).x + B(x,t).U
\tag{3}
$$

Where U the input to the System the following is a possible choice of the structure of a sliding mode controller

$$
U = U_{eq} + K.Sign\left[s(x,t)\right]
\tag{4}
$$

Where stands for equivalent control used when the System state is in the Sliding mode [2], K is a constant, being the maximal value of the controller output. S is switching function since the control action switches its sign on the two sides of the switching surface $S(x) = 0$, S is defined as [14]:

$$S(x) = \left(\frac{\partial}{\partial t} + \lambda\right)^{r-1} . e(x) \tag{5}$$

Where:

$$e(x) = x^* - x,$$

x^* Being the desired state. λ is a constant. Concerning the development of the control law, it is divided into two parts, the equivalent control Ueq and the attractively or reachability control Us. The equivalent control is determined off-line with a model that represents the plant as accurately as possible. If the plant is exactly is exactly identical to the model used for determining Ueq and there are no disturbances, there would be no need to apply an additional control Us. However, in practice there are a lot of differences between the model and the actual plant. Therefore, the control component Us is necessary which will always guarantee that the state is attracted to the switching surface by satisfying the condition [13], [14].

$$\dot{S}(x) . S(x) < 0$$

Therefore, the basic switching law is of the form:

$$U = U_{eq} + U_{sw} \tag{6}$$

U_{eq} is the equivalent control, and U_{sw} is the switching control. The function of U_{eq} is to maintain the trajectory on the sliding surface, and the function of U_{sw} is to guide the trajectory to this surface.

The surface is given by:

$$S_1 = z_1 = w_{md}^* - w_m \tag{7}$$

The derivative of the surface is:

$$\dot{S}_1 = \dot{z}_1 = \dot{w}_{md}^* - \dot{w}_m \tag{8}$$

In a conventional variable structure control, Un generates a high control activity. It was first taken as constant, a relay function, which is very harmful to the actuators and may excite the model dynamics of the System. This is known as a chattering phenomenon. Ideally, to reach the sliding surface, the chattering phenomenon should be eliminated [13], [14]. However, in practice, chattering can only be reduced.

The first approach to reduce chattering was to introduce a boundary layer around the sliding surface and to use a smooth function to replace the discontinuous part of the control action as follows:

$$\begin{cases} U_{sw} = \dfrac{K}{\varepsilon}.S(x) & if \quad |S(x)| < \varepsilon \\ U_{sw} = K.\operatorname{sgn}(S(x)) & if \quad |S(x)| > \varepsilon \end{cases} \tag{9}$$

The constant K is linked to the speed of convergence towards the sliding surface of the process (the reaching mode). Compromise must be made when choosing this constant, since if K is very small the time response is important and the robustness may be lost, whereas when K is too big the chattering phenomenon increases.

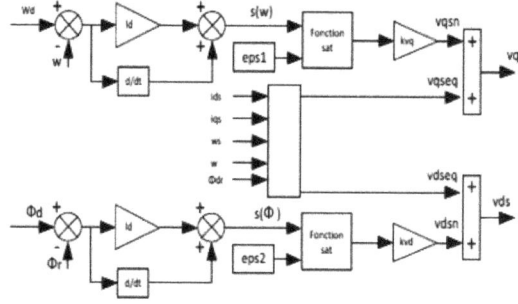

Figure 1. Block diagram speed control of IM Indirect field-oriented control (IFOC) of inductionby sliding mode

4. BACKSTEPPING CONTROL DESIGN

In this section, we use the Backstepping algorithm to develop the speed control law of the induction motor This speed will converge to the reference value from a wide set of initial conditions.

Step 1: Firstly we consider the tracking objective of the direct current (ϕ_{dr}). A tracking error $z_1 = w_{md} - w_m$ is defined, and the derivative becomes:

$$\dot{z}_1 = \dot{w}_{md} - \dot{w}_m \tag{10}$$

$$\dot{z}_1 = \dot{w}_{md} - a_6.i_{qs}.\phi_{dr} - a_7 w_m - a_8.C_r \tag{11}$$

The proposed virtual command is:

$$\left(a_6.i_{qs}.\phi_{dr}\right)^* = -a_7.w_m - a_8.C_r + c_1.z_1 + \dot{w}_{md} \tag{12}$$

$$\dot{v}(z_1) = z_1.\dot{z}_1 = z_1\left[\dot{w}_{md} - \left(-a_7.w_m - a_8.C_r + c_1.z_1 + \dot{w}_{md}\right) - a_7 w_m - a_8 C_r\right]$$
$$\dot{v}(z_1) = -c_1 z_1^2 \tag{13}$$

With $c_1 \succ 0$

Step 2: The derivative of the error variable

$$z_2 = \left(a_6.i_{qs}.\phi_{dr}\right)^* - a_6.i_{qs}.\phi_{dr} \tag{14}$$

$$z_2 = \left(-a_7.w_m - a_8.C_r + c_1.z_1 + \dot{w}_{md}\right) - a_6.i_{qs}.\phi_{dr} \tag{15}$$

$$\dot{z}_2 = \left[-a_7\left(a_6.i_{qs}.\phi_{dr} + a_7 w_m + a_8 C_r\right) - a_8\dot{C}_r + \ddot{w}_{md} + c_1\left(\dot{w}_{md} - a_6.i_{qs}.\phi_{dr} - a_7 w_m - a_8 C_r\right)\right]$$
$$- a_6\left[\phi_{dr}\left(a_1 i_{qs} - w_s i_{ds} + a_3\phi_{dr} w_m + bv_{qs}\right) + i_{qs}\left(a_4 i_{ds} + a_5\phi_{dr}\right)\right]$$

$$\dot{z}_2 = \Phi_1 + \Phi_2 v_{qs}$$

$$\Phi_1 = \left[\ddot{w}_{md} + c_1\dot{w}_{md} - a_8\dot{C}_r + \left(-c_1 - a_7\right)\left(a_6.i_{qs}.\phi_{dr} + a_7 w_m + a_8 C_r\right)\right]$$
$$- a_6\left[\phi_{dr}\left(a_1 i_{qs} - w_s i_{ds} + a_3\phi_{dr} w_m\right) + i_{qs}\left(a_4 i_{ds} + a_5\phi_{dr}\right)\right]$$

$$\Phi_2 = -a_6 b \phi_{dr} \tag{16}$$

$$\dot{v}(z_2) = z_2 . \dot{z}_2 = z_2 (\Phi_1 + \Phi_2 v_{qs}) \tag{17}$$

$$v_{qs} = \frac{-\Phi_1 - c_2 z_2}{\Phi_2}$$

$$\dot{v}(z_2) = -c_2 z_2^{\;2} \tag{18}$$

With $c_2 \succ 0$

Step 3:

$$z_3 = \phi_{dr}^* - \phi_{dr} \tag{19}$$

$$\dot{z}_3 = \dot{\phi}_{dr}^* - \dot{\phi}_{dr} = \dot{\phi}_{dr}^* - a_4 i_{ds} - a_5 \phi_{dr} \tag{20}$$

The proposed virtual command is:

$$(a_4 i_{ds})^* = -a_5 \phi_{dr} + \dot{\phi}_{dr}^* + c_3 z_3 \tag{21}$$

$$\dot{z}_3 = \dot{\phi}_{dr}^* - \left(- a_5 \phi_{dr} + \dot{\phi}_{dr}^* + c_3 z_3\right) - a_5 \phi_{dr}$$

$$\dot{v}(z_3) = z_3 . \dot{z}_3 = -c_3 z_3^{\;2} \tag{22}$$

With $c_3 > 0$

Step 4:

$$z_4 = (a_4 i_{ds})^* - a_4 i_{ds} \tag{23}$$

$$z_4 = (a_4 i_{ds})^* - a_4 i_{ds} = -a_5 \phi_{dr} + \dot{\phi}_{dr}^* + c_3 z_3 - a_4 i_{ds}$$

$$\dot{z}_4 = \Phi_3 + \Phi_4 v_{ds} \tag{24}$$

With,

$$\Phi_3 = \ddot{\Phi}_{dr}^* + c_3 \dot{\Phi}_{dr}^* - (a_5 + c_3)(a_4 i_{ds} + a_5 \phi_{dr}) - a_4 (a_1 i_{ds} + w_s i_{qs} + a_2 \Phi_{dr})$$

$$\Phi_4 = -a_4 b$$

$$v_{ds} = \frac{-\Phi_3 - c_4 z_4}{\Phi_4} \tag{25}$$

$$\dot{z}_4 = \Phi_3 + \left(\frac{-\Phi_3 - c_4 z_4}{\Phi_4}\right)\Phi_4$$

$$\dot{v}(z_4) = z_4.\dot{z}_4 = -c_4 z_4^2 \qquad (26)$$

With $c_4 > 0$.

5. ASSOCIATION BACKSTEPPING SLIDING MODE CONTROL

The control law obtained is:

$$\dot{z}_2 = \Phi_1 + \Phi_2 v_{qs} = -q_1 sign(z_2) - q_2 z_2$$

Then;

$$v_{qs} = \frac{(-\Phi_1 - q_1 sign(z_2) - q_2 z_2)}{\Phi_2} \qquad (27)$$

And,

$$\dot{z}_4 = \Phi_3 + \Phi_4 v_{ds} = -q_3 sign(z_4) - q_4 z_4$$

Then;

$$v_{ds} = \frac{(-\Phi_3 - q_3 sign(z_4) - q_4 z_4)}{\Phi_4} \qquad (28)$$

The Figure 2 shows the backstepping sliding control strategy scheme for each induction motor.

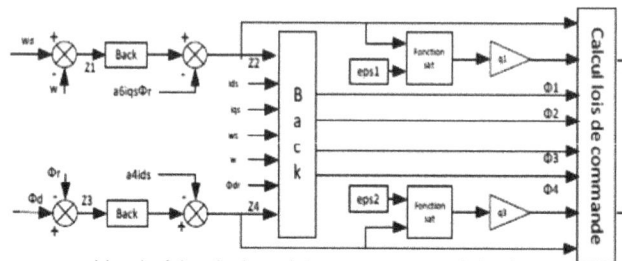

Figure 2. Block diagramspeed control of IM by a combination of the BACK-SMC command

6. SIMULATION RESULTS

The three controls adopted as PI, sliding mode, and Backstepping sliding mode are tested by numerical simulation for the values of these coefficients:

PI	$Rho_{pw}=10$	$Rho_{iw}=10$	$Rho_d=800$	$Rho_q=800$
SMC	Ld=800	Eps1=10^{-4}	Kvd=2500	
	Lq=500	Eps2=10^{-4}	Kvq=1500	
Back-SMC	C1=100	q1=100	q3=100	Eps1=10^{-6}
	C2=300	q2=100	q4=100	Eps2=10^{-6}

The simulation results are showing Figure 3-Figure 5. Figure 3 shows the speed with PI, SMC And Back-SMC, Figure 4 the torque and Figure 5 shows the current $I_{a,b,c}$.

Figure (3), (4) and (5) shows the evolution of electrical and mechanical parameters of the IM ideal voltage supplied to a load variation between 0.5 seconds and 1 second, and reverse speed set point at time 1.5 second 200 -200 [rad / s].

The results show a good response in IM alimented ideal tension, continuing with a very low response time and a static error to zero for Backstepping control mode by sliding the control input to the PI controller, and controller with sliding mode. The couple has a peak related to the start and fades during permanent regime. The load change has also allowed us to conclude on the rejection of the disturbance which is satisfactory

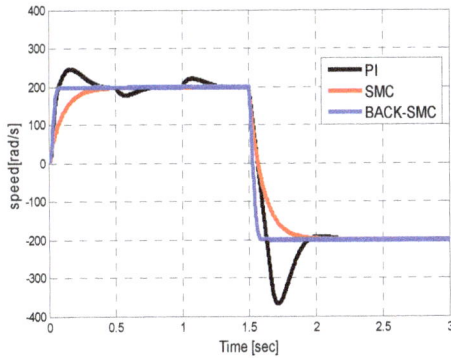

Figure 3. The speed of IM

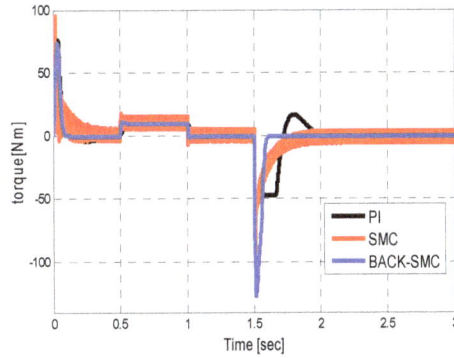

Figure 4. The torque

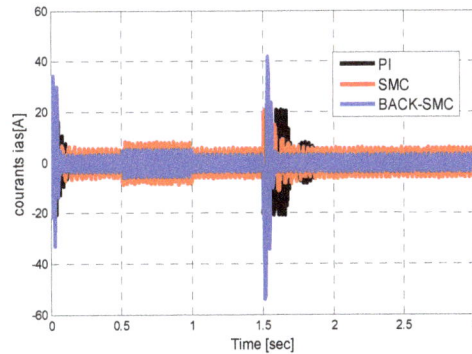

Figure 5. The Current Ia,b,c

7. CONCLUSION

The sliding mode control of the field oriented induction motor was proposed. To show the effectiveness and performances of the developed control scheme, simulation study was carried out. Good results were obtained despite the simplicity of the chosen sliding surfaces. The robustness and the tracking qualities of the proposed control system using sliding mode controllers appear clearly.

Furthermore, in order to reduce the chattering, due to the discontinuous nature of the controller, backstepping controllers were added to the sliding mode controllers.

These gave good results as well and simplicity with regards to the adjustment of parameters. The simulations results show the efficiency of the sliding mode controller technique, however the strategy of backstepping sliding mode controller brings good performances.

REFERENCES
[1] PC Krause, et al. Analysis of Electric Machinery and Drive Systems. John Wiley & Sons: 2002.
[2] PV Kokotovic, et al. Nonlinear and Adaptive control. John Wiley & Sons: 1995.
[3] Leonhard W. Control of Electrical Drives. Springer-Verlag, 1996.
[4] A Behal et al. An improved IFOC for the induction motor. IEEE Trans. Control Systems Technology. 2003 11(2): 248-252.
[5] D Casadei, et al. FOC and DTC: two viable schemes for induction motors torque control. *IEEE Trans. Power Electronics.* 2002; 17(5): 779-787.

[6]   J Chiasson. A new approach to dynamic feedback linearizationcontrol of an induction motor. *IEEE Trans. Automatic Control.* 1998; 42(3): 391-397.

[7]   R Marino et al. Adaptive input-output linearizing control of induction motors. *IEEE Tran. Automatic Control.* 1993; 38(2): 208-221.

[8]   GW Chang et al. Tuning rules for the PI gains of field-oriented controllers of induction motors. *IEEE Trans. Industrial Electronics.* 2000; 47(3): 592-602.

[9]   C Cecati et al. Torque and speed regulation of induction motors using passivity theory approach. *IEEE Tran. Industrial Electronics.* 1999; 46(1): 23-36.

[10] M Comanescu et al. Decoupled current control of sensorless induction-motor drives by integral sliding mode," *IEEE Trans. Industrial Electronics*, vol. 55, n°11, pp. 3836-3845, November 2008.

[11] A Derdiyok. Speed-sensorless control of induction motor using a continuous control approach of sliding-mode and flux observer. *IEEE Trans. Industrial Electronics.* 2005; 52(4): 1170-1176.

[12] Zhongze Chen Changhong Shan Huiling Zhu. Adaptive Fuzzy Sliding Mode Control Algorithm for a Non-Affine Nonlinear System » Energy Conversion, IEEE Transactions on. 2004; 19(2): 362 – 368

[13] M Zhiwen, T Zheng, F Lin, X You. *A New Sliding-Mode Current Controller for Field Oriented Controlled Induction Motor Drives.* IEEE Int. Conf. IAS. 2005: 1341-1346.

[14] RJ Wai. Adaptive Sliding-Mode Control for Induction Servomotor Drives. *IEE Proc. Elecrr. Power Appl.,* 2000; 147: 553-562.

[15] I Boldea et al. Vector Control of AC Drives. CRC Press: 1992.

14

Advanced Control of Wind Electric Pumping System for Isolated Areas Application

Mohamed Barara*, Abderrahim Bennassar*, Ahmed Abbou*, Mohamed Akherraz*, Badre Bossoufi**
* Laboratory of Power Electronic and Control, Mohamed V University Agdal Mohammadia School of Engineering, Rabat Morocco
** Laboratory of Electrical Engineering and Maintenance, Higher School of Technology, EST-Oujda, University of Mohammed I, Morocco

Keyword:	ABSTRACT
DC Voltage PWM Converter SIEG Wind Pump	The supply water in remote areas of windy region is one of most attractive application of wind energy conversion. This paper proposes an advanced controller suitable for wind-electric pump in isolated applications in order to have a desired debit from variation of reference speed of the pump also the control scheme of DC voltage of SIEG for feed the pump are presented under step change in wind speed. The simulation results showed a good performance of the global proposed control system.

Corresponding Author:

Mohamed Barara,
Laboratory of Power Electronic and Control,
Mohamed V University Agdal Mohammadia School of Engineering,
Rabat Morocco.
Email: Mohamed-barara@hotmail.fr

1. INTRODUCTION

The climate change is a complex global challenge its impacts on the environment and human health are now more understood requiring real solutions to reduce greenhouse gas emissions and accelerate the transition to a low-carbon future. Eventually, the world will run out of fossil fuels, or it will become too expensive to retrieve those that remain. Fossil fuels also causes air, water and soil pollution, and produce greenhouse gases that contribute to global warming.

In recent years, wind has become an increasingly attractive source of renewable energy. While wind power helps the environment by producing electricity without producing pollution, great resource to generate energy in remote locations, Efficient and reliable and they will never run out.

The wind energy can be used for pumping water and are particularly useful in remote locations where access to electrical utilities would be costly, different types of wind water pumps. Some are straight wind water pumps, such as the Aermotor windmill, while others are wind-electric pumps. In this case, the spinning of the wind turbine creates electricity that is used to power a water pump. Although mechanical windmills still provide a sensible, low-cost option for pumping water in low-wind areas, wind-electric pumping systems is an emerging technology that combines modern high-reliability small wind turbines and standard electric centrifugal pumps to provide a reliable and they can pump. In addition, mechanical windmills must be placed directly above the well, which may not take the best advantage of available wind resources. Wind-electric pumping systems can be placed where the wind resource is the best and connected to the pump motor with an electric cable and possibility to control the pumping system [24].

An overview of complete mechanical and electrical wind pumping systems is presented in Figure 1 and Figure 2.

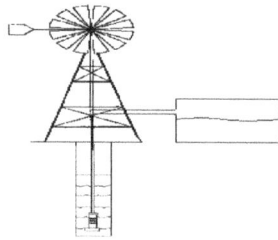

Figure 1. Mechanical Wind Pump

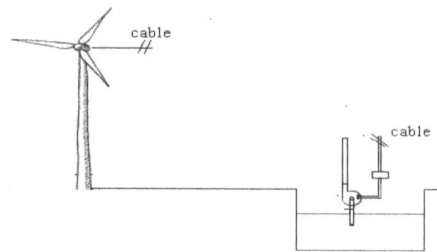

Figure 2. Wind electric pump systems

In this context our application aims to make good use of the wind electric pump system, including a good control of the pumping speed, and of course ensuring a control of the voltage source of the generator.

The induction machine is a very popular generator used wind turbine systems in isolated areas to generate electrical energy also for motor application because of its low price, mechanical simplicity, robust structure, as compared to other machine. However, the major draw back of the SIEG (Self Excited Induction Generator). A poor voltage regulation under change in load and speed in stand-alone system. In literature many researchers have proposed numerous control for regulating the terminal voltage [1]-[2], [4]-[5], [8], [10], [12]-[13].

Since the aims of the proposed system consists of a SEIG driven by an unregulated rotor speed supplies induction motor loaded with a centrifugal pump (non linear load). The proposed control should have keeps the DC bus voltage at a constant value for supplied inverter when the speed of the wind change and the application of pump, based in this regulation then we are more interested towards a desired state while varying the debit from the variation of the reference speed of the pump.

The indirect vector control using rotor flux orientation for two controls with fuzzy logic regulation applied in order to carry out DC voltage of SIEG and speed of the pump. Detailed Matlab/Simulink-based simulation studies are carried out to demonstrate the effectiveness of the proposed scheme.

2.   SYSTEM DESCRIPTION AND CONTROL SCHEME

A self-excited induction generator using three phase AC capacitors can start its voltage buildup only from a remnant magnetic flux in the core, the voltage buildup starts when the induction generator is driven at a given speed and an appropriate capacitance connected at its terminals, However, for a system with a single DC capacitor as proposed in this paper it cannot start the voltage buildup from the remnant flux in the core [2]. The proposed system starts its excitation process from a charging circuit an external battery. Since this paper focuses on modeling and behavior of the electrical part of the system, the turbine is not taken into account. Rotor speed is taken as an independent and variable input into the model. The main components of the proposed system are shown in Figure 3.

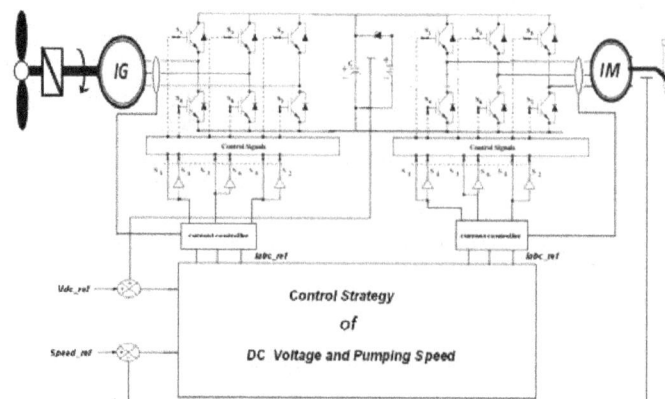

Figure 3. Control structure proposed

The components are Induction Generator, PWM rectifier PWM inverter, and Induction Motor coupled with a centrifugal pump those modelling are explained below.

2.1. Mathematical Model of Induction Machine

The following equations describe model of the squirrel-cage induction machine the stationary dq reference frame:

$$V_{sd} = R_s.i_{sd} + \frac{d\psi_{sd}}{dt} - \frac{d\theta_s}{dt}\psi_{sq} \tag{1}$$

$$V_{sq} = R_s.i_{sq} + \frac{d\psi_{sq}}{dt} + \frac{d\theta_s}{dt}\psi_{sd} \tag{2}$$

$$\psi_{rd} = L_r.i_{rd} + M.i_{sd} \tag{3}$$

$$\psi_{rq} = L_r.i_{rq} + M.i_{sq} \tag{4}$$

$$V_{rd} = R_r.i_{rd} + \frac{d\psi_{rd}}{dt} - \frac{d\theta_r}{dt}\psi_{rq} \tag{5}$$

$$V_{rq} = R_r.i_{rq} + \frac{d\psi_{rq}}{dt} + \frac{d\theta_r}{dt}\psi_{rd} \tag{6}$$

Electromagnetic torque is expressed as:

$$C_e = P\frac{M}{L_r}(\psi_{rd}.i_{sq} - \psi_{rq}i_{sq}) \tag{7}$$

2.2. Mathematical Modeling of the control scheme for induction generator

In order to model any field oriented control system, it is necessary to choose the synchronously rotating reference frame (d, q) In the RFO control system, the rotor flux vector is aligned with the d-axis Figure 4, which means:

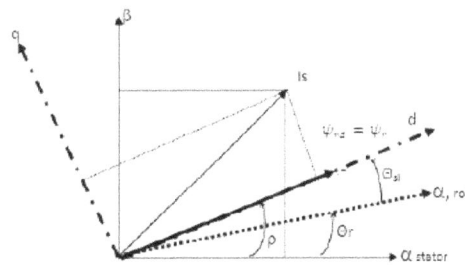

Figure 4. dq and alpha beta frame

$$\psi_{rd} = \psi_r \tag{8}$$

$$\psi_{rq} = 0 \tag{9}$$

From the desired value of the DC voltage, it is possible to express that the reference power by:

$$V_{dc\_ref}.i_{dc} = P_{ref} \tag{10}$$

The electromagnetic torque:

$$C_{em} = \frac{P_{ref}}{\Omega}$$
(11)

The control voltage V_{dc} can be done via the electromagnetic torque control, the derivative of rotor flux can be written as:

$$\frac{d\psi_r}{dt} = \frac{M}{\tau_r}i_{sd} - \frac{1}{\tau_r}\psi_r$$
(12)

The slip frequency can be written as:

$$\omega_r = \frac{M}{\tau_r}\frac{i_{sd}}{\psi_r}$$
(13)

And,

$$\omega_s = \omega + \omega_r$$
(14)

Then,

$$\omega_s = \omega + \frac{M}{\tau_r}\frac{i_{sd}}{\psi_r}$$
(15)

The field angle is calculated as:

$$\theta_s = \int \omega_s dt$$
(16)

The electromagnetic torque is expressed from the current i_{sq} by:

$$C_e = p.\frac{M}{L_r}.\psi_r.i_{sq}$$
(17)

The flux controlled by i_{sd} and electromagnetic torque controlled by i_{sq}.

$$\psi_r = p.\frac{M}{1 + p\tau_r}.i_{sd}$$
(18)

The several studies carried out shows that the fuzzy logic control provides good results for contributions to conventional regulation that was introduced for this type of regulator in order to have a good performance in our application.

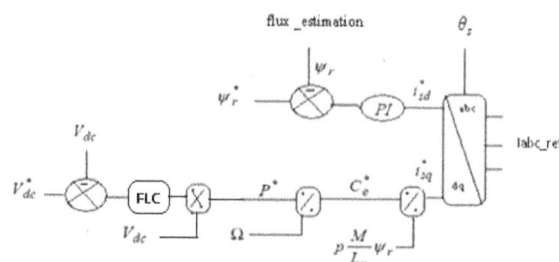

Figure 5. Block diagram control scheme of SIEG

3. FUZZY LOGIC CONTROL

The FLC consists of four major blocks, Fuzzification, knowledge base, inference engine and defuzzification.

There are two inputs, the voltage error $e(k)$ and the change of voltage error $ce(k)$. The two input variables are calculated at every sampling time as:

$$e(k) = V_{dc}(k)^* - V_{dc}(k) \tag{19}$$

$$ce(k) = V_{dc}(k)^* - V_{dc}(k-1) \tag{20}$$

Where $V_{dc}*(k)$ denotes the reference speed, $V_{dc}(k)$ is the actual speed and $e(k-1)$ is the value of error at previous sampling time.

3.1. Fuzzification

The crisp input variables are $e(k)$ and $ce(k)$ are transformed into fuzzy variables referred to as linguistic labels. The membership functions associated to each label have been chosen with triangular shapes. The following fuzzy sets are used, *NL* (Negative Large), *NM* (Negative Medium), *NS* (Negative Small), *ZE* (Zero), *PS* (Positive Small), *PM* (positive Medium), and *PL* (Positive Large). The universe of discourse is set between − 1 and 1. The membership functions of these variables are shown in Figures 6, 7 and 8.

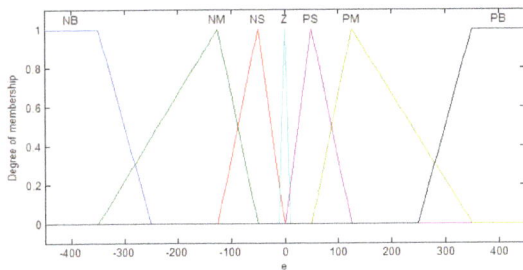

Figure 6. Membership function for input e

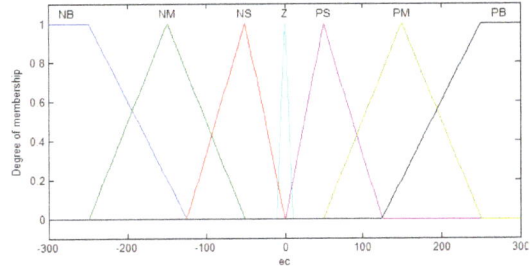

Figure 7. Membership function for input ce

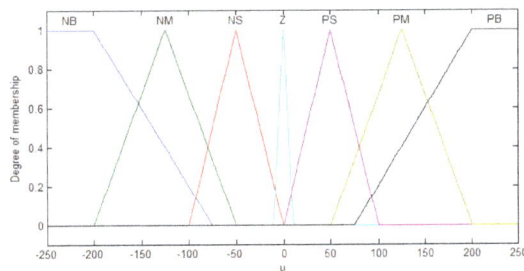

Figure 8. Membership function for output u

3.2. Knowledge Base and Inference Engine

The knowledge base consists of the data base and the rule base. The data base provides the information which is used to define the linguistic control rules and the fuzzy data in the fuzzy logic controller. The rule base specifies the control goal actions by means of a set of linguistic control rules [19]. The inference engine evaluates the set of IF-THEN and executes 7*7 rules as shown in Table 1. The linguistic rules take the form as in the following example:

IF e is NL AND ce is NL THEN u is NL

Table 1. Fuzzy Rules Base

ce/e	NL	NM	NS	ZE	PS	PM	PL
NL	NL	NL	NL	NL	NM	NS	ZE
NM	NL	NL	NL	NM	NS	ZE	PS
NS	NL	NL	NM	NS	ZE	PS	PM
ZE	NL	NM	NS	ZE	PS	PM	PL
PS	NM	NS	ZE	PS	PM	PL	PL
PM	NS	ZE	PS	PM	PL	PL	PL
PL	ZE	PS	PM	PL	PL	PL	PL

3.2.1. Defuzzification

In this stage, the fuzzy variables are converted into crisp variables. There are many defuzzification techniques to produce the fuzzy set value for the output fuzzy variable. In this paper, the centre of gravity defuzzification method is adopted here and the inference strategy used in this system is the Mamdani algorithm.

3.2.2. The Reference Rotor Flux Linkage Required

The reference rotor flux linkage required at any speed is calculated based on this maximum flux linkage, which corresponds to the minimum rotor speed hence at any rotor speed the reference rotor flux linkage is given by [2].

$$\psi_r^* = \frac{\omega_{min} \cdot \psi_{r\,max}}{\omega} \tag{21}$$

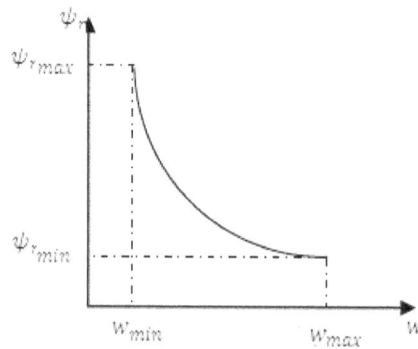

Figure 9. Relationship between rotor speed and rotor flux linkage

3.2.3. Mathematical model of PWM Converter

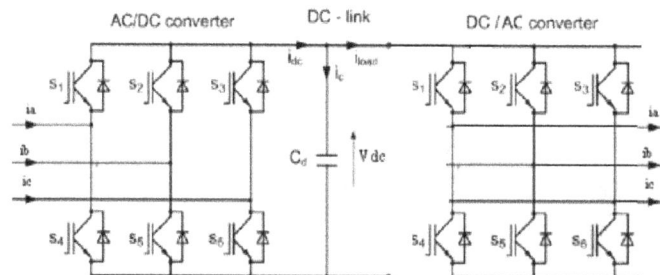

Figure 10. PWM Converter

The DC bus voltage reflects at the output of the inverter in the form of the three-phase PWM AC voltages V_{sa}, V_{sb} and V_{sc}. These voltages may be expressed as:

$$V_{sa} = \frac{1}{3}V_{dc}(2S_a - S_b - S_c) \tag{22}$$

$$V_{sb} = \frac{1}{3}V_{dc}(2S_b - S_c - S_a) \tag{23}$$

$$V_{sc} = \frac{1}{3}V_{dc}(2S_c - S_a - S_b) \tag{24}$$

The derivative of the DC bus voltage and when non linear load is present is defined as:

$$\frac{d}{dt}V_{dc} = \frac{1}{C}(S_a.i_a + S_b.i_b + S_c.i_c - i_{load}) \tag{25}$$

While i_{load} current drawn by the pump and S_a, S_b and S_c are the switching functions for the ON/OFF positions of the rectifier switches S_1-S_6.

The relation of the inverter input and output current are given by the following expression:

$$i_{load} = S_a.i_a + S_b.i_b + S_c.i_c \tag{26}$$

S_a, S_b and S_c are the switching functions for the ON/OFF positions of the inverter switches S_1-S_6.

3.2.4. Modeling of the control scheme for induction motor

Following the same procedure for the control of the generator, but in this case our regulation based to control rotor speed, the block diagram explain the control strategy as show in Figure 6.

It becomes possible to control the torque independently by the q-axis stator current, and the rotor flux can be controlled with the d-axis stator current with a delay. In this case, the torque can be expressed as:

$$C_e = p.\frac{M}{L_r}.\psi_r.i_{sq} \tag{27}$$

By keeping the rotor flux constant, the expression of the rotor flux can be given by:

$$\psi_r = \frac{M}{1 + p\tau_r}.i_{sd} \tag{28}$$

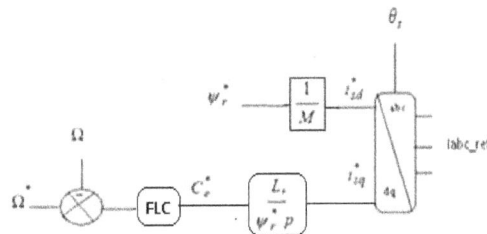

Figure 11. Block diagram control scheme of induction motor

In order to operate the motor at high efficiency, the inverter works on the principle of bang-bang control with three independent hysteresis controllers. The calculated values of the three-phase stator currents are compared with the reference values and the inverter elements are switched accordingly to impress the necessary terminal voltages to the motor phases [15].

4.   PUMP MODEL

The pump used is of centrifugal type which can be described by an aerodynamic load which is characterized by the following load equation:

$$C_r = K * \Omega^2 \tag{29}$$

Where K is the pump constant

5.   SIMULATION RESULTS AND DISCUSSION

The global of all circuit components. The system is implemented using MATLAB/SIMULINK .The dynamic performance of the whole system for different operating conditions is studied; the sequence of simulation is as follows:

a)   The simulation completed with in 30 seconds.
b)   The reference DC voltage is set at 600V.
c)   The voltage build up process is under no load condition.
d)   The pump applied to the induction generator at t=2s.
e)   The system was simulated for variable wind speed after connected the pump as show in Figure 12.
f)   Then reference pump speed is set to different value of 120 rad/s to 170 rad/s, and 140 rad/s, as show in Figure 17.

The SEIG output voltage is converted into DC voltage by using the controlled rectifier circuits. The output voltage of the rectifier is 600 volts. This DC voltage is given to the source inverter to produce required output voltage of the pump.

The Induction motor loaded with a centrifugal pump. suddenly is applied at t=2s it is observed that the value of the DC bus voltage is maintained at a constant value even if the wind speed changes at 14s and 18s and variation of pump speed at 12s and 20s. The fuzzy voltage controller provides a rapid and accurate response for the reference. The reference flux and estimate is shown in Figura 14. Also Figure 15 shows the variation in d-axis, q-axis stator currents in the rotating reference frame.

The Figure 16 shows the stator current at the terminals of the induction generator.

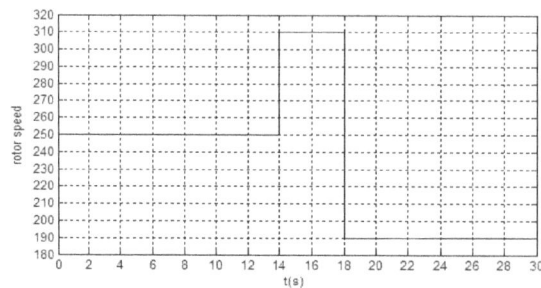

Figure 12. Variation of wind speed (rad/s)

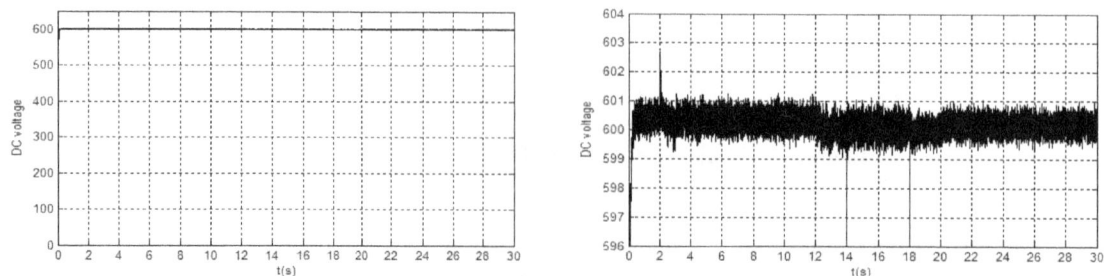

Figure 13. DC capacitor voltage profile of the SIEG

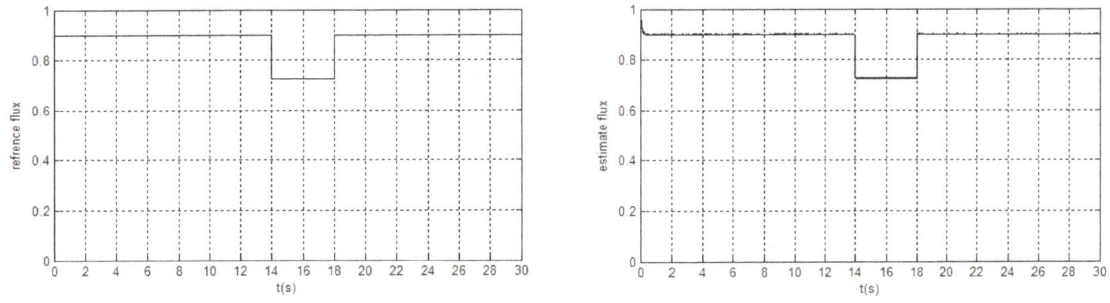

Figure 14. Reference and estimate rotor flux of the SIEG

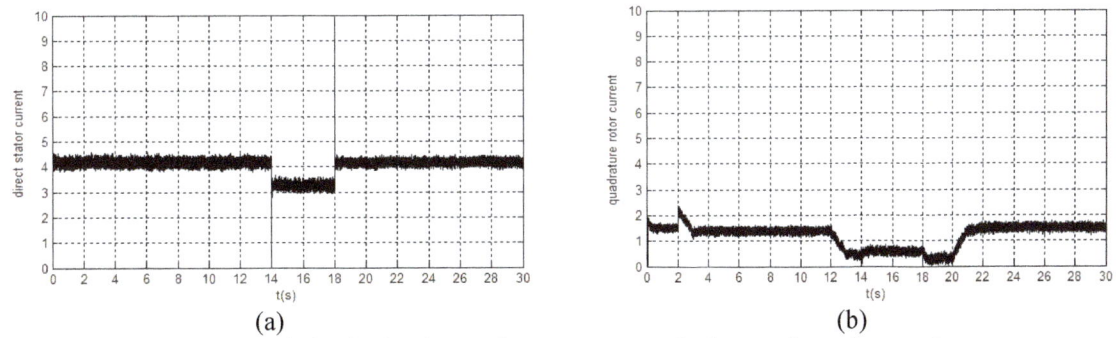

(a)   (b)

Figure 15. Variation in d-axis, q-axis stator currents in the rotating reference frame

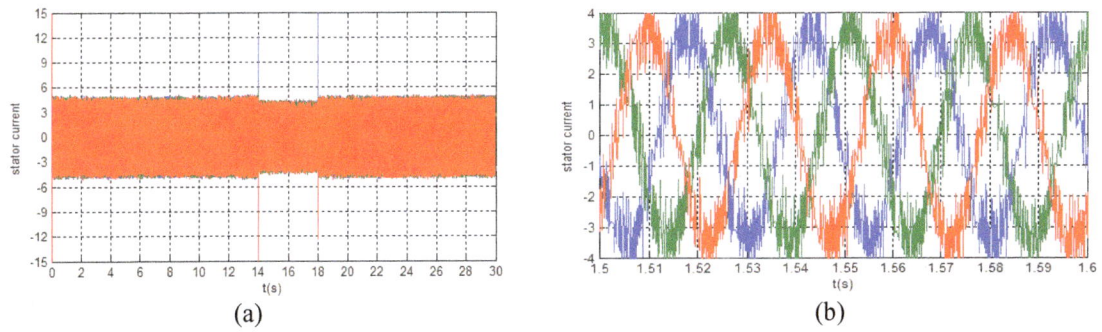

(a)   (b)

Figure 16. Stator current of the SIEG

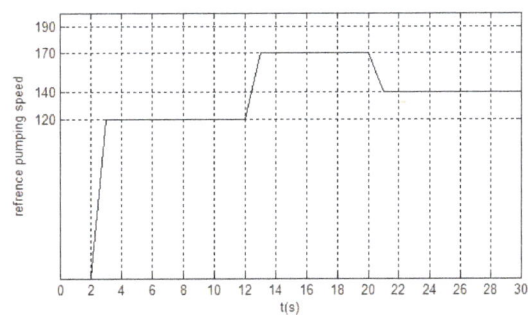

Figure 17. Reference pump speed ans pump speed

5.1. Regulation of the Pump Speed

After connecting the pump we can say that the pump speed follow the given reference as shown in Figure 18 at 12s and 20s ,also is observed that pump speed not affected by the variation of wind speed of the generator ,then the system became more stable and more robust. The Figure 19 shows the variation of stator currents of the pump.

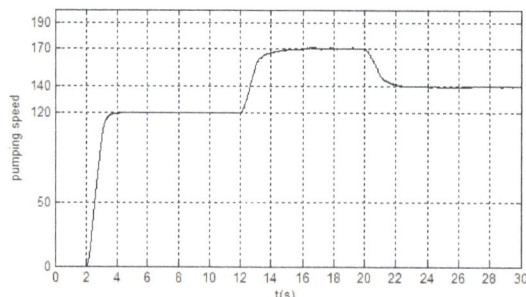

Figure 18. Reference pump speed ans pump speed

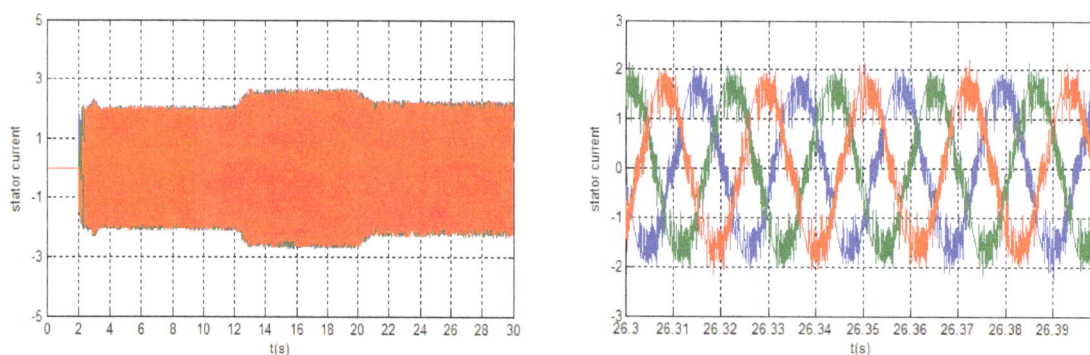

Figure 19. Stator Current of the Pump

6.   CONCLUSION

This paper introduces the modeling and simulation of the wind electric pumping systems using MatlabSimulink the studies are made by formulating the mathematical models and control for global system. It has been demonstrated that the system is able to feed pump system by regulated DC bus voltage and satisfactory desired debit from variation reference speed of the pump under variable wind speed. All results obtained confirm the effectiveness of the proposed controllers and it has been found to be satisfactory such wind electric pumping successfully in windy remote locations.

REFERENCES
[1]   Ahmed T, Nishida K, Nakaoka M, Tanaka. Advanced Control of a Boost ACDC PWM Rectifier for Variable-Speed Induction Generator. *K Applied Power Electronics Conference and Exposition. APEC '06.* Twenty First Annual IEEE. 2006; 956-962.
[2]   D Seyoum, MF Rahman C Grantham. *Terminal voltage control of a wind turbine driven isolated induction generator using stator oriented field control.* Proc. IEEE APEC'03, Miami Beach, FL, USA. 2003; 2: 846–852
[3]   Michał Knapczy, krzysztof Pienkowski. *Analysis of pulse width modulation techniques for AC/DC line-side converters.* Prace Naukowe Instytutu Maszyn, Napedów i Pomiarów Elektrycznych Politechniki Wrocławskiej. 2006
[4]   Raja Singh Khela, KS Sandhu. ANN Model for Estimation of Capacitance Requirements to maintain Constant Air-Gap Voltage of Self-Excited Induction Generator with Variable Load. *IJCST.* 2011; 2(4).
[5]   K Premalatha, S Sudha. Self-Excitation and Voltage Control of an Induction Generator in an Independent Wind Energy Conversion System. *International Journal of Modern Engineering Research (IJMER).* 2012; 2(2): 454-461.
[6]   ML Elhafyani, S Zouggar, Y Zidani, M Benkaddour. Permanent and Dynamic Behaviours of Selfexcited Induction Generator in Balanced Mode. *M. J. Condensed Mater.* 2006; 7: 49 - 53.

[7] Debta, Birendra Kumar, Mohanty KB. *Analysis on the effect of dynamic mutual inductance in voltage build up of a stand-alone brushless asynchronous generator.* Proc. of National Power Electronics Conference, IIT Roorkee. 2010.

[8] K Idjdarene. *Contrôle d'une génératrice asynchrone à cage dédiée à la conversion de l'énergie éolienne'.* JCGE'08 LYON. 2008.

[9] Mohammed Ali Elgendy, Bashar Zahawi. Comparison of Directly Connected and Constant Voltage Controlled Photovoltaic Pumping Systems. *IEEE Transactions on Sustainable Energy.* 2010; 1(3).

[10] Miranda MS, Lyra ROC, Silva SR. An alternative isolated wind electric pumping system using induction machines. *Energy Conversion, IEEE Transactions on.* 1999; 14(4): 1611 – 1616.

[11] Luc Moreau. modélisation, conception et commande de génératrices a reluctance variable basse vitesse, PhD Thesis, Université de Nantes France. 2005.

[12] Dinko Vukadinovi'cMateo Ba˘si''' A stand–alone induction generator whit improved stator flux oriented control'' *Journal of Electrical Engineering,* VOL. 62, NO. 2, 2011, 65–72.

[13] Karim H Youssef, Manal A Wahba, Hasan A Yousef, Omar A Sebakhy. *A New Method for Voltage and Frequency Control of Stand-Alone Self-Excited Induction Generator Using PWM Converter with Variable DC link Voltage.* American Control Conference Westin Seattle Hotel, Seattle, Washington, USA. 2008.

[14] K Idjdarene. Conttrubition à l'etude et la commande de génératrices asynchrones à cage dédiées à des centrales électriques éoliennes autonomes, thèse doctorat université Abderrahmane MIRA-Béjaia. Alegérie. 2010.

[15] Lamri LOUZE. Production décentralisée de l'énergie électrique: Modélisation et contrôled'une génératrice asynchrone auto excitée, thèse doctorat Université Menetouri Constantine. 2010.

[16] M Arrouf, N Bouguechal. Vector control of an induction motor fed by a photovoltaic generator. *Applied Energy.* 2003; 74: 159–167.

[17] A Abbou, M Barara, M Akherraz, H Mahmoudi, M Moutchou. *Fuzzy Controller for Self-Excited Induction Generator used in Wind energy.* International Renewable and Sustainable Energy Conference. Ouarzazate, Morocco. 2013.

[18] Mohamed Abdellatif Khalfa, Anis Sellami, Radhi M'hiri. *Sensorless Sliding Mode Control of. Induction Motor Pump fed by Photovoltaic.Generator. International.* Journal of Sciences and Techniques of Automatic control & computer engineering. IJ-STA. 2010; 4(2).

[19] V Peter. Sensorless vector and direct torque control, Oxford New York Tokyo, Oxford University Press, 1998.

[20] El Sousy FFM, Orabi M, Godah H. Maximum Power Point Tracking Control Scheme for Grid Connected Variable Speed Wind Driven Self-Excited Induction Generator. *Journal of Power Electronics (JPE).* 2006; 6(1): 52-66.

[21] Mohamed barara. Modeling and Control Voltage of Wind Pumping Systems using a Self Excited Induction Generator. *International Journal on Electrical Engineering and Informatics.* 2013; 5(2).

[22] Bennassar, A Abbou, M Akherraz, M Barara. *Fuzzy Logic Speed Control for Sensorless Indirect Field Oriented of Induction Motor Using an Extended Kalman Filter* International Review of Automatic Control (I.RE.A.CO.). 2013; 6(3).

[23] http://www.windpoweringamerica.gov/pdfs/small_wind/small_wind_guide.pdf

[24] BG Ziter. Electric Wind Pumping for Meeting Off-Grid Community Water Demands. *Guelph Engineering Journal.* 2009; 2: 14-23.

[25] Manel Ouali, Mohamed Ben Ali Kamoun, Maher Chaabene. Investigation on the Excitation Capacitor for a Wind Pumping Plant Using Induction Generator.*Smart Grid and Renewable Energy.* 2011; 2: 116-125.

[26] Gopal, M Mohanraj, P Chandramohan , P Chandrasekar. Renewable energy source water pumping systems A literature review. *Renewable and Sustainable Energy Reviews.* 2013; 25: 351–370.

[27] Abdelhafid SEMMAH, Ahmed MASSOUM, Habib HAMDAOUI. Patrice WIRA. Comparative Study of PI and Fuzzy DC Voltage Control for aDPC- PWM Rectifier. *Przegląd Elektrotechniczny* (Electrical Review). ISSN 0033-2097, R. 87 NR 10/2011

[28] M Sasikumar, S Chenthur Pandian. Performance Characteristics of Self-Excited Induction Generator fed Current Source Inverter for Wind Energy Conversion Applications. *International Journal of Computer and Electrical Engineering.* 2010; 2(6).

[29] B BOSSOUFI, MKARIM, A LAGRIOUI, M TAOUSSI. FPGA-Based Implementation nonlinear backstepping control of a PMSM Drive. *IJPEDS International Journal of Power Electronics and Drive System.* 2014; 4(1): 12-23.

15

FPGA-Based Implementation Direct Torque Control of Induction Motor

Saber Krim[1], Soufien Gdaim[2], Abdellatif Mtibaa[3], Mohamed Faouzi Mimouni[4]

[1,2,3]Laboratory of Electronics and Microelectronics (EuE), Faculty of Sciences of Monastir
University of Monastir, Tunisia
[4]Research Unit of industrial systems Study and renewable energy (ESIER),
University of Monastir, Tunisia.
[3,4]Department of Electrical Engineering, National Engineering School of Monastir, University of Monastir, Tunisia

---

ABSTRACT

Keyword:

Direct Torque Control
Induction Motor
Field Programmable Gate Array
Real Time
Sliding mode observer
VHDL
Xilinx System Generator

This paper proposes a digital implementation of the direct torque control (DTC) of an Induction Motor (IM) with an observation strategy on the Field Programmable Gate Array (FPGA). The hardware solution based on the FPGA is caracterised by fast processing speed due to the parallel processing. In this study the FPGA is used to overcome the limitation of the software solutions (Digital Signal Processor (DSP), Microcontroller...). Also, the DTC of IM has many drawbacks such as for example; The open loop pure integration has from the problems of integration especially at the low speed and the variation of the stator resistance due to the temperature. To tackle these problems we use the Sliding Mode Observer (SMO). This observer is used estimate the stator flux, the stator current and the stator resistance. The hardware implementation method is based on Xilinx System Generator (XSG) which a modeling tool developed by Xilinx for the design of implemented systems on FPGA; from the design of the DTC with SMO from XSG we can automatically generate the VHDL code. The model of the DTC with SMO has been designed and simulated using XSG blocks, synthesized with Xilinx ISE 12.4 tool and implemented on Xilinx Virtex-V FPGA.

Corresponding Author:

Saber KRIM,
Laboratory of Electronics and Microelectronics (EuE), Faculty of Sciences of Monastir,
National Engineering School of Monastir, University of Monastir, Tunisia
E-mial: krimsaber@hotmail.fr

---

1. INTRODUCTION

With technological advancement in the field of microelectronics new digital solutions such as FPGAs (Field Programmable Gate Array) or ASIC (Application Specific Integrated Circuit) are available and can be used as numerical targets for the implementation of algorithms command. The inherent parallelism of these digital solutions and their high calculation capacity make the calculation time is negligible in spite of the complexity of the algorithms to be implanted. These hardware solutions can meet the new demands of modern controls, such as reduction of the calculation time, the processing parallelism of these hardware solutions allows integrating on a single target several algorithms that provide various features and which can work independently of each other. For the control of the variable speed electrical machines, various control algorithms can be used. These algorithms often have several nested control loops. In our case we use the DTC that contains a speed control loop, stator flux regulator, electromagnetic torque regulator and the sliding mode observer; this is why we are interested in the implementation on FPGA of Direct Torque Control based on the sliding mode observer for controlling an induction motor. During the past few years several researchers use the FPGA for controlling electrical system [1]-[7]. Most of them develop the algorithm on a VHDL hardware description language. For the hardware implementation of the Directe Torque Control with

Sliding Mode Observer of an induction motor on the FPGA we use Xilinx System Generator (XSG) toolbox developed by Xilinx and added to matlab/simulink. The XSG advantages are the rapid time to market, real time and portability. Once the design and simulation of the proposed algorithme is completed we can automatically generate the VHDL code in Xilinx ISE.

The DTC of IM is based on the orientation of the stator flux by a direct action on the states of the switches of the inverter [8]-[11]. The DTC control based on an open loop estimator of stator flux having well-known problems of integration especially at a low speed [12]-[14]; also, it is sensitive to the variation of the machine parameters such as stator resistance [15]. To solve these problems many observation methods are used, such as the Extended Kalman Filter [16] but the drawback of this observer that the knowledge of load dynamics is not usually possible, Model Reference Adaptive System (MRAS) [17], [18]; the drawback of this algorithm that it is sensitive to uncertainties of the induction motor parameters, the luenberger Observer is used for state estimation of IM [19]. In this work, we propose to use the adaptive sliding mode observer for the estimation of the stator flux, stator current and the adaptation of the variation of the stator resistance. That is a powerful observer that can estimate simultaneously the stator flux, stator current, rotor speed and motor parameters. It is introduced to replace the open-loop estimator of stator flux. Furthermore it has been provided with an adaptation mechanism of the stator resistance. Thus, the aim of this paper is first, to give a fair comparison between a DTC with an open loop estimator and DTC with sliding mode observer at the stage of adjustment of the stator resistance. Secondly, the proposed model is developed using Xilinx System Generator for implementation on FPGA, to enjoy the performances of FPGAs in the field of digital control of electrical machines in real time. The performance of the proposed model is proved by simulation results, Resources used and execution time.

2. DIRECT TORQUE CONTROL OF AN INDUCTION MOTOR

2.1. Induction Machine Model

The state model of an induction machine can be expressed as follows:

$$\frac{dX}{dt} = [A]X + [B]U \tag{1}$$

Where A, B, X and U are the evolution matrix, the control matrix, state vector X and the stator voltage respectively.

$$[A] = \begin{pmatrix} -\dfrac{\frac{R_S}{L_S}+\frac{R_r}{L_r}}{\sigma} & -\omega & \dfrac{R_r}{\sigma L_r L_S} & \dfrac{\omega}{\sigma L_S} \\ \omega & -\dfrac{\frac{R_S}{L_S}+\frac{R_r}{L_r}}{\sigma} & -\dfrac{\omega}{\sigma L_S} & \dfrac{R_r}{\sigma L_r L_S} \\ -R_S & 0 & 0 & 0 \\ 0 & -R_S & 0 & 0 \end{pmatrix}, \quad [B]=\begin{pmatrix} \dfrac{1}{\sigma * L_S} & 0 \\ 0 & \dfrac{1}{\sigma * L_S} \\ 1 & 0 \\ 0 & 1 \end{pmatrix}, \quad X=\begin{pmatrix} i_{s\alpha} \\ i_{s\beta} \\ \varphi_{s\alpha} \\ \varphi_{s\beta} \end{pmatrix}, \quad U=\begin{pmatrix} V_{s\alpha} \\ V_{s\beta} \end{pmatrix}$$

The state vector X is composed by stator current and flux components. The vector command U is constituted by the stator voltage components.

2.2. Direct Torque Control Principle

Direct torque control of an induction machine is based on the direct determination of the control sequence applied to the switches of a voltage inverter. The choice of sequences is based on the two hysteresis comparators of the stator flux and electromagnetic torque [20]. The voltage vector Vs is the output of a three-phase voltage inverter whose the state of the inverter switches are controlled by three Boolean variables Sj (j = a, b, c). The voltage vector can be written as:

$$\begin{cases} V_S = \sqrt{\dfrac{2}{3}} U \left(S_a + S_b . e^{j\frac{2\pi}{3}} + S . e^{j\frac{4\pi}{3}} \right) \\ \overline{V}_s = V_{s\alpha} + jV_{s\beta} \end{cases} \tag{3}$$

The components of the stator voltage vector $V_S\left(V_{S\alpha}, V_{S\beta}\right)$ and the stator flux vector $\varphi_S\left(\varphi_{S\alpha}, \varphi_{S\beta}\right)$ in Concordia reference are given by Equation (4) and (5). The calculation of the position and module of the stator flux are based on the use of components $\left(\varphi_{S\alpha}, \varphi_{S\beta}\right)$. The module of the stator flux and its position are given by Equation (6).

$$\begin{cases} \varphi_{s\alpha} = \int_0^t (v_{s\alpha} - R_s i_{s\alpha})dt \\ \varphi_{s\beta} = \int_0^t (v_{s\beta} - R_s i_{s\beta})dt \end{cases} \tag{4}$$

$$\begin{cases} V_{S\alpha} = \sqrt{\dfrac{2}{3}}U\left(Sa - \dfrac{1}{2}(Sb + Sc)\right) \\ V_{S\beta} = \sqrt{\dfrac{2}{3}}U(Sb - Sc) \end{cases} \tag{5}$$

$$\begin{cases} \varphi_s = \sqrt{\varphi_{s\alpha}^2 + \varphi_{s\beta}^2} \\ \theta_S = \arg \varphi_s = arctg\,\dfrac{\varphi_{s\beta}}{\varphi_{s\alpha}} \end{cases} \tag{6}$$

The electromagnetic torque is expressed in terms of the components of stator flux vector and the components of stator current vector as:

$$C_e = \frac{3}{2}p(\varphi_{s\alpha}i_{s\beta} - \varphi_{s\beta}i_{s\alpha}) \tag{7}$$

The estimated values of the stator flux and electromagnetic torque are compared with their reference values Φsref, Teref respectively. Switching states are selected by the switching table, where E_C is the error of electromagnetic torque after hysteresis block and E_φ is the error of the stator flux after hysteresis block, $S_i(i = 1...6)$ means the sector (Table 1) [21]:

Table 1. Switching table for direct torque control

$E\varphi$	Ec	S1	S2	S3	S4	S5	S6
1	1	V2	V3	V4	V5	V6	V1
	0	V7	V0	V7	V0	V7	V0
	-1	V6	V1	V2	V3	V4	V5
0	1	V3	V4	V5	V6	V1	V2
	0	V0	V7	V0	V7	V0	V7
	-1	V5	V6	V1	V2	V3	V4

The structure of DTC of an induction motor is given, as shown by the Figure 1:

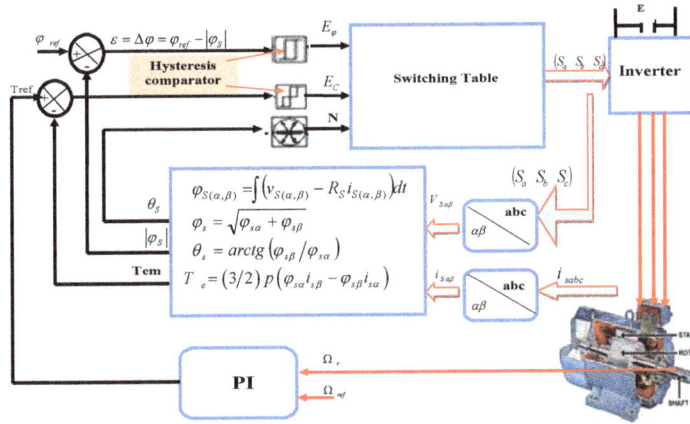

Figure 1. Schematic of conventional DTC

3. DIRECT TORQUE CONTROL OF AN INDUCTION MOTOR WITH SLIDING MODE OBSERVER

The sliding mode observer (SMO) is wedely used for non linear systems due to its robustness to the parameter variations. The SMO is used to contstruct the state variables and the stator resistance. The diagram of the SMO is shown in Figure 2.

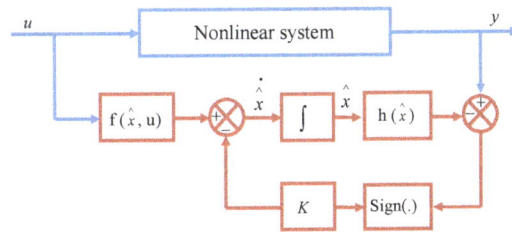

Figure 2. Principle of the sliding mode observer

The mathematical model of the observed stator current is presented as:

$$
\begin{pmatrix} \dot{\hat{i}}_{S\alpha} \\ \dot{\hat{i}}_{S\beta} \end{pmatrix} = \begin{pmatrix} -\dfrac{1}{\sigma}\left(\dfrac{R_S}{L_S}+\dfrac{R_r}{L_r}\right) & -\omega \\ \omega & -\dfrac{1}{\sigma}\left(\dfrac{R_S}{L_S}+\dfrac{R_r}{L_r}\right) \end{pmatrix}\begin{pmatrix} \hat{i}_{S\alpha} \\ \hat{i}_{S\beta} \end{pmatrix} + \begin{pmatrix} \dfrac{R_r}{\sigma L_r L_S} & \dfrac{\omega}{\sigma L_S} \\ -\dfrac{\omega}{\sigma L_S} & \dfrac{R_r}{\sigma L_r L_S} \end{pmatrix}\begin{pmatrix} \hat{\varphi}_{S\alpha} \\ \hat{\varphi}_{S\beta} \end{pmatrix} + \begin{pmatrix} \dfrac{1}{\sigma * L_S} & 0 \\ 0 & \dfrac{1}{\sigma * L_S} \end{pmatrix}\begin{pmatrix} v_{S\alpha} \\ v_{S\beta} \end{pmatrix} + \begin{pmatrix} A_{11} & A_{12} \\ A_{21} & A_{22} \end{pmatrix}\begin{pmatrix} I_{S1} \\ I_{S2} \end{pmatrix}
$$
(8)

The matimatical model of the observed stator flux is given by the following system:

$$
\begin{pmatrix} \dot{\hat{\varphi}}_{S\alpha} \\ \dot{\hat{\varphi}}_{S\beta} \end{pmatrix} = \begin{pmatrix} -R_S & 0 \\ 0 & -R_S \end{pmatrix}\begin{pmatrix} \hat{i}_{S\alpha} \\ \hat{i}_{S\beta} \end{pmatrix} + \begin{pmatrix} \dfrac{1}{\sigma * L_S} & 0 \\ 0 & \dfrac{1}{\sigma * L_S} \end{pmatrix}\begin{pmatrix} v_{S\alpha} \\ v_{S\beta} \end{pmatrix} + \begin{pmatrix} A_{31} & A_{32} \\ A_{41} & A_{42} \end{pmatrix}\begin{pmatrix} I_{S1} \\ I_{S2} \end{pmatrix}
$$
(9)

Where:
$\begin{pmatrix} A_{11} & A_{12} \\ A_{21} & A_{22} \end{pmatrix}, \begin{pmatrix} A_{31} & A_{32} \\ A_{41} & A_{42} \end{pmatrix}$ and $\begin{pmatrix} I_{S1} \\ I_{S2} \end{pmatrix} = \begin{bmatrix} signe(S_1) \\ signe(S_2) \end{bmatrix}$: are the matrixes of gains of the observed stator current, matrix of gains of the observed stator flux, and the sign vector of the sliding mode surface respectively.

3.1.   Determining of the SMO Characteristics

The sliding surface is based on the error between the real stator current $i_{s\alpha}$ and $i_{s\beta}$, and the observed stator current $\hat{i}_{s\alpha}$ and $\hat{i}_{s\beta}$ as follows:

$$
S = \begin{bmatrix} S_1 \\ S_2 \end{bmatrix} = \frac{1}{\sigma L_s (\frac{1}{T_r^2} + \omega^2)} \begin{bmatrix} \frac{1}{T_r} & -\omega \\ \omega & \frac{1}{T_r} \end{bmatrix} \begin{bmatrix} i_{s\alpha} - \hat{i}_{s\alpha} \\ i_{s\beta} - \hat{i}_{s\beta} \end{bmatrix}
\tag{10}
$$

Where: $i_s = \begin{bmatrix} i_{s\alpha} & i_{s\beta} \end{bmatrix}^T$ and $\hat{i}_s = \begin{bmatrix} \hat{i}_{s\alpha} & \hat{i}_{s\beta} \end{bmatrix}^T$ are the real and observed stators currents vectors respectively.

$T_r = \dfrac{L_r}{R_r}$: Rotor time constant.

The matrix of gains related to the current observer is as follows:

$$
A_i = \begin{bmatrix} A_{i1} & A_{i2} \\ A_{i3} & A_{i4} \end{bmatrix} = \Gamma \begin{bmatrix} \delta_1 & 0 \\ 0 & \delta_2 \end{bmatrix}
\tag{11}
$$

Where $\delta_1 \ and \ \delta_2$ are two positive constants, which are determined by applying the stability conditions defined by the Lyapunov approach.

The gain matrix of the stator flux is as follows:

$$
A_\varphi = \begin{bmatrix} A_{\varphi1} & A_{\varphi2} \\ A_{\varphi3} & A_{\varphi4} \end{bmatrix} = \begin{bmatrix} q_1\delta1 & \dfrac{\omega}{\sigma L_S} \\ -\dfrac{\omega}{\sigma L_S} & q_2\delta_2 \end{bmatrix}
\tag{12}
$$

Where $q_1 \ and \ q_2$ are two positive constants.

The stability of the SMO depends on its convergence towards its sliding surface. To study the stability of this observer the following Lyapunov function is used:

$$
V = \frac{1}{2} S^T S
\tag{13}
$$

When the sliding surface S=0, the error between the real and observed stator current must bee zero, $i_{s\alpha} - \hat{i}_{s\alpha} = 0$ and the derivate of the lyapunov fuction is strictly negative ($\dot{V} \prec 0$).

$$
\dot{V} = S^T \dot{S} \prec 0
\tag{14}
$$

The major drawback of the SMO observer is the chattering phenomenon. To weaken this phenomenon a saturation function is used to replace the Bang-Bang function. The saturation function is given by the system (15).

$$
sign \ (S_i) = \begin{cases} 1 & si & S_i > \lambda \\ -1 & si & S_i < -\lambda & ; i = 1,2 \\ \dfrac{S_i}{\lambda} & si & |S_i| < \lambda \end{cases}
\tag{15}
$$

Where λ is a positive constant with a low value.

3.2. Mechanism of Adaptation of the Stator Resistance

During operation the stator résistance vary, due to the temperature and the low speed operation. To online estimate of the stator resistance another term added to the lyapunov fuction.

$$V = \frac{1}{2} S^T S + \frac{\lambda}{2} \left(\hat{R}_S - R_S \right)^2 \tag{16}$$

$$\dot{V} = S^T \dot{S} + \lambda \left(\hat{R}_S - R_S \right) \dot{\hat{R}}_S \prec 0 \tag{17}$$

To satisfy the condition of the Equation (17), the estimated stator resistance can be expressed as follow:

$$\dot{\hat{R}}_S = -k \left(\hat{i}_{S\alpha} * (i_{S\alpha} - \hat{i}_{S\alpha}) + \hat{i}_{S\beta} * (i_{S\beta} - \hat{i}_{S\beta}) \right) \tag{18}$$

With k is a positive constant.

4. USE OF XILINX SYSTEM GENERATOR (XSG) IN THE CONTROLLER DESIGN

4.1. Description of Xilinx System Generator

Xilinx System Generator (XSG) is a modeling tool developed by Xilinx for the design of implemented systems on FPGA. It has a library of varied blocks, which can be automatically compiled into an FPGA [22]. In this work, Xilinx System Generator (XSG) is used to implement the architecture of the Direct Torque Control of induction Motor with sliding mode observer on FPGA. In the first step, we begin by implementing of the proposed architectures using the XSG blocks available on the Simulink library. Once the Design of the system is completed and gives the desired simulation results, the VHDL code can be generated by the XSG tool [23]. The design flow of the Xilinx System Generator is given by figure 3. After generation of VHDL code and the synthesis, we can generate the bitstream file. Then we can move this configuration file to program the FPGA [24].

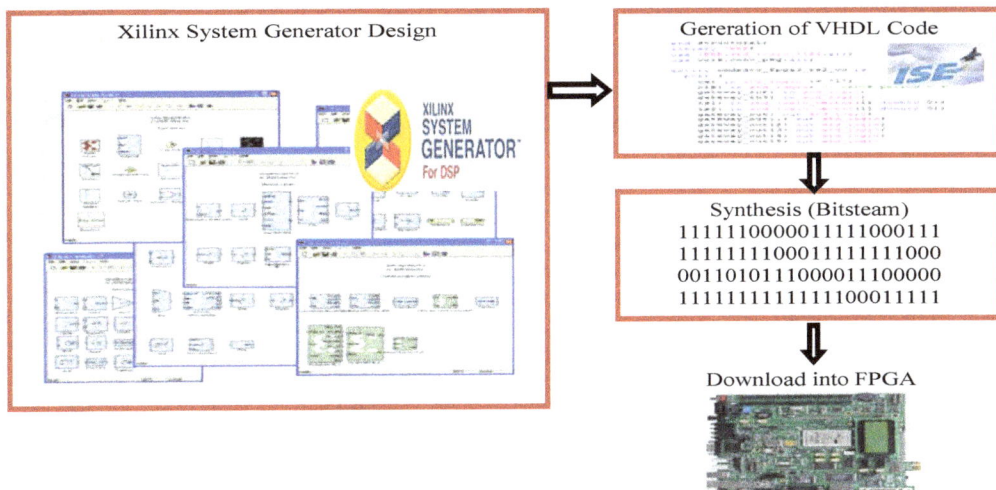

Figure 3. Configuring an FPGA

4.2. Design of the Sliding Mode Observer using XSG

4.2.1. Design of the Currents Observed

The Design of the direct component of the observed stator current vector introduced into the equation system (8) given by Figure 4.

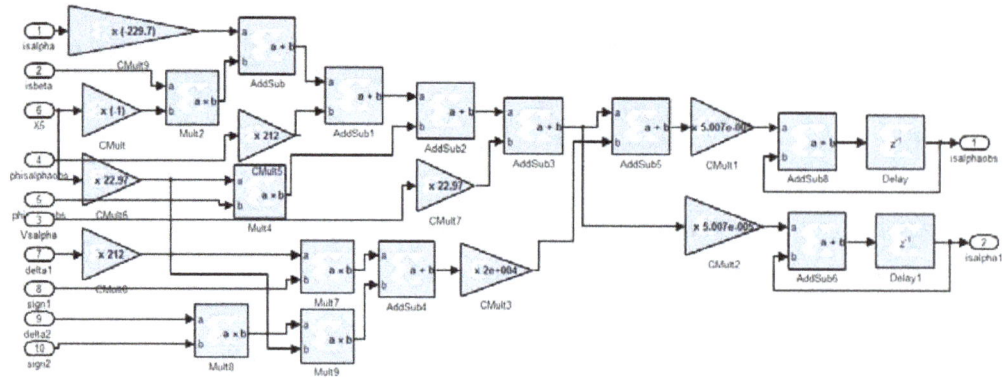

Figure 4. Design of the Component $\hat{i}_{S\alpha}$ from the XSG

4.2.2. Design of Sliding Surface, gain Matrix and Sign Function

The sliding surface, the saturation function and the gain matrix are given in equations (10), (11), (12) and (15), are illustrated using XSG as shown in Figure 5.

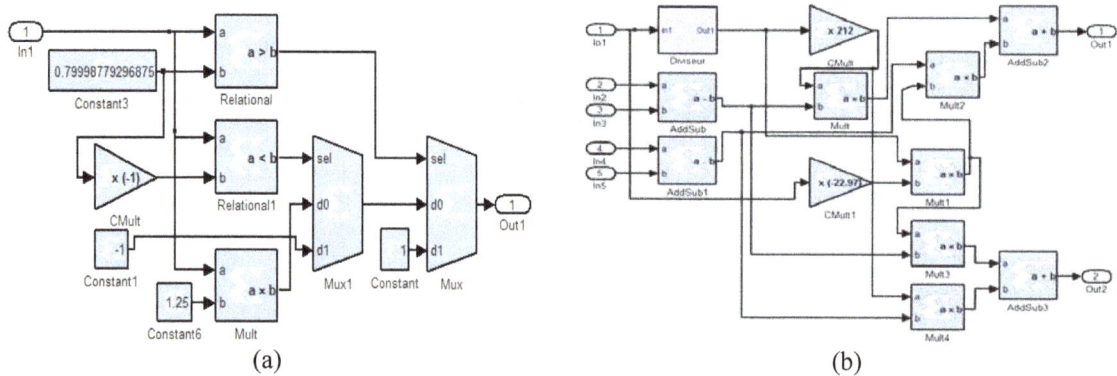

(a)

(b)

Figure 5. Design of sliding surface, gain matrix and sign function from the XSG

4.3. Simulation Results using Xilinx System Generator and Discussions

The structure of the direct torque control with the adaptative sliding mode observer of an induction motor is shown in Figure 6.

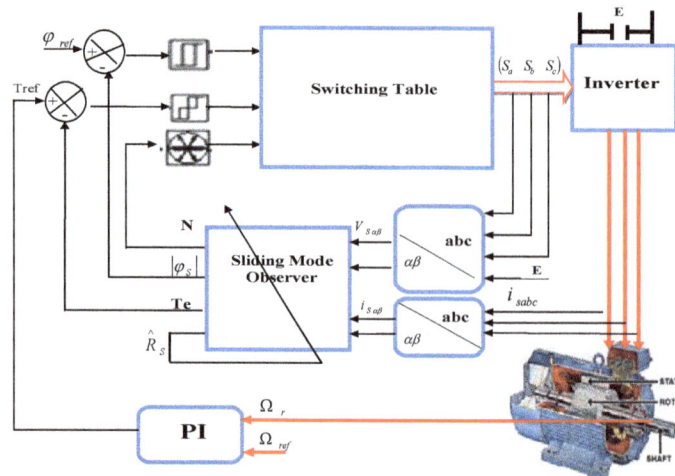

Figure 6. Schematic of a conventional DTC with an adaptative sliding mode observer

4.3.1. The Stator Resistance is Constant (Rs=5.717Ω)

The simulation of the conventional DTC with sliding mode observer is achieved using the XSG. The speed and flux references used in the simulation results are 150rad/s and 0.91wb respectively. The electromagnetic torque reference presents the output of PI controller of speed. At time t = 0.5sec a load torque of 10 Nm is applied.

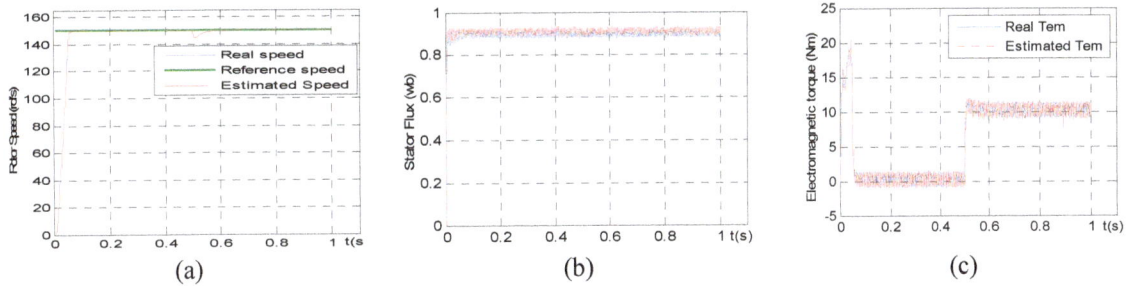

(a)

(b)

(c)

Figure 7(a). Evolution of real and estimated speed

Figure 7(b). Evolution of the real and observed stator flux

Figure 7(c). Evolution of the real and estimated electromagnetic torque

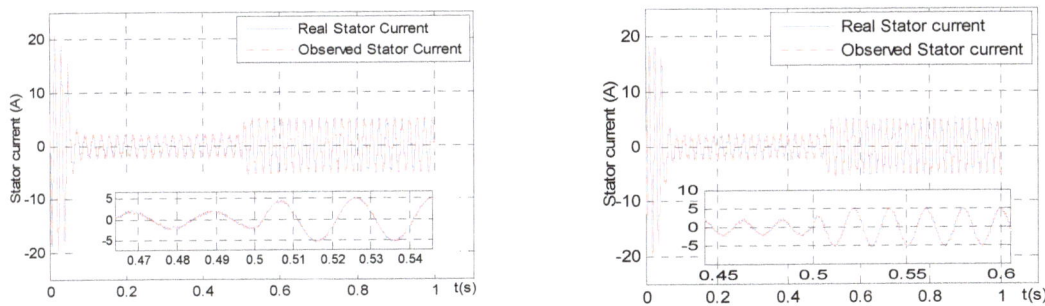

Figure 8(a). Variation of the real and observed stator current $\hat{i}_{S\alpha}$

Figure 8(b): Variation of the observed and real stator current $\hat{i}_{S\beta}$

4.3.2. The Stator Resistance Varies from 100% (Rs=2*5.717=11.434Ω)

The simulation of the conventional DTC with an open loop estimator and the conventional DTC with a sliding mode observer is achieved using the XSG at a low speed. The rotor speed and stator flux references used in the simulation results are 31.4 rad/s and 0.91 wb, respectively. At time t = 0.2sec a load torque of 10 Nm is applied. At t = 0.4sec the stator resistance increases by 100%.

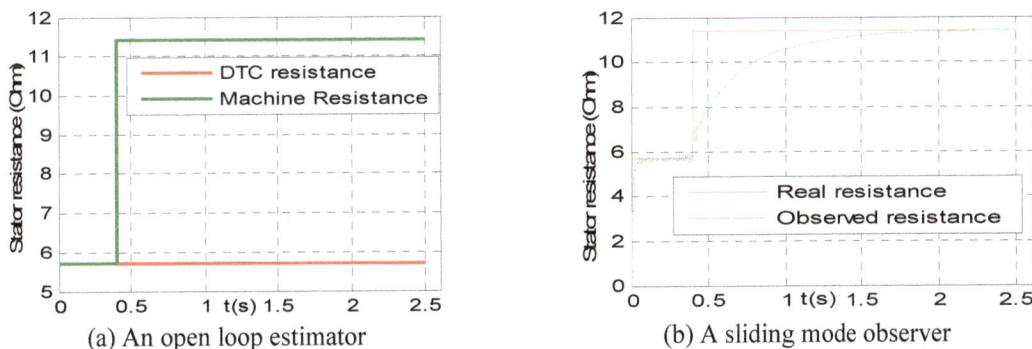

(a) An open loop estimator

(b) A sliding mode observer

Figure 9. Variation of Rs for DTC with

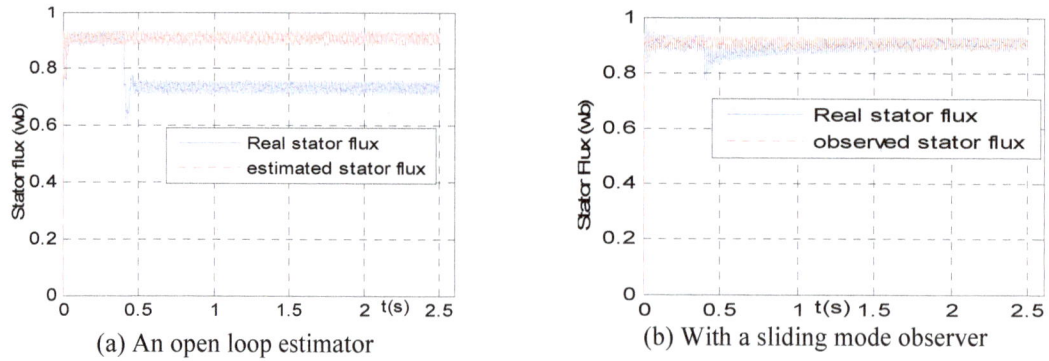

(a) An open loop estimator   (b) With a sliding mode observer

Figure 10. Evolution of the stator flux for DTC with

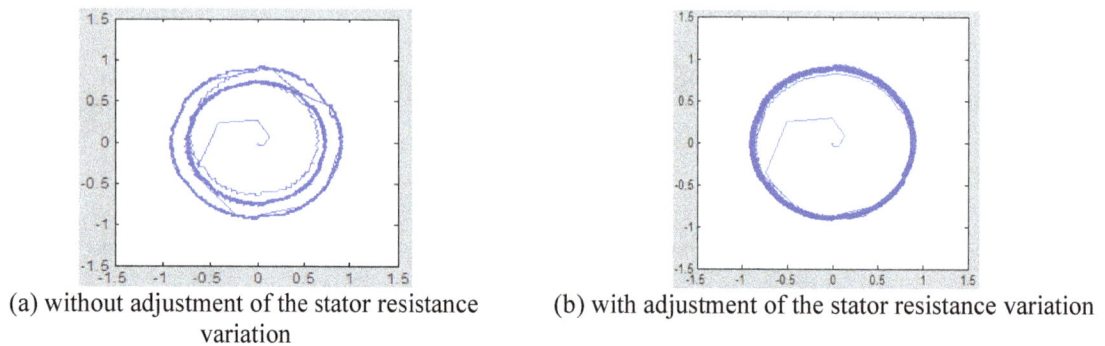

(a) without adjustment of the stator resistance   (b) with adjustment of the stator resistance variation
                    variation

Figure 11. Evolution of the stator flux trajectories

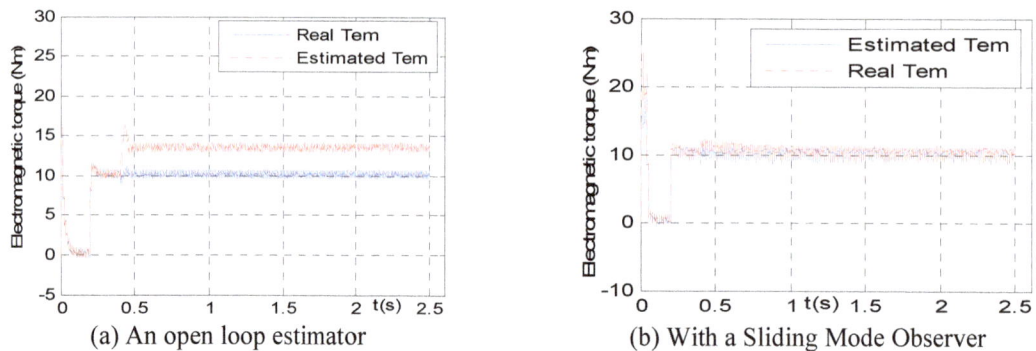

(a) An open loop estimator   (b) With a Sliding Mode Observer

Figure 12. Variation of electromagnetic torque for DTC with

The simulation results of the direct torque control with sliding Mode Observer given by Figure 7 and 8 show that the real and the estimated variables are similar. In Figure 7(b) and 7(c) it can be seen that the stator flux and the electromagnetic torque are characterized by high ripples due to the use of the hysteresis comparator.

Figure 9 show the sensitivity of the Direct Torque Control of an Induction Motor. In Figure 9(a) we can see at t=0.4sec that the stator resistance increases 2 times the nominal stator resistance due to temperature, it can be seen that in the case of the DTC with an open loop estimator, the stator resistance used in the DTC kept constant. By contrast, in Figure 9(b) the observed stator resistance converges rapidly to the nominal value, this is due to the on line adaptation of the stator resistance by the sliding mode observer. The simulation results demonstrates the robustness of the Sliding Mode Observer against the abruptly variation of the machine parametres.

In Figure 10(a), at t = 0.4sec the real stator flux is affected by de variation of the stator resistance, it decreases abruptly to 0.72 wb, the error between the real and the reference stator flux kept constant. Yet, in the Figure 10(b), the static error gradually vanishes due to the presence of the adaptive online mechanism of the stator resistance using the Sliding Mode Observer.

In Figure 11(a), we can notice that the stator flux trajectories increase due to the variation of the stator resistance at t=0.4sec. By contrast, in Figure 11(b) the stator flux trajectory is kept constant due to the presence of the adaptive online mechanism of the stator resistance using the Sliding Mode Observer.

In Figure 12(a), at t = 0.4sec the electromagnetic torque increases, and the error between the electromagnetic torque and the load torque remains constant. However, in Figure 12(b), for the sliding mode observer; the electromagnetic torque is kept constant.

4.4. FPGA Implementation Results of the Proposed Approach and Discussions

Once the simulation is completed and gives the desired results, we can generate the VHDL code and synthesized the hardware block. The implementation results are given by Figure 13 and Table 2.

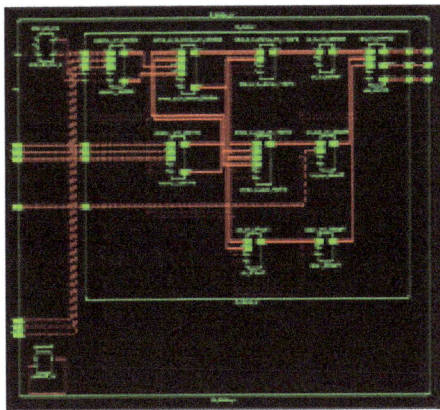

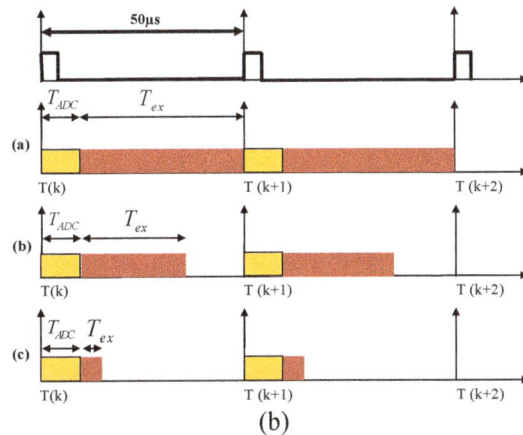

Figure 13(a). RTL schematic of the conventional DTC with a sliding mode observer from Xilinx ISE

Figure 13(b). Timing Diagram for the Imlementation on (a): Microcontroller, (b): Digital Signal Processor (DSP), (c): Field Programmable Gate Array (FPGA)

Table 2. Resources Utilisation for the DTC with SMO

Resourses	Used resources	Available resources
Number of bonded I/O	68	640
Number of Slice LUTs	2087	44800
Number of Slices Registers	478	44800
Number of DSP48Es	8	128

Execution time: $T_{ex} = 0.94$ µs

The Table 2 shows the implementation results in term of the used resources and the execution time of the Direct Torque Control with SMO using the FPGA Virtex-5 Device. The RTL schematic of the CDTC with SMO is given by Figure 13(a). The Figure 13(b), shows the performance of computing time for the hardware implementation on FPGA compared to software solutions (Microcontroller, Digital Signal Processor). T_{ADC} and T_{ex} are the analogue to digital conversion time and the execution time respectively.

In this work the execution time equal to 0.94 µs. But in papers [25] and [26], the sampling time equal to 100 µs using the dSPACE 1104 (digital signal processing and control engineering). In paper [27], the sampling time equal to 500 µs. It can be seen that the execution time is too important relative to the FPGA due to the sequential processing of the dSPACE.

5. CONCLUSION

In this paper, the digital implementation of the Direct Torque Control with Sliding Mode Observer using the FPGA has been presented. The Sliding mode Observer has been proposed to improve the performences of the Direct Torque Control of an Induction Motor. Using the Sliding Mode Observer consists to remplacing the open loop pure integration. The simulations results using Xilinx System Generator have shown that the proposed observation strategy has better performances than the open loop pure integration especially in term of variation of the stator resistance. The obtained design of the Direct Torque Control with Sliding Mode Observer from XSG can be translated automatically into a VHDL (VHSIC Hardware Description Language) from Xilinx Integrated Software Environment tool (ISE) and can be embedded into the Xilinx Virtex-V FPGA.

Induction Machine Parameters

Number of pairs of poles: 2	Rotor resistance: 4,282 Ω	Mutual inductance :441,7mH
Rated frequency: 50 Hz	Stator inductance :464mH	Moment of inertia :0.0049 kg.m²
Rated voltage :220/380 V	Rotor inductance :464mH	Viscous friction coefficient: 0.0029kg.m^2/s
Stator resistance :5,717 Ω		

REFERENCES

[1] HM Hasanien. *FPGA implementation of adaptative ANN controller for speed regulation of permanent stepper motor drives*. Energy Conversion and Management. Elsevier Publisher. 2011

[2] E Monmasson, L Idkhajine, MN Cirstea, I Bahri, A Tisan, MW Naouar. FPGAs in Industrial Control Applications. *IEEE Transactions on Industrial Informatics*. 2011; 7(2): 224-243.

[3] M Dagbagi, L Idkhajine, E Monmasson, I.Slama-Belkhodja. *FPGA Implementation of Power Electronic Converter Real-Time Model*. International Symposium on Power Electronics, Electrical Drives, Automation and Motion. 2012; 658 – 663.

[4] E MONMASSON, I BAHRI, L IDKHAJINE, A MAALOUF, MW NAOUAR. *Recent Advancements in FPGA-based controllers for AC Drives Applications*. 13th International Conference on Optimization of Electrical and Electronic Equipment (OPTIM), IEEE. 2012; 8-15.

[5] M Shahbazi, P Poure, S Saadate, MR Zolghadri. FPGA-Based Reconfigurable Control for Fault-Tolerant Back-to-Back Converter without Redundancy. *IEEE Transactions on Industrial Electronics*. 2013; 60(8): 3360 - 3371.

[6] K Jezernik, J Korelic, R Horvat. PMSM Sliding Mode FPGA-Based Control for Torque Ripple Reduction. *IEEE Transactions on Power Electronics*. 2013; 28(7): 3549 - 3556.

[7] T Sutikno, NR Idris, A Jidin, MN Cirstea. An Improved FPGA Implementation of Direct Torque Control for Induction Machines. *IEEE Transactions on Industrial Informatics*. 2013; 9(3): 1272–1279.

[8] P Bibhu, P Dinkar, S Sabyasachi. A Simple hardware realization of switching table based direct torque control of induction motor. *Electric Power Systems research. Elsevier Publisher*. 2007; 77(2): 181-190.

[9] GS Buja, MP Kazmierkowski. Direct torque control of PWM inverter-fed AC motors-A survey. *IEEE Transactions on Industrial Electronics, 2004*; 51(4): 744-757.

[10] P Vas. Sensor less Vector and Direct Torque Control. Oxford University Press, London. 1998.

[11] I Takahashi, Y Ohmori. High-performance Direct Torque Control of an Induction Motor. *IEEE Transactions on Industry Applications*. 1989; 25(2): 257-264.

[12] R Rajendran, Dr N Devarajan. A Comparative Performance Analysis of Torque Control Schemes for Induction Motor Drives. *International Journal of Power Electronics and Drive System (IJPEDS)*. 2012, 2(2): 177~191.

[13] M Barut, S Bogosyan, M Gokasan. Speed sensorless direct torque control of IMs with rotor resistance estimation. *International Journal Energy Conv. and* Manag. 2005; 46(3): 335-349.

[14] D Casadei, F Profumo, G Serra, A Tani, "FOC and DTC: Two Viable Schemes for Induction Motors Torque Control. *IEEE Transaction on Power Electronics*, 2002; 17(5): 779 – 787.

[15] S Meziane, R Toufouti, H Benalla. Speed Sensorless Direct Torque Control and Stator Resistance   Estimator of Induction Motor Based MRAS Method. *International Journal of Applied Engineering Research (IJAER)*. 2008; 3(6): 733-747.

[16] M Messaoudi, H Kraiem, M Ben Hamed, L Sbita, MN Abdelkrim. A Robust Sensorless Direct Torque Control of Induction Motor Based on MRAS and Extended Kalman Filter. *Leonardo Journal of Sciences*, 2008; 7(12): 35-56.

[17] Cirrincione M, Pucci M. Sensorless direct torque control of an induction motor by a TLS-based MRAS observer with adaptive integration. *Automatica*, 2005; 41(11): 1843-1854.

[18] MK Metwally. Control of Four Switch Three Phase Inverter Fed Induction Motor Drives Based Speed and Stator Resistance Estimation. *International Journal of Power Electronics and Drive System (IJPEDS)*. 2014; 4(2): 192-203.

[19] Sbita L, Ben Hamed M. An MRAS–based full order Luenberger observer for sensorless DRFOC of induction motors. *Int. J. ACSE*, 2007; 7(1): 11-20.

[20] MM Rezaei, M Mirsalim. Improved Direct Torque Control for Induction Machine Drives based on Fuzzy Sector Theory. *Iranian Journal of electrical and Electronic Engineering*, 2010; 6(2): 110-118.

[21] A Mahfouz, WM Mamadouh. Intelligent DTC for PWSM Drive using ANFIS technique. *International Journal of Engineering Science and Technology (IJEST)*, 2012; 4(3): 1208-1222.

[22] XSG, 1998. Xilinx system generator v2.1 basic tutorial. Printed in USA.

[23] JG Mailloux. Prototypage Rapide de la Commande Vectorielle sur FPGA à l'Aide des Outils SIMULINK SYSTEM GENERATOR, l'Université de Québec, Mars 2008.

[24] White paper: Using System Generator for Systematic HDL Design, Verification, and Validation WP283. 2008. (v1.0)

[25] Bhoopendra Singh, Shailendra Jain, Sanjeet Dwivedi. Direct Torque Control Induction Motor Drive with Improved Flux Response. *Hindawi Publishing Corporation Advances in Power Electronics.* 2012, Article ID 764038: 1-11.

[26] A ELBACHA, Z BOULGHASOUL, E ELWARRAKI. A Comparative Study of Rotor Time Constant Online Identification of an Induction Motor Using High Gain Observer and Fuzzy Compensator. *WSEAS TRANSACTIONS on SYSTEMS and CONTROL.* 2012; 7(2): 37-53.

[27] M Boussak, K Jarray. A High-Performance Sensorless Indirect Stator Flux Orientation Control of Induction Motor Drive. *IEEE Transactions on industrial electronics.* 2006; 53(1): 14-49.

16

A Fuzzy Logic Control Strategy for Doubly Fed Induction Generator for Improved Performance under Faulty Operating Conditions

G. Venu Madhav*, Y. P. Obulesu**

* Department of Electrical and Electronics Engineering, Padmasri Dr. B. V. Raju Institute of Technology
** Department of Electrical and Electronics Engineering, LakiReddy BaliReddy College of Engineering

ABSTRACT

Keyword:

Wind turbine
Doubly fed induction generator
Fuzzy logic control
PI controller
Synchronous generator

In this paper, decouple PI control for output active and reactive powers which is the common control technique for power converter of Doubly Fed Induction Generator (DFIG) is presented. But there are some disadvantages with this control method like uncertainty about the exact model, behavior of some parameters or unpredictable wind speed and tuning of PI parameters. To overcome the mentioned disadvantages a fuzzy logic control of DFIG wind turbine is presented and is compared with PI controller. To validate the proposed scheme, simulation results are presented, these results showed that the performance of fuzzy control of DFIG is excellent and it improves power quality and stability of wind turbine compared to PI controller. The Fuzzy logic controller is applied to rotor side converter for active power control and voltage regulation of wind turbine. The entire work is carried out in MATLab/Simulink. Different faulty operating conditions are considered to prove the effective implementation of the proposed control scheme.

Corresponding Author:

G. Venu Madhav,
Department of Electrical and Electronics Engineering,
Padmasri Dr. B. V. Raju Institute of Technology,
Vishnupur, Narsapur, Medak Dist., 502313, AP, India.
Email: venumadhav.gopala@gmail.com

1. INTRODUCTION

Wind energy is one of the extra ordinary sources of renewable energy due to its clean character and free availability. Moreover, because of reducing the cost and improving techniques, the growth of wind energy in Distributed Generation (DG) units has developed rapidly.

In terms of wind power generation technology, because of numerous technical benefits (higher energy yield, reducing power fluctuations and improving var supply) the modern MW-size wind turbines always use variable speed operation which is achieved by a converter system [1]. These converters are typically associated with individual generators and they contribute significantly to the costs of wind turbines. Between variable speed wind turbine generators, Doubly Fed Induction Generators (DFIGs) and Permanent Magnet Synchronous Generators (PMSGs) with primary converters are emerging as the preferred technologies [2].

Doubly Fed Induction Generator (DFIG) is one of the most popular wind turbines which include an induction generator with slip ring, a partial scale power electronic converter and a common DC-link capacitor. Power electronic converter which encompasses a back to back AC-DC-AC voltage source converter has two main parts; Grid Side Converter (GSC) that rectifies grid voltage and Rotor Side Converter (RSC) which feeds rotor circuit. Power converter only processes slip power therefore it's designed in partial scale and just about 30% of generator rated power [3] which makes it attractive from economical point of view.

Many different structure and control algorithm can be used for control of power converter. In this paper, decouple PI control of output active and reactive power to improve dynamic behavior of wind turbine which is one of the most common control techniques is presented. But due to uncertainty about the exact model and behavior of some parameters such as wind, wind turbine, etc and also parameters values differences during operation because of temperature, events or unpredictable wind speed, tuning of PI parameters is one of the main problems in this control method. Based on the analysis, fuzzy logic controller has been designed to improve the dynamic performance of DFIG.

In fuzzy logic control there is no need of a detailed mathematical model of the system and just using the knowledge of the total operation and behavior of system is enough in designing the controller. The performance of PI control is compared with that of fuzzy logic controller and it is investigated that the dynamic performance of fuzzy logic controller is quite good in comparison with PI controller.

In this paper, the dynamic performance of DFIG under different fault conditions is investigated.

2. THE SAMPLE TEST SYSTEM

Sample test system is shown in Figure 1. It consists of three main feeders, two DG units and five local loads. The two DG units are a DFIG and a synchronous generator. In the proposed system, different cases of abnormal conditions are considered, when there is a single phase line to ground fault near DFIG, a single phase line to ground fault on the grid and three phase line to ground fault near DFIG etc. The configurations and parameters of the DFIG and synchronous generator system are extracted from [4]. Main grid is represented by a three phase 69 kV voltage source with 1000MVA short circuit capacity and X/R ratio of 22.2.

Connection point of main and micro-grid systems is called Point of Common Coupling (PCC). 2MVA DFIG wind turbine consists of power electronic converter control unit which feeds generator's rotor and grid. Power electronic converter unit is to control active and reactive power of generator separately and to improve power quality and stability of the network. The parameters of 5MVA synchronous generator are given in Table 1.

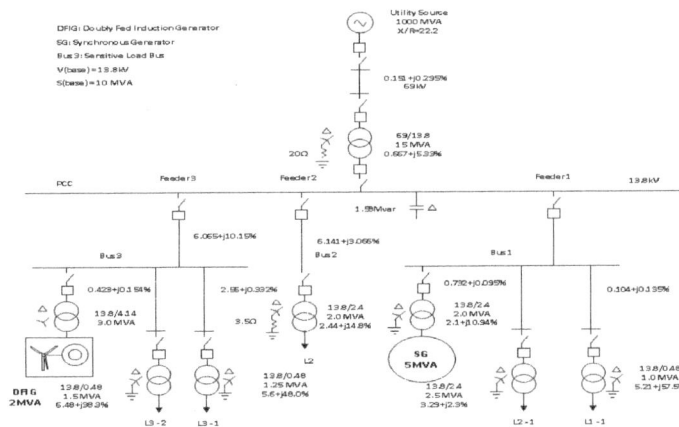

Figure 1. Sample test system

3. MODELING OF BASIC COMPONENTS
3.1. Wind and Wind Turbine

Wind effect plays a fundamental rule in wind turbine modeling especially for interaction analysis between wind turbines and the power system to which they are connected. Wind model describes wind fluctuation in wind speed which causes power fluctuation in generator. For wind model four components can be considered, as describe in (1) [5]:

$$V_{wind} = V_{bw} + V_{gw} + V_{rw} + V_{nw} \tag{1}$$

Where, V_{bw} = Base wind component (*m/s*); V_{gw} = Gust wind component (*m/s*); V_{rw} = Ramp wind component (*m/s*); V_{nw} = Noise wind component (*m/s*).

The base component is a constant speed; wind gust component may be expressed as a sine or cosine wave function or their combination [6]; a simple ramp function will be used for ramp component and a triangle wave for noise function which it's frequency and amplitude will be accordingly adjusted. The simple block diagram for generation of wind speed is illustrated in Figure 2 and which includes all of four components mentioned above.

For electrical analysis, a simplified aerodynamic model of wind turbine is normally used. Accordingly wind blade torque from wind speed will be produced which is as follows:

$$\lambda = \frac{R\omega_{rot}}{V_{wind}} \tag{2}$$

$$P_w = \frac{1}{2}\rho\pi R^2 C_p\left(\lambda,\theta\right)V_{wind}^3 \tag{3}$$

$$T_w = \frac{P_w}{\omega_{rot}} = \frac{\rho\pi R^2 C_p\left(\lambda,\theta\right)V_{wind}^3}{2\lambda} \tag{4}$$

Where T_w is an aerodynamic torque extracted from the wind (Nm), ρ is the air density (kg/m^3), R is the wind turbine rotor radius (m), Vwind is the equivalent wind speed (m/s), θ is the pitch angle of the rotor (deg), λ is the tip speed ratio, ω_{rot} is the mechanical speed of the generator (rad/s) and C_p is the power coefficient.

C_p can be expressed as a function of the Tip Speed Ratio (TSR) and pitch angle which is given by (5) [7], [8]:

$$C_p\left(\lambda,\theta\right) = 0.22\left(\frac{116}{\lambda_i} - 0.4\theta - 5\right)e^{\frac{-12.5}{\lambda_i}}$$

$$\lambda_i = \frac{1}{\left(\frac{1}{\lambda+0.08\theta} - \frac{0.035}{\theta^3+1}\right)} \tag{5}$$

By increasing pitch angle, power coefficient and therefore torque decreases moreover C_p growth rate changes in different speed by λ.

3.2. DFIG Model

As illustrated in Figure 3, DFIG system is a wound rotor induction generator with slip ring, with stator directly connected to the grid and with rotor interfaced through a back to back partial scale power converter. The converter consists of two conventional voltage source converters that are called Rotor Side Converter (RSC) and Grid Side Converter (GSC) and a common DC-link [3]. Consequently the DFIG can be regarded as a traditional induction machine with a nonzero rotor voltage.

Using the Concordia and Park transformation allows to write a dynamic model in a d-q reference frame from the traditional a-b-c frame as follows [9]:

Electromagnetic torque:

$$T_{em} = \frac{3}{2}\left(\psi_{ds}i_{qs} - \psi_{qs}i_{ds}\right) \tag{6}$$

Active and reactive power of stator:

$$P_s = \frac{3}{2}\left(V_{ds}i_{ds} + V_{qs}i_{qs}\right) \tag{7}$$

$$Q_s = \frac{3}{2}\left(V_{ds}i_{ds} - V_{qs}i_{qs}\right) \tag{8}$$

Table 1. Synchronous generator parameters

Rated Power	5 MVA	Rated Voltage	13.8 kV
R_a	0.0052 p.u	X_{ls}	0.2 p.u
X_d	2.86 p.u	X_q	2.0 p.u
X_d	0.7 p.u	X_q	0.85 p.u
X_d	0.22 p.u	X_q	0.2 p.u
T_{do}	3.4 s	T_{qo}	0.05 s
T_{do}	0.01 s	H	2.9 s

Table 2. Induction generator parameters of wind turbine (DFIG)

Rated Power	2 MVA
Rated Voltage	0.69 kV
Stator/rotor ratio	0.4333
Angular moment of inertia (J=2H)	1.8293 p.u
Mechanical damping	0.02 p.u
Stator resistance	0.0183 p.u
Rotor resistance	0.0205 p.u
Stator leakage inductance	0.2621 p.u
Rotor leakage inductance	0.3152 p.u
Mutual inductance	5.572 p.u

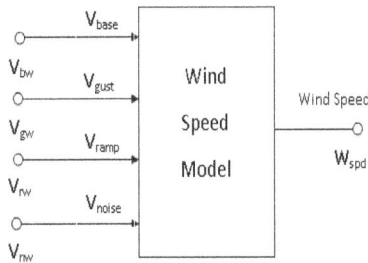

Figure 2. Model of wind speed

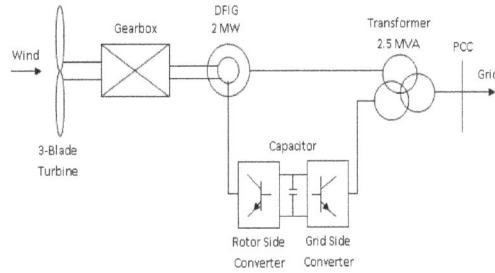

Figure 3. Schematic representation of a DFIG wind turbine

Table 2 shows the parameters of the DFIG which is used in this proposed system. The rotor side converter operates at the slip frequency. The power converter processes only the slip power, thus if the DFIG to be varied within about ±30% slip, the rating of power converter is only about 30% of rated power of the wind turbine [10].

Setting the stator flux vector to align with *d*-axis and assuming the per phase stator resistance negligible, we have:

$$\psi_s = \psi_{ds}, V_s = V_{qs} \tag{9}$$

$$|\psi_s| \angle \theta_s = \int (V_s - r_s i_s) dt \tag{10}$$

Substitution (9) in (7) and (8), the active and reactive power of stator flow into the grid can be expressed as:

$$P_s = -\frac{3}{2} \frac{L_m}{L_m + L_s} V_s i_{qr} \tag{11}$$

$$Q_s = \frac{3}{2} \frac{V_s}{L_m + L_s} \left(L_m i_{dr} - \frac{V_s}{\omega_s} \right) \tag{12}$$

Where, i_{qr} and i_{dr} are rotor current (A) in d- and q-axis respectively, L_{ls} and L_m are stator leakage and mutual inductance (H), ω_s is the electrical angular velocity (rad/s) and V_s is the magnitude of the stator phase voltage (V). This means that using vector control with d-axis oriented stator flux vector in rotor side converter, active and reactive power can be controlled separately. This will be achieved by regulating i_{qr} and i_{dr} respectively.

Grid side converter is presented for keeping DC link voltage of capacitor constant regardless to the magnitude and direction of rotor power. Neglecting power losses in the converter, capacitor current can be described as follow:

$$i_{dc} = C\frac{dV_{dc}}{dt} = \frac{3}{4}mi_{gcd} - i_{dcr} \tag{13}$$

Where i_{gcd} stands for the *d*-axis current flowing between grid and grid side converter (*A*), i_{dcr} is the rotor side DC current (*A*), C is the DC-link capacitance (*F*) and m is the PWM modulation index of the grid side converter.

The reactive power flow into the grid from GSC can be expressed as:

$$Q_g = \frac{3}{2}V_g i_{gcq} \tag{14}$$

Where V_g is the magnitude of grid phase voltage (*V*) and i_{gcq} is *q*-axis current of grid side converter (*A*). Therefore it is seen from (13) and (14), by adjusting i_{gcd} and i_{gcq}, DC-link voltage and Q_g can be controlled respectively.

3.3. Pitch Control

To produce a maximum energy, the blade angle must be tuned with wind straightforward using pitch angle control of wind turbine blades. It is worth noticing that we can use this characteristic in abnormal conditions such as grid faults to protect generator from over speeding. In two different cases, an increasing rotor speed may be occurred; a wind speed as input power and an abnormal case due to a fault existence.

These must be distinguished first, before a control takes place. When the output terminal voltage falls under 0.9 p.u and the rotor speed is increased, it means a fault is happened.

To actuate the event and to decrease the rotor speed, the pitch angle must be manipulated. An emergency pitch angle should be added with rate of $+10(deg/s/1000rpm)$ for over speed protection.

4. A FUZZY LOGIC AND PI CONTROL STRATEGY

The four main components of fuzzy logic controller are fuzzification, fuzzy inference engine, rule base and defuzzification. Inputs are fuzzified, then based on rule base and inference system, outputs are produced and finally the fuzzy outputs are defuzzified and applied to the main control system. Error of inputs from their references and error deviations in any time interval are chosen as inputs. Mamdani type fuzzy logic control is considered here.

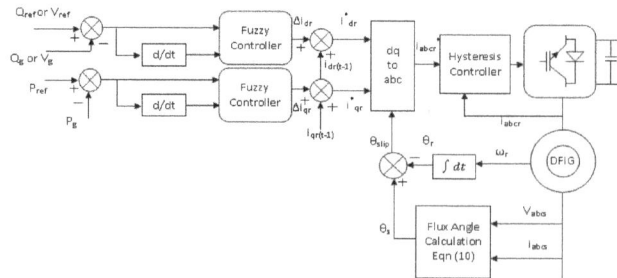

Figure 4. Rotor side converter fuzzy controller unit structure

Figure 4 shows the block diagram of rotor side converter with fuzzy controllers. Similarly, PI controllers are used in place of fuzzy controllers. The main objectives of this part are active power control and voltage regulation of DFIG wind turbine using output reactive power control. As illustrated in Figure 6 rotor side converter manages to follow reference active (P_{ref}) power and voltage (V_{ref}) separately using fuzzy controllers, hysteresis current controller converter and vector control algorithm. Based on (11), (12) and Figure 6, inputs of fuzzy controller are error in active and reactive power or voltage and the rate of changes in errors in any time interval. After the production of reference d- and q-axis rotor currents, they converted to a-b-c reference frame using flux angle, rotor angle and finally slip angle calculation and Concordia and Park transformation matrix. Then they applied to a hysteresis current controller to be compared with actual currents and produce switching time intervals of converter.

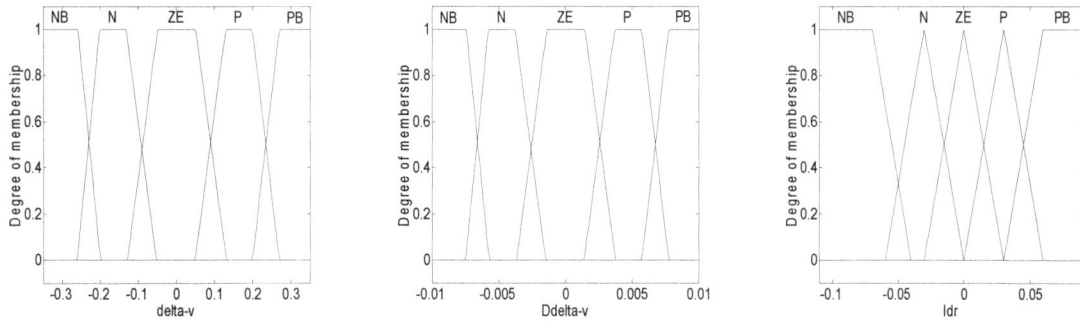

Figure 5. Input and output membership functions of voltage controller

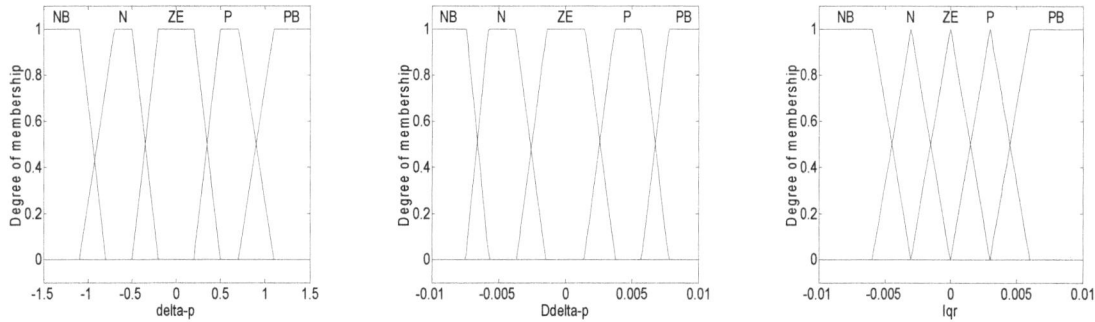

Figure 6. Input and output membership functions of active power controller

Table 3. Rule bases of voltage fuzzy controller

ΔI_{dr}		ΔE (V)				
		NB	N	ZE	P	PB
E (V)	NB	NB	NB	N	N	ZE
	N	NB	N	N	ZE	P
	ZE	N	N	ZE	P	P
	P	N	ZE	P	P	PB
	PB	ZE	P	P	PB	PB

Table 4. Rule bases of active power fuzzy controller

ΔI_{qr}		ΔE (P)				
		NB	N	ZE	P	PB
E (P)	NB	NB	NB	N	N	ZE
	N	NB	N	N	ZE	P
	ZE	N	N	ZE	P	P
	P	N	ZE	P	P	PB
	PB	ZE	P	P	PB	PB

Figure 5 and 6 shows inputs and output membership functions. To avoid miscalculations due to fluctuations in wind speed and the effects of noise on data, trapezoidal membership functions are chosen to have smooth and constant region in the main points. Rule bases are shown in Table 3 and 4. NB, N, ZE, P and PB represents negative big, negative, zero, positive and positive big respectively. For instance when E (P), the error of active power and ΔE (P), the rate of change of active power error in a time interval, are NB mean the output voltage is more than reference and is increasing dramatically therefore reference q-axis rotor current which controls active power should decrease rapidly that represents NB.

In this paper, Proportional and Integral (PI) controllers are used in place of fuzzy controllers as shown in Figure 4 and the results of both the controllers are compared. PI controller blocks operate in the feed forward path of both active power (P) and reactive power (Q) feedback loops. PI controller gains are tuned by using the Simulink Control Design software which makes the control systems design and analyze in Simulink environment.

5.   RESULTS AND DISCUSSION

For investigation of dynamic behavior of proposed system with fuzzy logic and PI controller, different situations and events are considered. Based on different fault locations and severity, the system has different responses. In each condition, different parameters such as voltage, active and reactive power, rotor currents and dc link voltage are taken to prove the capability of the proposed controller.

(a) Single line to ground fault near synchronous generator:

A single line to ground short circuit fault with duration of 0.1s is occurred near the synchronous generator. The fault duration is from 5s to 5.1s. Figure 7 shows different responses of the synchronous generator and DFIG in test systems. During the fault, there is little variation in active and reactive power of wind turbine and in AC and DC-link voltages because the fault is far from the wind turbine and near the synchronous generator, so, variation in active and reactive power of synchronous generator is high.

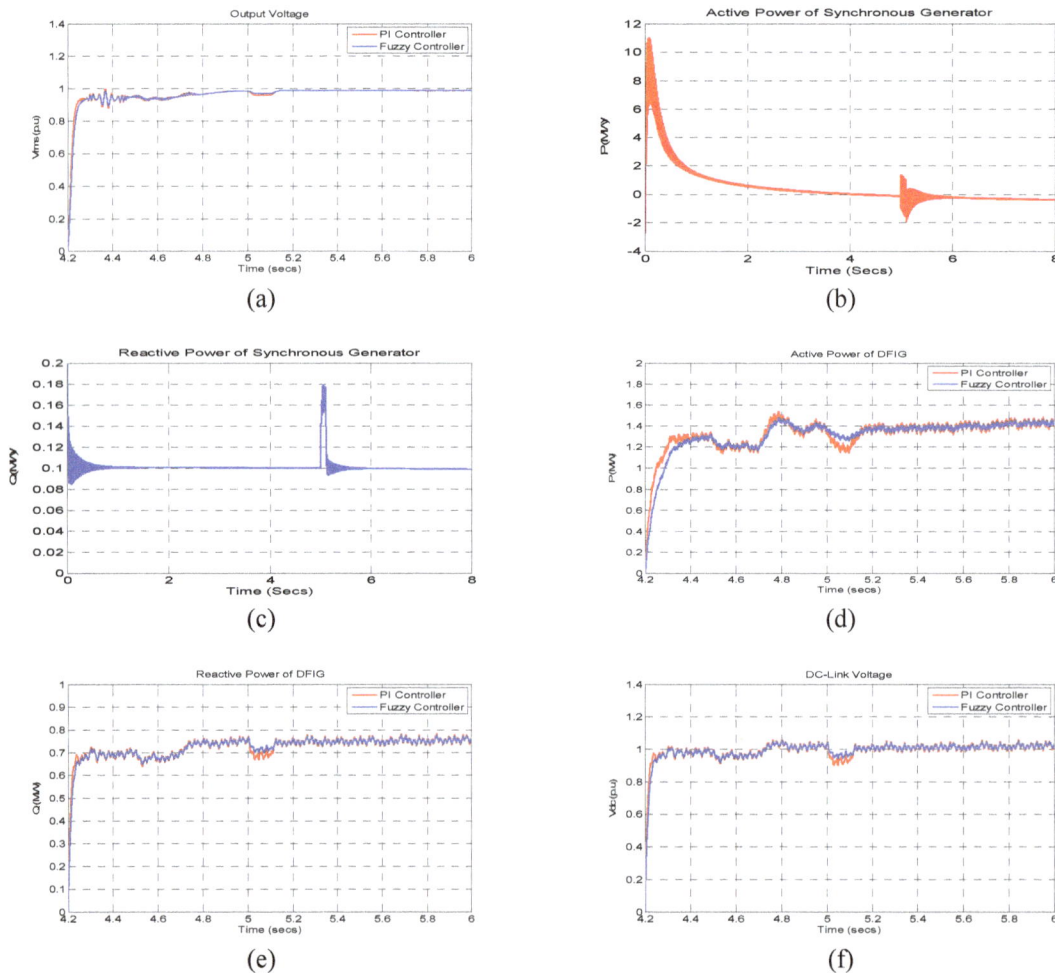

Figure 7. Single line to ground fault near synchronous generator at 5s with duration of 0.1s (a) output voltage (b) active power of synchronous generator (c) reactive power of synchronous generator (d) active power of DFIG (e) reactive power of DFIG (f) dc-link voltage

GSC generally controls the dc bus voltage of the back-to-back converter and the exchange of reactive power to the grid. The proposed controllers produce the necessary values of direct and quadrature axis rotor currents which are converted into three phase currents to maintain control on the machine stability. The power delivered from RSC will be increased due to increase of rotor currents and voltages which in turn increase the dc bus voltage.

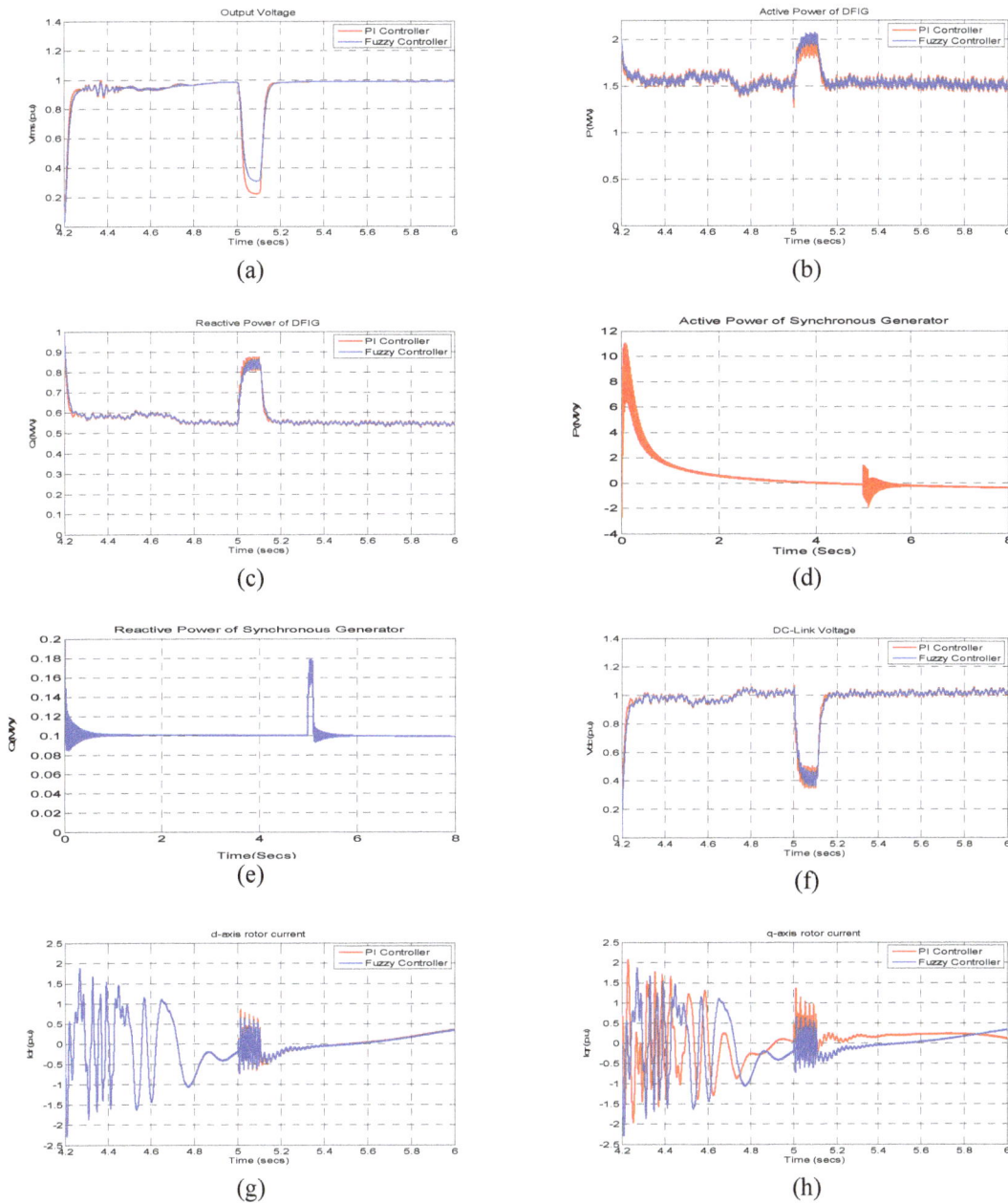

(a)

(b)

(c)

(d)

(e)

(f)

(g)

(h)

Figure 8. Single line to ground fault near DFIG wind Turbine (a) output voltage (b) active power of DFIG (c) reactive power of DFIG (d) active power of synchronous generator (e) reactive power of synchronous generator (f) dc-link voltage (g) d-axis rotor current (h) q-axis rotor current

(b) Three line to ground fault near DFIG:

To prove performance of fuzzy logic controller in comparison with decouple PI control and to investigate dynamic behavior of doubly fed induction generator in one of the worst case situations, a severe three line to ground short circuit fault is considered near the wind turbine. Figure 9 shows the waveforms, there is reduction in voltage and it reduces to near zero. In addition, active and reactive deviations in DFIG are the most severe. Rotor current reaches to its limit and crowbar protection unit short circuits the rotor and rotor side converter but still stator is connected to the network and due to super synchronous operation of wind turbine it can produce active power. The proposed controller maintains the rotor currents under their safety limits without high over currents. Due to mitigation of the over currents of the rotor the back-to-back converter is less affected by this perturbation which produces short dc bus voltage oscillations.

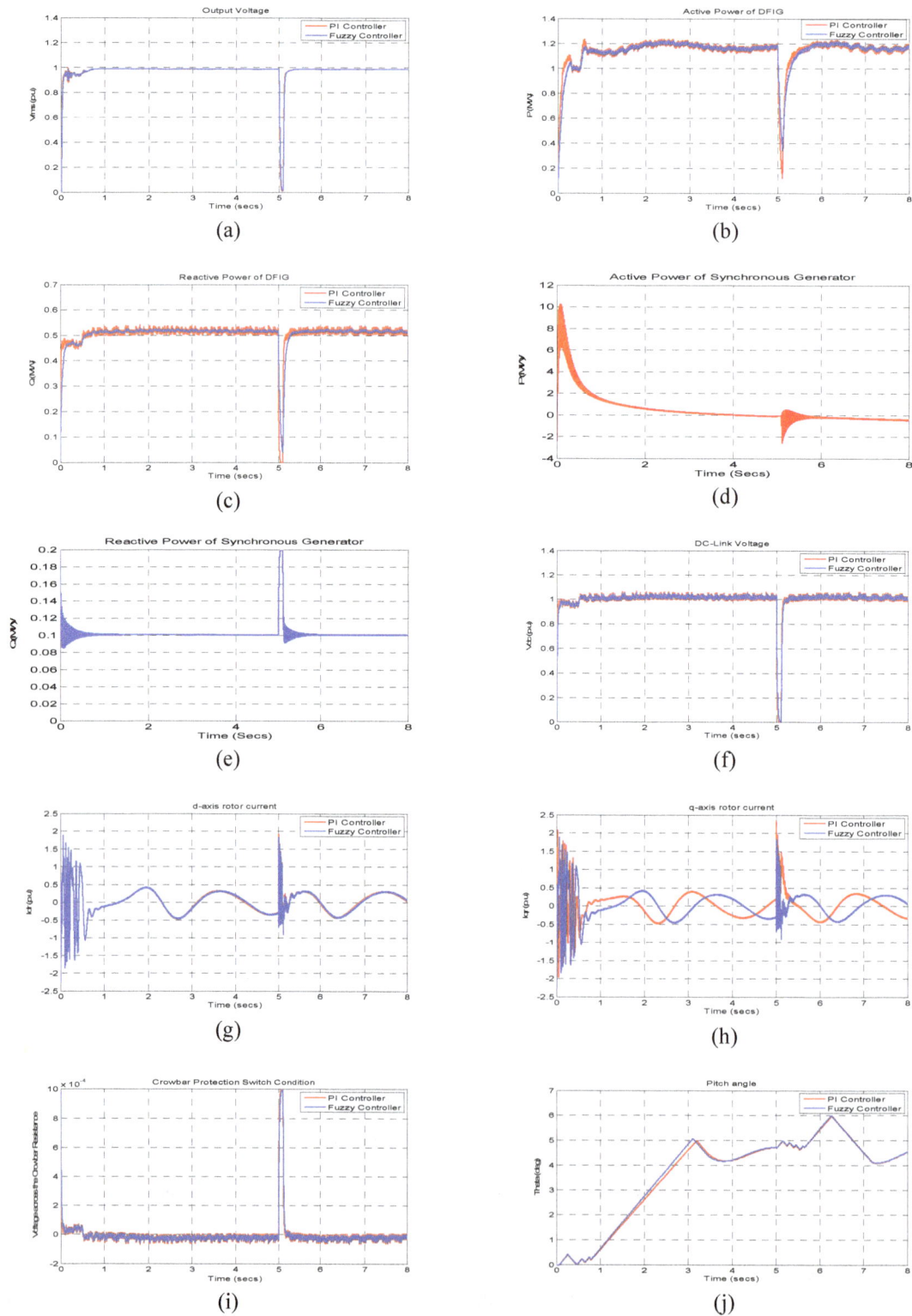

(a)

(b)

(c)

(d)

(e)

(f)

(g)

(h)

(i)

(j)

Figure 9. Three line to ground short circuit fault near DFIG wind turbine (a) output voltage (b) active power of DFIG (c) reactive power of DFIG (d) active power of synchronous generator (e) reactive power of synchronous generator (f) dc-link voltage (g) d-axis rotor current (h) q-axis rotor current (i) voltage across the crowbar resistance (j) pitch angle

Furthermore, beside electrical protection, an emergency pitch angle is introduced with slope of ± 10 (deg/s). When voltage drops under 0.8 p.u and wind speed is constant, emergency pitch angle due to external fault activates to protect DFIG from over speeding and keep output power below rated value. As soon as voltage and speed come back to normal situation it starts to decrease and returns to normal situation. Grid side converter acts as STATCOM and tries to restore voltage. After rotor current returns under the limit and a constant time delay, crowbar switch opens and rotor side converter continues to operate. As illustrated in Figure 9, fuzzy control unit of wind turbine maintains good stability and restores parameters to their predefined values as well in comparison with PI controller.

6. CONCLUSION

In this paper, dynamic performance of DFIG under different fault conditions with PI controller and fuzzy logic control has been investigated. The PI controller and fuzzy logic controller has been designed and implemented in MATLab/Simulink. To prove the performance of controller unit, the abnormal situations of single line to ground fault near and away from DFIG and three phase line to ground fault near DFIG are exerted on proposed system. The output voltage, active and reactive powers, dc-link voltage, direct and quadrature axis totor currents are improved for fuzzy logic controller compared to PI controller for different cases of fault near and away from DFIG. The performance of fuzzy logic controller is found quite satisfactory in improving stability and power quality of wind turbine compared to PI controller. Closer fault location to the wind turbine causes more severe effect and a three line to ground short circuit fault near the wind turbine as the worst case in which voltage decreases until zero and rotor current exceeds its limit.

REFERENCES

[1] Datta R, Ranganathan VT. Variable-Speed Wind Power Generation Using Doubly Fed Wound Rotor Induction Machine - A comparison With Alternative Schemes. *IEEE Transactions on Energy Conversion*. 2002; 17(3): 414–421.

[2] Li G, M Yin, M Zhou, C Zhao. Decoupling control for multi terminal VSCHVDC based wind farm interconnection. *IEEE. Power Engineering Society General Meeting*. 2007: 1-6.

[3] Holdsworth L, XG Wu, JB Ekanayake and N Jenkins. Comparison of fixed speed and doubly-fed induction wind turbines during power system disturbances. *IEE Proc. Gener. Transm. Distrib.*, 2003; 150 (3): 343-352.

[4] Katiraei F, MR Iravani and PW Lehn. Micro-Grid Autonomous Operation During and Subsequent to Islanding Process. *IEEE Trans. Power Delivery*. 2005: 20(1).

[5] Youjie Ma, Haishan Yang, Xuesong Zhou, Li Ji , *2009. The dynamic modeling of wind farms considering wake effects and its optimal distribution.* World Non-Grid-Connected Wind Power and Energy Conference, 2009. WNWEC 2009; 22 (2): 1 - 4.

[6] Reynolds MG. Stability of wind turbine generators to wind gusts. Purdue University Report TR-EE 79-20.

[7] Heier S. Grid integration of wind energy Conversion systems. Chichester: John Wiley and Sans Ltd. 1998; 35-302.

[8] Slaotweg G, H Polindcr, WL Kling. *Dynamic modeling of a wind turbine with direct drive synchronous generator and back to back voltage source converter and its control.* Proceedings of the European Wind Energy Conference, Copenhagen, Denmark. 2001; 1014-1017.

[9] Bose BK. 1986. Power electronics and AC drives. New Jersey: Prentice-Hall. 1986; 46-52.

[10] Jang J, Y Kim, D Lee. Active and reactive power control of DFIG for wind energy conversion under unbalanced grid voltage. Proc. IEEE Power Electronics and Motion Control Conference. 2006; 3.

17

Proposed Voltage Vector to Optimize Efficiency of Direct Torque Control

Goh Wee Yen*, Ali Monadi*, Nik Rumzi Nik Idris*, Auzani Jidin**, Tole Sutikno**

*Department of Electrical Power Engineering, Universiti Teknologi Malaysia (UTM), Johor Bahru, Malaysia
** Department of Power Electronics and Drives, Universiti Teknikal Malaysia Melaka (UTeM), Malacca, Malaysia
*** Department of Electrical Engineering, Universitas Ahmad Dahlan (UAD), Yogyakarta, Indonesia

ABSTRACT

Keyword:

Direct torque control
Induction machine
Proposed voltage vector
Search controller
Torque response

Compared to field-oriented control (FOC) system, direct torque control (DTC) system has gained attractiveness in control drive system because of its simpler control structure and faster dynamic control. However, supplying the drive system with rated flux at light load will decrease the power factor and efficiency of the system. Thus, an optimal flux has been applied during steady-state in order to maximize the efficiency of drive system. But when a torque is suddenly needed, for example during acceleration, the dynamic of the torque response would be degraded and it is not suitable for electric vehicle (EV) applications. Therefore, a modification to the voltage vector as well as look-up table has been proposed in order to improve the performance of torque response.

Corresponding Author:

Nik Rumzi Nik Idris,
Department of Electrical Power Engineering,
Universiti Teknologi Malaysia (UTM),
81310 UTM Skudai, Johor, Malaysia.
Email: nikrumzi@fke.utm.my

1. INTRODUCTION

A simple control structure of DTC that has been proposed by Takahashi [1] has gained popularity in industrial motor drive applications. Due to its simpler control structure and faster dynamic control compared to FOC system, the popularity of DTC system is increased rapidly in the past decades [1-3]. In FOC, the torque and flux are controlled based on stator current components whereas in DTC, the torque and flux are controlled directly and independently via an optimized selection of voltage vectors using look-up table.

As illustrated in Figure 1, the simple control structure of DTC consists of three-phase voltage source inverter (VSI), hysteresis comparators, stator flux and torque estimators, as well as look-up table. By using two-level and three-level hysteresis comparators, the stator flux and electromagnetic torque can be controlled independently. Based on look-up table, an appropriate voltage vector is selected to satisfy its flux and torque requirement. The selected voltage vector is then applied to activate the VSI in which it will in turn operate the induction machine.

A fast instantaneous control of torque and flux occurs because of de-coupled control of torque and flux, in which it leads to the faster dynamic control of DTC system compared to the FOC system. To achieve more accurate flux estimation, the current model is applied during low speed operations whereas the voltage model is employed in high speed operations. Only stator resistance and terminal quantities such as stator voltages and currents are required in the estimation of voltage model.

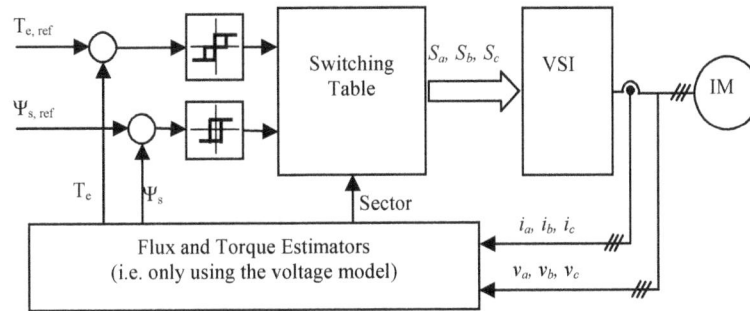

Figure 1. Simple Control Structure of DTC

In order to fully utilized the power and lengthen the life span of induction motor, an optimal efficiency of the drive system is an important factor to be implemented in EV applications. Usually, induction motors are operated at light load and thus, supplying the motor at its rated flux will decrease the power factor and efficiency of the drive [4]. Therefore, researchers have been working on the efficiency optimization of drive system in recent years but there is still no suitable method to achieve the fast instantaneous torque response of DTC drive.

Two main methods have been proposed to maximize efficiency in DTC drive system. These methods are known as flux search controller (SC) [4-12] and loss model controller (LMC) [13-16]. The former method measures input power or stator current of the system while decreasing the flux value in a consecutive step. When the input power or stator current is at its minimized value, the optimal flux is obtained. Meanwhile, by applying loss model equations, the optimal flux of latter method is determined. When copper losses are approximately equal to iron losses, the optimize efficiency of drive system is achieved.

Instead of just concentrated on searching for the optimal flux, improving the dynamic torque is also an essential factor to be considered in order to optimize the efficiency of DTC system. This is because supplying the drive system at its optimum flux will cause the torque response to be degraded when a rated torque is suddenly needed. Therefore, a modification to look-up table as well as DTC algorithm has been done so that the dynamic torque is achieved during transient state.

2. EFFECTS OF VOLTAGE VECTOR

The effect of voltage vector on torque response has been studied in order to improve the performance of torque response, as shown in Figure 2. Based on Figure 2 (b), in sector 4, the voltage vectors, $v_{s,5}$ and $v_{s,6}$, are applied to increase and decrease the stator flux, respectively. In Figure 2 (a), $v_{s,5}$ is activated to increase the stator flux and at the same time, it is capable to increase the output torque dynamically. But activating $v_{s,6}$ to decrease the flux causes the output torque to increase slightly, and thus, it degraded the torque performance. This case is worsening when the flux is set to its optimized value for efficiency purposes.

In sector 4, $v_{s,5}$ is considered as the most optimized voltage vector compared to $v_{s,6}$ because it has larger tangential to the stator flux and consequently, it can produce dynamically torque. Note that, at the very beginning of sector 4, the response of torque is more dynamic when $v_{s,5}$ is activated for a longer time because this voltage vector is tangential to stator flux. Therefore, the voltage vector that is applied to decrease the stator flux has to be modified so that the proposed voltage vector can produce larger tangential to stator flux in order to improve torque performance.

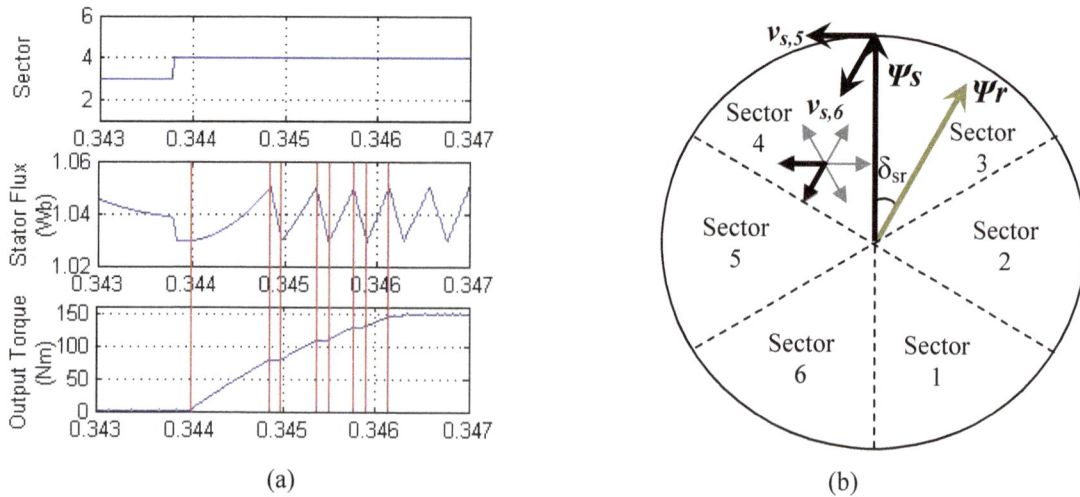

Figure 2. Voltage Vector (a) Effects on Torque Response, and (b) Controlling Stator Flux

3. RESEARCH METHOD

A proposed voltage vector is produced between two conventional voltage vectors by applying vector's parallelogram law. Compared to the conventional voltage vector, the proposed voltage vector has longer amplitude and an angle of 30° adjacent to the conventional voltage vector. For instance, addition of $v_{s,5}$ to $v_{s,6}$ will obtain $v_{s,5-6}$, as shown in Figure 3, and their respective equations are calculated in (1).

$$v_{s,5-6} = [(v_{sd,5} + v_{sd,6}), (v_{sq,5} + v_{sq,6})]\tag{1}$$

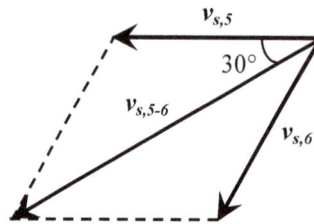

Figure 3. Vector's Parallelogram Law

In Figure 4, the red line indicates the proposed voltage vector whereas the black line represents the conventional voltage vector. In proposed method, $v_{s,5-6}$ is activated instead of $v_{s,6}$ when the flux is required to be reduced. The proposed voltage vector has a larger tangential to stator flux in which it is believed to improve the torque performance when decreasing the flux. In conventional DTC system, the switching of voltage vector is more regular in the middle of a sector compared to the beginning and end of a sector. Consequently, activating the proposed voltage vector also reduces the switching of voltage vector when it is in the middle of a sector and thus, it increases the torque dynamically.

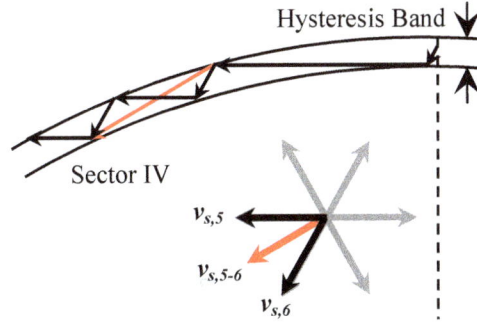

Figure 4 Proposed Voltage Vector

By applying the proposed method, the amplitude of proposed voltage vector is almost twice of the amplitude of conventional method. The increasing amplitude of voltage vector will cause the value of stator flux and output torque to be increased as well. Therefore, the amplitude of proposed voltage vector can be reduced by estimating a ratio between proposed and conventional voltage vector, as follows:

$$Ratio = \frac{V_{s,k-k}}{V_{s,k}} \tag{2}$$

where $V_{s,k-k}$ is the amplitude of proposed voltage vector and $V_{s,k}$ is the amplitude of conventional voltage vector.

After introducing the estimated ratio, it can be seen that the d-q axis of proposed voltage vector is exchanged with the d-q axis of conventional voltage vector. Therefore, the d-q axis of proposed voltage vector is given in (3) and (4):

$$\boldsymbol{v_{sd,k-k}} = \frac{2}{3}V_{DC}(0.866S_b - 0.866S_c) \tag{3}$$

$$\boldsymbol{v_{sq,k-k}} = \frac{2}{3}V_{DC}(S_a - 0.5S_b - 0.5S_c) \tag{4}$$

The switching state of VSI can be implemented in the look-up table with modified DTC algorithm since d-q axis of proposed voltage vector is exchanged with d-q axis of conventional voltage vector. The proposed look-up table with modified DTC algorithm is shown in Table 1. In order to improve the dynamic of output torque, the modified look-up table with DTC algorithm is implemented only during transient state. Meanwhile, the conventional look-up table is applied during steady-state.

Table 1 Modified Look-Up Table

Stator Flux Error Status, Ψ_s^+	Torque Error Status, T_{stat}	Sector 1	Sector 2	Sector 3	Sector 4	Sector 5	Sector 6
1	1	$v_{s,2}$	$v_{s,3}$	$v_{s,4}$	$v_{s,5}$	$v_{s,6}$	$v_{s,1}$
	0	$v_{s,0}$	$v_{s,7}$	$v_{s,0}$	$v_{s,7}$	$v_{s,0}$	$v_{s,7}$
	-1	$v_{s,5-6}\,(v_{s,6})$	$v_{s,1-6}\,(v_{s,5})$	$v_{s,1-2}\,(v_{s,4})$	$v_{s,2-3}\,(v_{s,3})$	$v_{s,3-4}\,(v_{s,2})$	$v_{s,4-5}\,(v_{s,1})$
0	1	$v_{s,2-3}\,(v_{s,3})$	$v_{s,3-4}\,(v_{s,2})$	$v_{s,4-5}\,(v_{s,1})$	$v_{s,5-6}\,(v_{s,6})$	$v_{s,1-6}\,(v_{s,5})$	$v_{s,1-2}\,(v_{s,4})$
	0	$v_{s,7}$	$v_{s,0}$	$v_{s,7}$	$v_{s,0}$	$v_{s,7}$	$v_{s,0}$
	-1	$v_{s,5}$	$v_{s,6}$	$v_{s,1}$	$v_{s,2}$	$v_{s,3}$	$v_{s,4}$

As shown in Table 1, the respective proposed voltage vector can be obtained when the voltage vector in bracket is activated. In other words, the voltage vector in bracket indicates the switching state of the respective proposed voltage vector. For example, in sector 1, $v_{s,2}$ is applied to increase the flux and $v_{s,2-3}$ is activated to decrease the flux. Both of these voltage vectors are capable to increase the output torque. But in order to activate $v_{s,2-3}$, the switching state of $v_{s,3}$ has to be implemented.

4.   RESULTS AND ANALYSIS

The simulation of DTC drive system has been constructed using MATLAB's SIMULINK blocks, as shown in Figure 5. The specifications and parameters of induction machine used in the simulation are given in Table 2. The modified look-up table with DTC algorithm has been attached in parallel to the conventional look-up table with DTC algorithm. The modified look-up table with DTC algorithm has been activated for 5ms only during transient state whereas during steady-state, the conventional look-up table with DTC algorithm has been implemented.

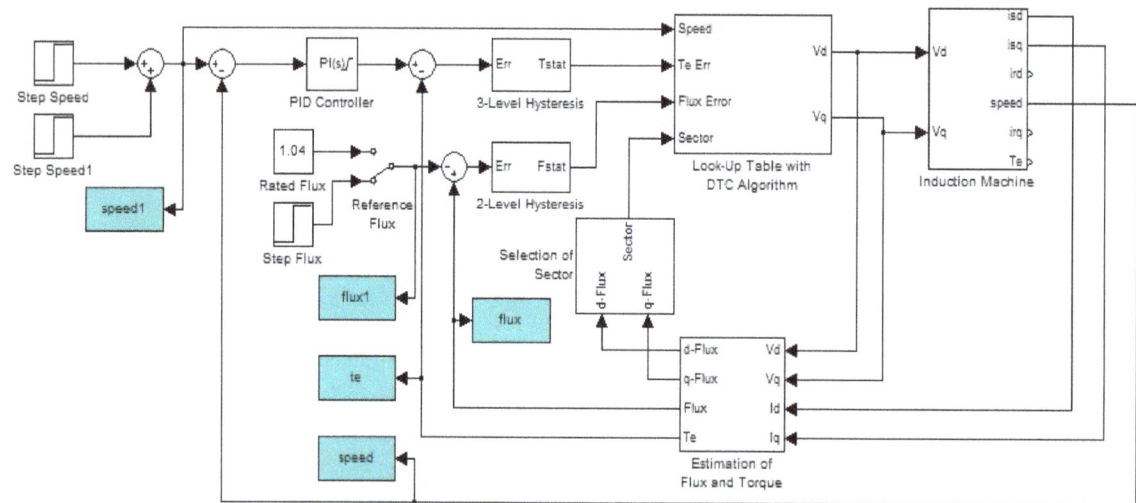

Figure 5. Simulation of DTC Drive System

Table 2 Specifications and Parameters of Induction Motor

Parameters	Values
DC voltage	340V
Stator resistance, R_s	0.25Ω
Rotor resistance, R_r	0.2Ω
Stator inductance, L_s	0.0971H
Rotor inductance, L_r	0.0971H
Mutual inductance, L_m	0.0955H
Frequency, f	50Hz
Inertia motor, J	0.046kgm^2
Pole pairs, p	2
Sampling time	50μs
Rated flux	1.04Wb
Rated torque	150Nm

In order to optimize the efficiency of drive system, the SC method has been implemented in DTC drive system. The SC is activated at t=1s with step flux of 0.043Wb, sample time of 0.1s and at its rated speed. Basically, the rated flux is applied to the system and after the system has reached its steady-state, the corresponding current value is measured. Then, the flux is decreased with step flux and the corresponding current value is measured again. When the new stator current ($I_{s, k}$) is smaller than the previous stator current ($I_{s, k-1}$), the flux value is decreased with step flux, and vice-versa. The SC method is continuing until it reaches its optimum flux value. From Figure 6, the optimal flux is obtained after 0.3s with flux value of 0.89Wb.

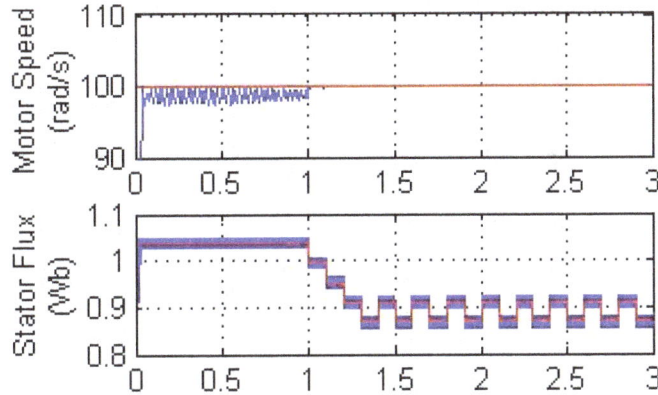

Figure 6. Flux Search Controller

3.1. Voltage Vector

As discussed earlier, the conventional and proposed voltage vector has been proven in Figure 7. The blue line indicates the conventional voltage vector whereas the red line denotes the proposed voltage vector. Based on this figure, the proposed voltage vector has same amplitude as the conventional voltage vector and it is 30° adjacent to the conventional voltage vector.

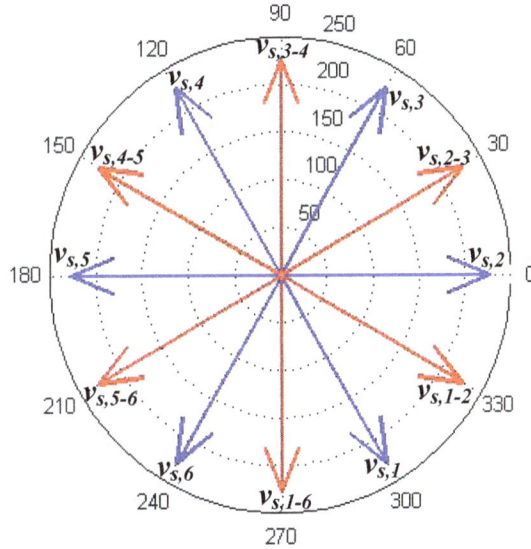

Figure 7. Voltage Vector

3.2. Torque Performance

In order to examine the effectiveness of proposed voltage vector towards the dynamic torque at light load, a step flux and torque is applied from 0.89Wb to 1.04Wb and from 0Nm to 150Nm respectively at beginning, middle and end of a sector, as shown in Figure 8. In Figure 8 (a) and (b), the stator flux of proposed method is increased beyond rated flux for a short duration because d-axis stator flux is slightly decreased and q-axis stator flux is slightly increased compared to conventional method. However, the output torque is not affected by the slightly increased of stator flux. Meanwhile, in Figure 8 (a), the output torque is decreased to 120Nm at t=0.505s because the flux is decreased to its rated value.

At beginning of a sector, the conventional and proposed method achieved the rated torque in 2ms. But when a step torque is applied at the middle of a sector, the conventional method requires 2.8ms to achieve its rated torque whereas the proposed method needs 2.4ms to attain its rated torque. The proposed method has improved the performance of torque response by 0.4ms compared to conventional method. Meanwhile, at the end of a sector, the conventional and proposed method requires 3.5ms and 3.0ms to reach their rated torque, respectively. By implementing the proposed method, the torque response can be improved by 0.5ms.

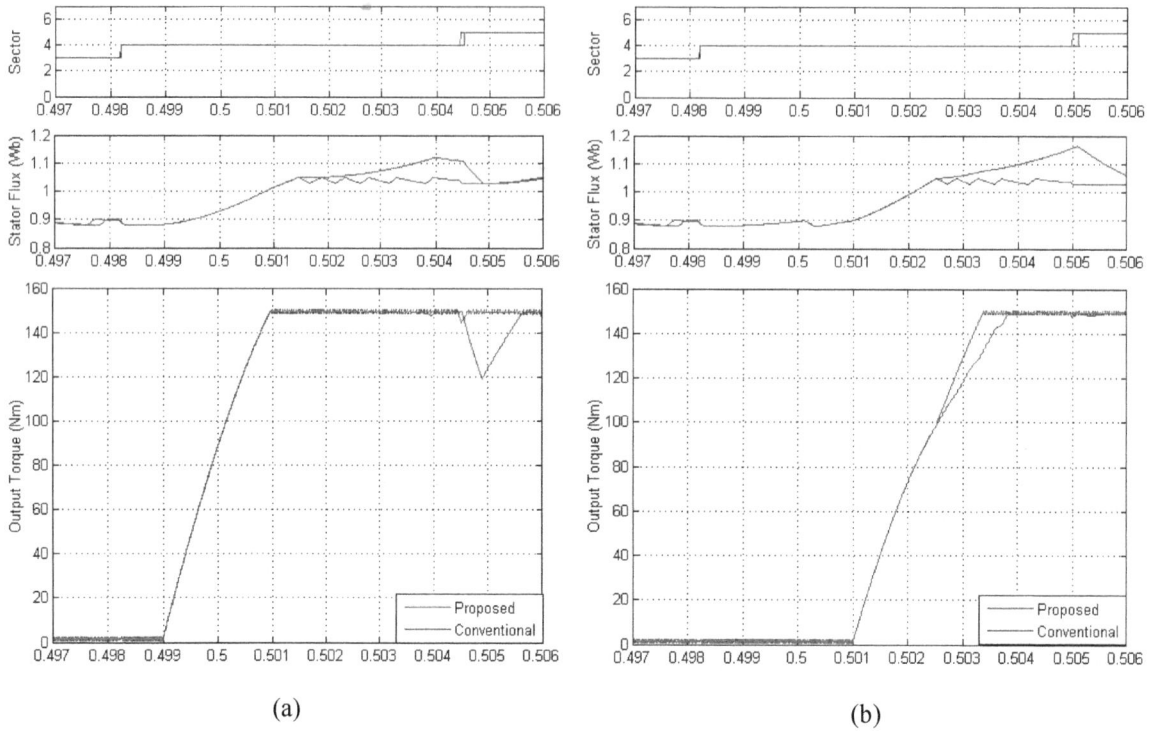

(a)   (b)

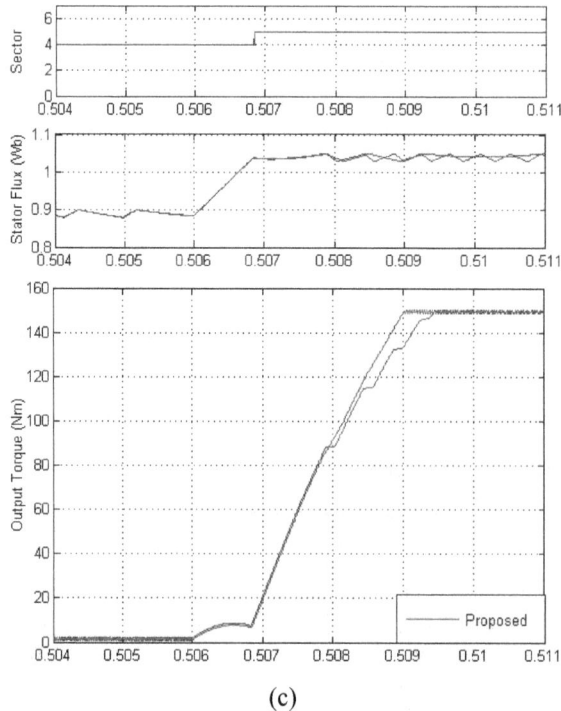

(c)

Figure 8 Torque Performance on (a) Beginning, (b) Middle, and (c) End, of a Sector

From Figure 8 (a), the applied voltage vector is activated for a longer period since the voltage vector is tangential with respect to stator flux. As mentioned earlier, only the voltage vector that is used to decrease the stator flux is modified; hence, the conventional and proposed method applied the same voltage vector to increase flux in which it caused the conventional and proposed method to reach rated torque at the same time. Based on Figure 8 (b) and (c), it can be seen that the torque performance is improved when the voltage vector is activated for a longer period, in which it can be controlled by generating a larger tangential to the stator flux. Besides that, voltage vector that has larger tangential respective to stator flux is able to reduce the switching of voltage vector. Consequently, it is necessary to modify the angle of voltage vector so that a dynamic torque can be produced during transient state.

5. CONCLUSION

During steady-state, the flux has to be set to an optimum value in order to optimize the efficiency of DTC drive system. However, the dynamic of output torque would be degraded when a torque is suddenly needed and it is not suitable to be implemented in EV applications. Therefore, an adjustment to the look-up table as well as DTC algorithm has been constructed by modifying the angle of voltage vector so that a larger tangential with respect to the flux is yielded. Based on the results, it can be concluded that the proposed voltage vector improves the performance of torque response during transient state. Therefore, this method is believed to optimize the efficiency of DTC drive system and at the same time, the dynamic of torque response is improved.

ACKNOWLEDGEMENTS

The authors would like to express their appreciation to Universiti Teknologi Malaysia (UTM) for providing Zamalah's Scholarship and Ministry of Education for fund research grant (R.J130000.7823.4F380) in this research.

REFERENCES

[1]   I. Takahashi and T. Noguchi, "A New Quick-Response and High-Efficiency Control Strategy of an Induction Motor," *Industry Applications, IEEE Transactions on,* vol. IA-22, pp. 820-827, 1986.

[2]   T. Abe, T. G. Habetler, F. Profumo, and G. Griva, "Evaluation of a high performance induction motor drive using direct torque control," in *Power Conversion Conference, 1993. Yokohama 1993., Conference Record of the,* 1993, pp. 444-449.

[3]   I. Takahashi and T. Noguchi, "Take a look back upon the past decade of direct torque control [of induction motors]," in *Industrial Electronics, Control and Instrumentation, 1997. IECON 97. 23rd International Conference on,* 1997, pp. 546-551 vol.2.

[4]   S. Kaboli, E. Vahdati-Khajeh, M. R. Zolghadri, and A. Homaifar, "A fast optimal flux search controller with improved steady state behavior for DTC based induction motor drives," in *Electric Machines and Drives, 2005 IEEE International Conference on,* 2005, pp. 1732-1736.

[5]   S. Kaboli, M. R. Zlghadri, and A. Emadi, "A Fast Flux Search Controller for DTC Based Induction Motor Drives," in *Power Electronics Specialists Conference, 2005. PESC '05. IEEE 36th,* 2005, pp. 739-744.

[6]   S. Kaboli, M. R. Zolghadri, and E. Vahdati-Khajeh, "A Fast Flux Search Controller for DTC-Based Induction Motor Drives," *Industrial Electronics, IEEE Transactions on,* vol. 54, pp. 2407-2416, 2007.

[7]   S. Kaboli, E. Vahdati-Khajeh, M. R. Zolghadri, and A. Homaifar, "on the Performance of Optimal Flux Search Controller for DTC Based Induction Motor Drives," in *Electric Machines and Drives, 2005 IEEE International Conference on,* 2005, pp. 1752-1756.

[8]   S. Kaboli, M. R. Zolghadri, D. Roye, and A. Emadi, "Online optimal flux controller for DTC based induction motor drives," in *Industrial Electronics Society, 2004. IECON 2004. 30th Annual Conference of IEEE,* 2004, pp. 1391-1395 Vol. 2.

[9]   N. Sadati, S. Kaboli, H. Adeli, E. Hajipour, and M. Ferdowsi, "Online Optimal Neuro-Fuzzy Flux Controller for DTC Based Induction Motor Drives," in *Applied Power Electronics Conference and Exposition, 2009. APEC 2009. Twenty-Fourth Annual IEEE,* 2009, pp. 210-215.

[10]   S. Vamsidhar and B. G. Fernandas, "Design and development of energy efficient sensorless direct torque controlled induction motor drive based on real time simulation," in *Industrial Electronics Society, 2004. IECON 2004. 30th Annual Conference of IEEE,* 2004, pp. 1349-1354 Vol. 2.

[11]   G. Calzada-Lara, F. Pazos-Flores, and R. Alvarez-Salas, "A new Direct Torque Control for a better efficiency of the induction motor," in *Power Electronics Congress (CIEP), 2010 12th International,* 2010, pp. 78-83.

[12]   I. Kioskeridis and N. Margaris, "Loss minimization in scalar-controlled induction motor drives with search controllers," *Power Electronics, IEEE Transactions on,* vol. 11, pp. 213-220, 1996.

[13]   D. Gan and O. Ojo, "Efficiency Optimizing Control of Induction Motor Using Natural Variables," *Industrial Electronics, IEEE Transactions on,* vol. 53, pp. 1791-1798, 2006.

[14] X. Zhang, H. Zuo, and Z. Sun, "Efficiency optimization of direct torque controlled induction motor drives for electric vehicles," in *Electrical Machines and Systems (ICEMS), 2011 International Conference on*, 2011, pp. 1-5.

[15] G. Bhuvaneswari and A. P. Satapathy, "ANN based optimal flux determination for efficiency improvement in Direct Torque controlled induction motor drives," in *Power and Energy Society General Meeting, 2010 IEEE*, 2010, pp. 1-6.

[16] I. Kioskeridis and N. Margaris, "Loss minimization in induction motor adjustable-speed drives," *Industrial Electronics, IEEE Transactions on*, vol. 43, pp. 226-231, 1996.

18

Performance Analysis of a DTC and SVM Based Field-Orientation Control Induction Motor Drive

Md. Rashedul Islam[*]**, Md. Maruful Islam**[**]**, Md. Kamal Hossain**[***] **and Pintu Kumar Sadhu**[****]

[*,**,***]Department of Electrical and Electronic Engineering, Dhaka University of Engineering & Technology, Gazipur-1700, Bangladesh

[****]Service Operation Center, Grameenphone Limited, Bangladesh

ABSTRACT

Keyword:

Direct Torque Control
Electric Drive
Field-orientation control
Induction Motor
Space Vector Modulation

This study presents a performance analysis of two most popular control strategies for Induction Motor (IM) drives: direct torque control (DTC) and space vector modulation (SVM) strategies. The performance analysis is done by applying field-orientation control (FOC) technique because of its good dynamic response. The theoretical principle, simulation results are discussed to study the dynamic performances of the drive system for individual control strategies using actual parameters of induction motor. A closed loop PI controller scheme has been used. The main purpose of this study is to minimize ripple in torque response curve and to achieve quick speed response as well as to investigate the condition for optimum performance of induction motor drive. Depending on the simulation results this study also presents a detailed comparison between direct torque control and space vector modulation based field-orientation control method for the induction motor drive.

Corresponding Author:

Md. Kamal Hossain,
Assistant Professor, Departement of Electrical and Electronic Engineering,
Dhaka University of Engineering & Technology,
Gazipur-1700, Bangladesh
Email: mkhossain87@gmail.com

1.   INTRODUCTION

The most common type of ac motor being used throughout the world today is the induction motor (IM). Three phase induction motors are widely used in various industries as prime workhorses to produce rotational motions and forces. Traditionally, it has been used in constant and variable-speed drive applications that do not cater for fast dynamic processes [1]. Due to the requirements of the load and the need for economy have resulted in developments of several types of induction motor drives and these induction motor drives require a great attention in controlling speed. Because of the marriage of power electronics with motors and recent development of several new control technologies this situation is changing rapidly. Such control technologies are direct torque control (DTC), space vector modulation (SVM) and field-orientation control (FOC) technique. These technologies are widely used in high performance motion control of induction motors [2].

This work presents a comparative study on three most popular control strategies for induction motor (IM) drives: direct torque control (DTC), space vector modulation (SVM) and field-oriented control (FOC) [3]-[5]. Here fixed value of the proportional and integral gain of PI controller is used to achieve quick speed response. In DTC it is possible to control the stator flux and the torque is controlled by selecting the appropriate inverter state [6]. But conventional DTC scheme has two main disadvantages [7]: Current and torque distortions caused by the sector changes and starting and low - speed operation problems. To overcome these problems SVM technique is implemented. For applying SVM scheme power is taken from dc source and converts it to three phase ac using dc-to-ac converter [8]. In order to achieve fast speed response and improved torque characteristic, field-orientation control (FOC) technique is used [9]. The methodology of field-orientation control is normally developed based on estimation of induction motor fluxes.

2.   RESEARCH METHOD
2.1. Induction Motor Model under Field-Orientation Control Principle

To study the transient and dynamic conditions generally mutually perpendicular stationary and synchronously rotating fictitious coils are considered. For the induction motor considered will have the following assumptions: symmetrical two-pole, three phases windings, slotting effects are neglected, permeability of the iron parts is infinite, the flux density is radial in the air gap, iron losses are neglected, stator and the rotor windings are simplified as a single, multi-turn full pitch coil situated on the two sides of the air gap.

From stationary two axis model and synchronously rotating two axis model the fifth order non-linear state space model of induction motor is represented in the synchronous reference frame (d-q) as follows:

$$v_{ds} = (R_s + pL_s)i_{ds} - L_s\omega_e i_{qs} + pL_m i_{dr} - L_m\omega_e i_{qr} \tag{1}$$

$$v_{qs} = \omega_e L_s i_{ds} + (R_s + pL_s)i_{qs} + L_m\omega_e i_{dr} + pL_m i_{qr} \tag{2}$$

$$0 = pL_m i_{ds} - \omega_{sl} L_m i_{qs} + (R_r + pL_r)i_{dr} - L_r\omega_{sl} i_{qr} \tag{3}$$

$$0 = L_m\omega_{sl} i_{ds} + pL_m i_{qs} + (R_r + pL_r)i_{qr} + L_r\omega_{sl} i_{dr} \tag{4}$$

$$T_e = Jp\omega_m + B\omega_m + T_L \tag{5}$$

From the developed electromagnetic torque in terms of d- and q- axes components is given by:

$$T_e = \frac{3}{2}P_p L_m\left(i_{qs}i_{dr} - i_{ds}i_{qr}\right) \tag{6}$$

Components of rotor flux are:

$$\psi_{dr} = L_r i_{dr} + L_m i_{ds} \tag{7}$$

$$\psi_{qr} = L_r i_{qr} + L_m i_{qs} \tag{8}$$

From (7) and (8),

$$i_{dr} = \frac{1}{L_r}(\psi_{dr} - L_m i_{ds}) \tag{9}$$

$$i_{qr} = \frac{1}{L_r}(\psi_{qr} - L_m i_{qs}) \tag{10}$$

Substituting from (7) to (10) into (3) and (4) yields:

$$\frac{d\psi_{dr}}{dt} + \frac{R_r}{L_r}\psi_{dr} - \frac{L_m}{L_r}R_r i_{ds} - \omega_{sl}\psi_{qr} = 0 \tag{11}$$

$$\frac{d\psi_{qr}}{dt} + \frac{R_r}{L_r}\psi_{qr} - \frac{L_m}{L_r}R_r i_{qs} + \omega_{sl}\psi_{dr} = 0 \tag{12}$$

If the field orientation is established such that q-axis rotor flux is set zero, and d-axis rotor flux is maintained constant then Equation (11), (12), (9), (10) and (6) becomes $\psi_{dr} = L_m i_{ds}$; $\omega_{sl} = \dfrac{1}{\tau_r}\dfrac{i_{qs}}{i_{ds}}$; $i_{dr} = 0$

; $i_{qr} = -\dfrac{L_m}{L_r}i_{qs}$ $T_e = \dfrac{3}{2}P_p\dfrac{L_m}{L_r}\psi_{dr}i_{qs}$; where $\tau_r (= L_r / R_r)$ is the time constant of the rotor.

2.2 Direct Torque Control Technique

Direct torque control technique is used in variable frequency drives to control the torque (and thus finally the speed) of three-phase ac electric motors. In direct torque it is possible to control directly the stator flux and the torque by selecting the appropriate state [10]. The way to impose the required stator flux is by means of choosing the most suitable Voltage Source Inverter (VSI) state. Decoupled control of the stator flux modulus and torque is achieved by acting on the radial and tangential components respectively of the stator flux-linkage space vector. Figure 1. shows the possible dynamic locus of the stator flux, and its different variation depending on the VSI states chosen [10]. The possible global locus is divided into six different sectors signaled by the discontinuous line. In accordance with Figure 1, the general table I can be written.

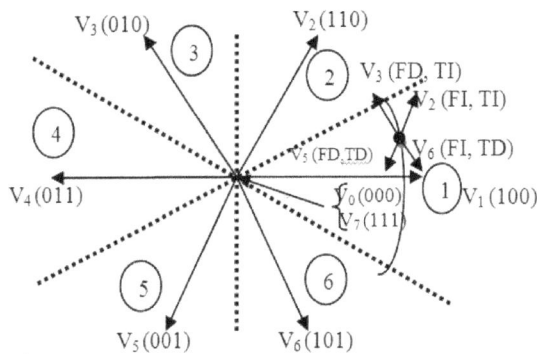

Figure 1. Stator flux vector locus and possible switching.

Table 1. Selection Table for DTC

Voltage Vector	Increase	Decrease
Stator Flux	V_k, V_{k+1}, V_{k-1}	V_{k+2}, V_{k-2}, V_{k+3}
Torque	V_{k+1}, V_{k+2}	V_{k-1}, V_{k-2}

The sectors of the stator flux space vector are denoted from S1 to S6. Stator flux modulus error after the hysteresis block ($\Delta\psi$) can take just two values. The zero voltage vectors V0 and V7 are selected when the torque error is within the given hysteresis limits, and must remain unchanged.

Table 2. Lookup Table for DTC

Torque error ($\Delta\psi$)	Torque error (ΔT)	S_1	S_2	S_3	S_4	S_5	S_6
FI	TI	V_2	V_3	V_4	V_5	V_6	V_1
FI	TE	V_7	V_0	V_7	V_0	V_7	V_0
FI	TD	V_6	V_1	V_2	V_3	V_4	V_5
FD	TI	V_3	V_4	V_5	V_6	V_1	V_2
FD	TE	V_0	V_7	V_0	V_7	V_0	V_7
FD	TD	V_5	V_6	V_1	V_2	V_3	V_4

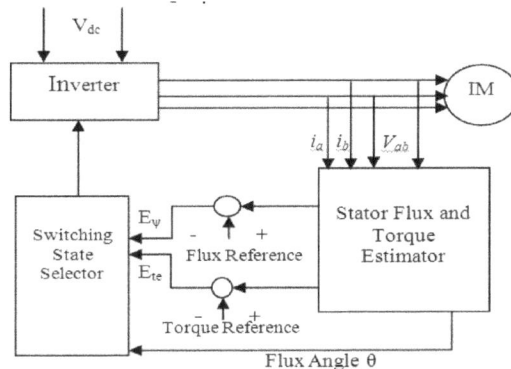

Fig. 2. DTC Block diagram.

2.3 Space Vector Modulation Technique

Space Vector Modulation principle [4] is shown in Fig.3. The reference vector u^* is sampled at the fixed clock frequency $2f_s$. The reference voltage vector u^* can be generated from the machine command $-\alpha$ and $-\beta$ axes voltages $v_{\alpha s}^*$ and $v_{\beta s}^*$ as:

$$u^* = \sqrt{v_{\alpha s}^{*2} + v_{\beta s}^{*2}} \tag{13}$$

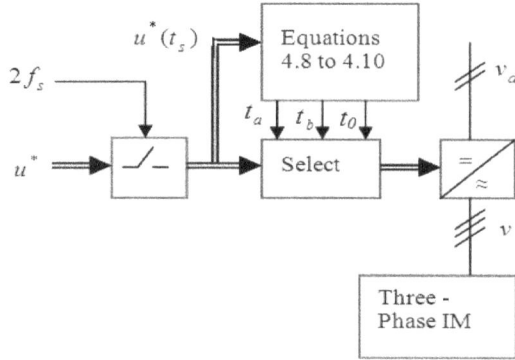

Figure 3. SVM Signal flow diagram

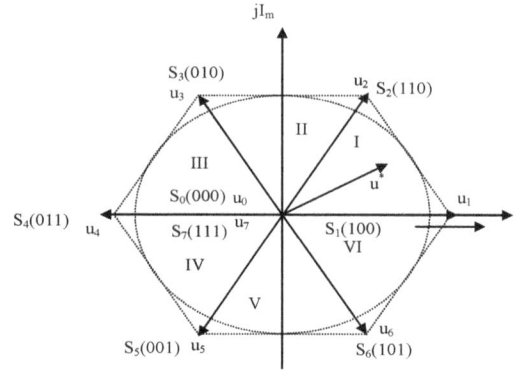

Figure 4. All voltage space vectors

If T_s is the sampling time then the sampled value reference voltage vector $u^*(t_s)$ is used to solve the equations.

$$\frac{2}{T_s}\left(t_a u_a + t_b u_b\right) = u^*(t_s) \tag{14}$$

$$t_0 = t_7 = \frac{1}{2}(T_s - t_a - t_b) \tag{15}$$

$$t_a = T_s u^*(t_s)\frac{3}{\pi}(\cos\alpha - \frac{1}{\sqrt{3}}\sin\alpha) \tag{16}$$

$$t_b = T_s u^*(t_s)\frac{2\sqrt{3}}{\pi}\sin\alpha \tag{17}$$

$$t_0 = t_7 = \frac{1}{2}(T_s - t_a - t_b) \tag{18}$$

2.4. Field-Orientation Control Method

The field-orientation control consists of controlling the stator currents represented by a vector. This control is based on projections which transform a three phase time and speed dependent system into a two co-ordinate (d and q co-ordinates) time invariant system. Figure 5. Shows the Basic scheme of FOC for IM drive [5].

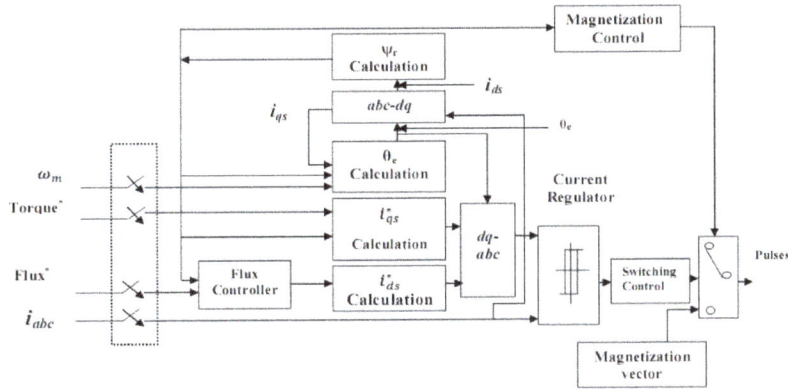

Figure 5. Basic scheme of FOC for IM drive

3. RESULTS AND ANALYSIS

The simulation scenarios shown in this thesis paper cover the following situations: Generation of pulses for inverter, transient and steady state behavior of 3-phase current, speed and torque response, a step change in load torque, a step change in speed reference [11] and condition for optimum performance.

3.1. Rotor and Stator flux

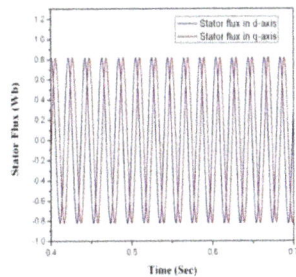 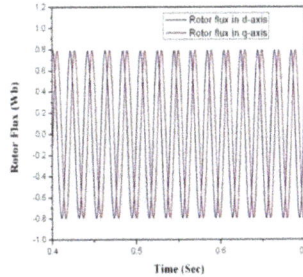 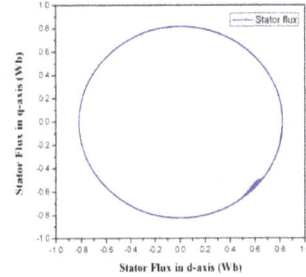

Figure 6. *d-q* axis stator fluxes   Figure 7. *d-q* axis rotor fluxes   Figure 8. Locus of stator fluxes in Stationary reference frame

Figure 6 and Figure 7. shows the stator and rotor fluxes. It is noticed that 90° phase difference is obtained and are approximately sinusoidal. Figure 8. indicates the locus of the stator flux and it is noticed that flux follows a circular shape. The components of stator fluxes in stationary reference frame are sinusoidal and 90° phase displacement to each other.

3.2. Simulation Results for DTC based FOC method

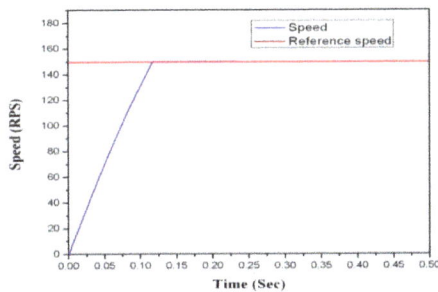 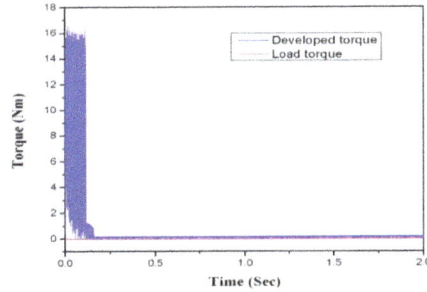

Figure 9. Speed response in DTC using FOC   Figure 10. Torque developed in DTC using FOC

With the help of field-orientation control method quick speed response is achieved and catches the reference speed within 0.10 sec. as indicated in Figure 9 and Figure 10 shows the actual torque developed curve for DTC based FOC method. It can be said that, region of torque distortion is smaller than DTC.

3.3. Simulation Results for SVM based FOC Method

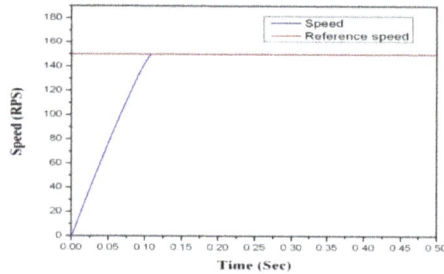

Figure 11. Speed response in SVM using FOC

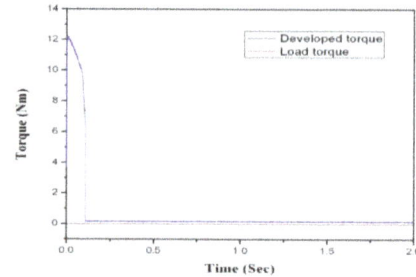

Figure 12. Torque developed in SVM using FOC

By using field-orientation principle in SVM technique overshoot problem is eliminated as well as quick speed response is achieved as shown in Figure 11. Figure 12. shows the torque response when the motor is unloaded, it is evident that the torque response is better than DTC based FOC method.

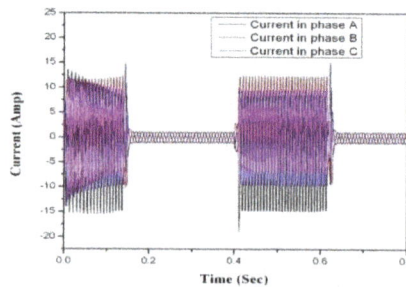

Figure 13. Three-phase currents under load condition

3.4. Effect of Change of Load for DTC Technique

Sudden application of load torque from 0 Nm to 6 Nm at t = 0.4 sec. to t = 0.6 sec. causes a change in three-phase current and transient phenomenon is occurred and after few seconds latter steady state condition has been reached when load adjustment is done. Figure 14. indicates the simulated response of the motor speed when it is suddenly loaded from 0 Nm to 6 Nm. It is observed that sudden application of load torque causes a non-uniform dip and overshoot in the speed curve. Sudden application of load torque from t = 0.4 sec. to t = 0.6 sec. causes torque ripple at that particular time in the developed torque curve as shown in Figure 15.

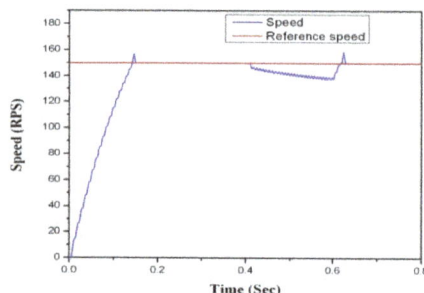

Figuure 14. Effect of change of load torque on speed response

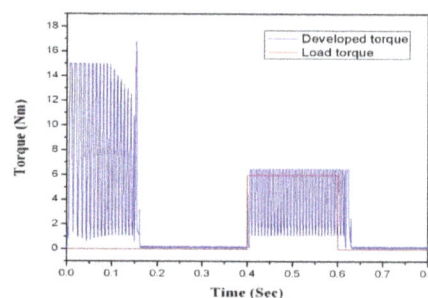

Figure 15. Torque developed when load torque is increased suddenly

3.5. Simulation Results for SVM Technique under Load Condition

Effect of change of load torque on three-phase current is illustrated in Figure 16. Sudden application of load torque from 0 Nm to 6 Nm causes a change in three-phase currents but the oscillation of current is smaller than that of DTC technique. Figure 17. indicates the motor speed when it is suddenly loaded. It is observed that sudden application of load torque causes a very little dip and there is no overshoot in the speed curve.

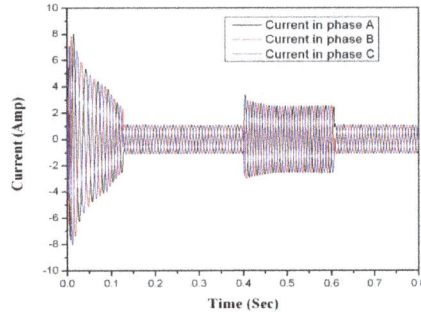

Figure 16. Three-phase currents under load condition

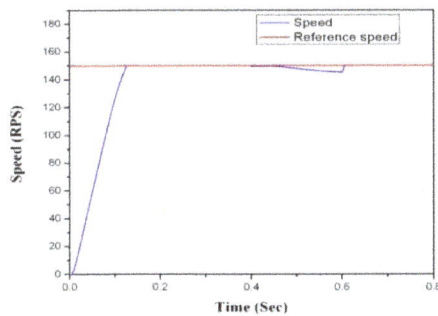

Figure 17. Effect of change of load torque on speed response

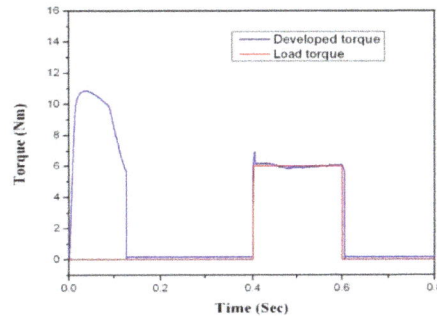

Figure 18. Torque developed when load torque is increased suddenly

Hence, better speed response curve has been achieved by using SVM than that of DTC technique. The oscillation found in the torque developed curve under load condition is also reduced as shown in Figure 18.

4.   CONCLUSION

In this paper, main characteristics of direct torque control and space vector modulation based field-orientation control scheme for induction motor drives are studied and performance analysis has been investigated with the help of simulation results with a view to highlighting the advantages of each schemes. For achieving high performance IM drive, a suitable mathematical model is used. Performance analysis of individual scheme is carried out by changing the load torque at a particular time interval and by changing the reference speed. Lower value of reference stator flux during the simulation causes greater torque distortion and bad speed response curve. That is why the reference stator flux is determined from the IM parameters. From the simulation results for DTC and SVM based field-orientation control method; it can be concluded that the SVM based field-orientation control method shows better performance for induction motor drives because of its quick speed response and elimination of the overshoot problem unlike DTC based field-orientation control. When field-orientation principle is introduced in SVM technique; there is a reduction of ripple in the developed torque. Moreover, SVM based FOC method is capable to follow the reference speed quickly and has practically thus will find many applications.

REFERENCES

[1]   Muhammad H Rashid. *Power Electronics, Circuits, Devices and Applications*, 3rd edition.

[2]   AF Puchstein, TC Lioyd, AG Conrad. *Alternating Current Machines*. Edition 2006-2007.

[3]   M Vasudevan, R Arumugam, S Paramasivam. *High – Performance Adaptive Intelligent Direct Torque Control Schemes for Induction Motor Drives. Serbian Journal of Electrical Engineering*. 2005; 2(1); 93 – 116.

[4]   Bimal K Bose. *Power Electronics and Variable Frequency Drive*. IEEE Press.

[5]   Jae Ho Chang, Byung Kook Kim. *Minimum-Time Minimum Loss Speed Control of Induction Motors Under Field-Oriented Control. IEEE Trans. on Ind. Elecron.* 1997; 44(6);809-815.

[6]   Tejavathu Ramesh, Anup Kumar Pandl, Y Suresh, Suresh Mikkili. *Direct Flux and Torque Control of Induction Motor Drive for Speed Regulator using PI and Fuzzy Logic Controllers. IEEE- International Conference On Advances In Engineering, Science And Management (ICAESM -2012)*. 2012; 31; 288-295.

[7]   Manoj Datta, Md Abdur Rafiq, BC Ghosh. *Genetic Algorithm Based Fast Speed Response Induction Motor Drive without Speed Encoder. POWERENG 2007, Setubal, Portugal*. 2007.

[8]   Marcin Żelechowski. *Space Vector Modulated – Direct Torque Controlled (DTC–SVM) Inverter – Fed Induction Motor Drive, Warsaw, Poland*.2005.

[9]   Donald Grahame Holmes, Brendan Peter McGrath, Stewart Geoffrey Parker.*Current Regulation Strategies for Vector-Controlled Induction Motor Drives. IEEE Transactions on Industrial Electronics*. 2012; 59(10); 3680-3689.

[10]   A Jidin, NRN Idris, AHM Yatim, T Sutikno, Malik E Elbuluk. Extending switching frequency for torque ripple reduction utilizing a constant frequency torque controller in dtc of induction motors. *Journal of Power Electronics*. 2011; 11(2); 148-155.

[11]   Elwy E El-kholy. *High Performance Induction Motor Drive Based on Adaptive Variable Structure Control. Journal of Electrical Engineering*. 2005; 56(3-4); 64–70.

19

Speed Sensorless Vector Control of Induction Motor Drive with PI and Fuzzy Controller

R. Gunabalan*, V. Subbiah**

* Departement of Electrical and Electronics Engineering, Dr. Sivanthi Aditanar College of Engineering, Tiruchendur
** Departement of Electrical and Electronics Engineering, PSG College of Technology, Coimbatore

ABSTRACT

Keyword:

Fuzzy controller
Induction motor
Natural observer
Sensorless control
Simulation

This paper directed the speed-sensorless vector control of induction motor drive with PI and fuzzy controllers. Natural observer with fourth order state space model is employed to estimate the speed and rotor fluxes of the induction motor. The formation of the natural observer is similar to and as well as its attribute is identical to the induction motor. Load torque adaptation is provided to estimate the torque and rotor speed is estimated from the load torque, rotor fluxes and stator currents. There is no direct feedback in natural observer and also observer gain matrix is absent. Both the induction motor and the observer are characterized by state space model. Simple fuzzy logic controller and conventional PI controllers are used to control the speed of the induction motor in closed loop. MATLAB simulations are made with PI and fuzzy controllers and the performance of fuzzy controller is better than PI controller in view of torque ripples. The simulation results are obtained for various running conditions to exhibit the suitability of this method for sensorless vector control. Experimental results are provided for natual observer based sensorless vector control with conventional PI controller.

Corresponding Author:

R. Gunabalan,
Departement of Electrical and Electronics Engineering,
Chandy College of Engineering, Thoothukudi 628005
Anna University, Chennai, TamilNadu, INDIA.
Email: gunabalanr@yahoo.co.in

1. INTRODUCTION

Induction motors are preferred for most of the industry applications because of the limitations of commutation and rotor speed in DC drives. The induction motor is in fact 'brushless' and can operate with simple control methods not requiring a shaft position transducer. With no shaft position feedback, the motor remains stable only as long as the load torque does not exceed the breakdown torque. At low speeds it is possible for oscillatory instabilities to develop. To overcome these limitations 'field-oriented' or 'vector' control has been developed in which the phase and magnitude of the stator currents are regulated so as to maintain the optimum angle between stator mmf and rotor flux. This control is based on transforming a three phase time and frequency dependent system into a two co-ordinate (d and q axes) time invariant system. These projections lead to a structure similar to that of a separately excited DC motor control. Field orientation, however, requires either a shaft position encoder or an in-built control model whose parameters are specific to the motor.

Generally, two types of field oriented control schemes are available. 1. Direct field oriented control 2. Indirect field oriented control. In the direct scheme, the instantaneous position of rotor flux (θ_e) has to be measured using flux sensors. This adds to the cost and complexity of the drive system. In the indirect scheme, a model of the induction motor is required to calculate the reference angular slip frequency that has to be added to the measured rotor speed. The sum is integrated to calculate the instantaneous position of the rotor flux. Rotor time constant (L_r/R_r) is used to calculate the slip frequency and is sensitive to temperature and flux level. To avoid these complications, different algorithms are projected, to estimate both the rotor flux vector and/or rotor shaft speed. The induction motor drives without mechanical speed sensors have the attractions of low cost, high reliability, smaller in size and lack of additional wiring for sensors or devices mounted on the shaft. Nowadays, a number of estimation techniques are available for speed and flux calculation. The standard speed estimators are Extended Kalman Filter (EKF) [1]–[6], Luenberger observer [7]–[9] and Model Reference Adaptive System (MRAS) [10]. The initial selection of noise covariance matrices is not easy in EKF and subsequently the algorithm is complicated. The selection of the observer gain constant is difficult in Luenberger observer. The number of inputs to the estimators mentioned above is different to the number of inputs to the induction motor since they utilize output feedback. To overcome the difficulties of the above estimators, natural observer proposed in [11] is used in this paper. In natural observer, the dynamic behavior is exactly the same as the motor and there is no external feedback. The load torque adaptation is used to estimate the load torque from the active power error. Fifth order state space model was used in [11] whereas fourth order induction motor model is used in this paper to reduce the computational burden and the equations are similar to Luenberger observer.

Recent developments in the application of control theory are such that the conventional techniques for the design of controllers are being replaced by artificial intelligence based controllers. The main purpose of using artificial intelligence based controllers is to reduce the tuning efforts associated with the conventional PI controllers and also to obtain the improved responses. PID controllers are commonly intended for linear systems and they provide a preferable cost/benefit ratio [12]. However, the presence of nonlinear effects limits their performances.

Fuzzy logic controllers (FLC's) have the following advantages over the conventional controllers [13]: they are cheaper to develop, they cover a wider range of operating conditions, and they are more readily customizable in natural language terms. Application of PI-type fuzzy controller increases the quality factor. In contrast with traditional linear and nonlinear control theory, a FLC is not based on a mathematical model and is widely used to solve problems under uncertain and vague environments, with high nonlinearities.

In this paper, natural observer with reduced order state space model is proposed to estimate the speed of the induction motor and fuzzy controller is employed instead of conventional PI controller for speed control. Mean value of the rotor flux is maintained constant by employing PI controller in the rotor flux feedback path. Simulations are performed for different running conditions to study the performance of fuzzy controller over PI controller. Experimental results are provided with PI controller to validate the proposed method.

2.   NATURAL OBSERVER

The arrangements and the characteristics of the natural observer are similar to the induction motor for the specified input voltage and load torque condition. The major difference between the natural observer and the conventional observer is that feedback is employed only in the adaptation algorithm and no direct feedback. So, the convergence rate of the natural observer is faster than that of the motor in reaching the steady state behaviour. To estimate the rotor speed, fourth order induction motor model in stator flux oriented reference frame is used in this paper, whereas fifth order state space model is used in [11]. The dq-axes stator currents and rotor fluxes are considered as state variables. The state space representation of the three-phase induction motor is as follows:

$$\frac{dX}{dt} = AX + BV_s \qquad (1)$$
$$Y = CX \qquad (2)$$

Where,

$$A = \begin{bmatrix} \dfrac{-\left(R_s + R_r\left(\frac{L_m}{L_r}\right)^2\right)}{\sigma L_s} & 0 & \dfrac{L_m}{\sigma L_s L_r \tau_r} & \dfrac{\omega_r L_m}{\sigma L_s L_r} \\[4mm] 0 & \dfrac{-\left(R_s + R_r\left(\frac{L_m}{L_r}\right)^2\right)}{\sigma L_s} & \dfrac{-\omega_r L_m}{\sigma L_s L_r} & \dfrac{L_m}{\sigma L_s L_r \tau_r} \\[4mm] \dfrac{L_m}{\tau_r} & 0 & \dfrac{-1}{\tau_r} & -\omega_r \\[4mm] 0 & \dfrac{L_m}{\tau_r} & \omega_r & \dfrac{-1}{\tau_r} \end{bmatrix}$$

$$B = \begin{bmatrix} \dfrac{1}{\sigma L_s} & 0 \\ 0 & \dfrac{1}{\sigma L_s} \\ 0 & 0 \\ 0 & 0 \end{bmatrix} C = \begin{bmatrix} 1 & 0 & 0 & 0 \\ 0 & 1 & 0 & 0 \end{bmatrix}$$

$\sigma = 1 - \dfrac{L_m^2}{L_s L_r}$ - leakage coefficient

$X = \begin{bmatrix} i_{ds}^s & i_{qs}^s & \varphi_{dr}^s & \varphi_{qr}^s \end{bmatrix}^T$

$Y = \begin{bmatrix} i_{ds}^s & i_{qs}^s \end{bmatrix} = i_s$

$V_s = \begin{bmatrix} V_{ds}^s & V_{qs}^s \end{bmatrix}^T$

L_s, L_r – stator and rotor self inductance respectively (H)

L_m- mutual inductance (H)

τ_r-rotor time constant $= \dfrac{L_r}{R_r}$

ω_r -motor angular velocity (rad/s)

Figure 1 shows the structure of the natural observer and the system described by Equation (1) and Equation (2) are exactly the same form of the induction motor model and no external feedback [11]. Estimation of the stator currents and the rotor fluxes can be written by the following equations:

$$\frac{d\hat{X}}{dt} = \hat{A}\hat{X} + BV_s \tag{3}$$

$$\hat{Y} = C\hat{X} \tag{4}$$

$\hat{X} = \begin{bmatrix} \hat{i}_{ds}^s & \hat{i}_{qs}^s & \hat{\varphi}_{dr}^s & \hat{\varphi}_{qr}^s \end{bmatrix}^T$

$\hat{Y} = \begin{bmatrix} \hat{i}_{ds}^s & \hat{i}_{qs}^s \end{bmatrix}^T = \hat{i}_s$

Where, "^" represents the estimated quantities.

The load torque is estimated from the active power error by the following equation [11]:

$$\hat{T_L} = K_P e_P + K_I \int e_P \, dt \tag{5}$$

$$e_P = V_{ds}^s (\hat{i}_{ds}^e - i_{ds}^e) + V_{qs}^s (\hat{i}_{qs}^e - i_{qs}^e) \tag{6}$$

Rotor speed is estimated from the estimated stator current, rotor flux and the estimated load torque and it is represented as follows [14]:

$$\hat{\omega_r} = \left(\frac{3}{2}\right)\left(\frac{n_p}{J}\right)\left(\frac{L_m}{L_r}\right)\left[\hat{\varphi}_{dr}^s \hat{i}_{qs}^s - \hat{\varphi}_{qr}^s \hat{i}_{ds}^s\right] - \frac{\hat{T_L}}{J} \tag{7}$$

Where n_p is the no. of pole pairs and J is of inertia of motor load system (kg.m^2).

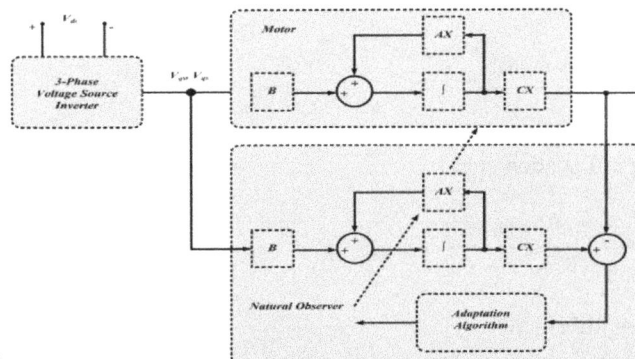

Figure 1. Structure of a natural observer

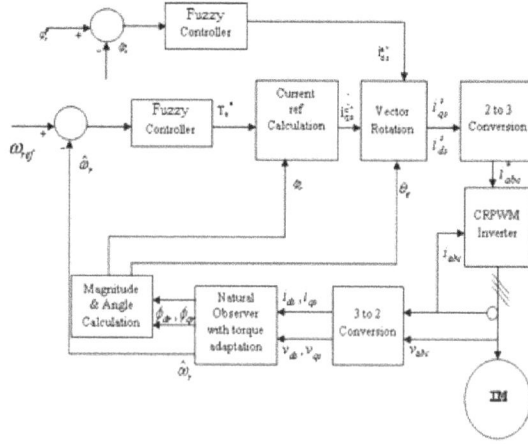

Figure 2. Closed loop sensorless speed control of induction motor drive with fuzzy controller

The closed loop structure of the natural observer is shown in Figure 2. The main components are: natural observer with adaptive load torque estimation, calculation blocks of reference current values, PI/fuzzy controllers and current regulated pulse width modulated (CRPWM) voltage source inverter. The space vector model of the induction motor is used to derive the equations for i_{ds}^{e*} and i_{qs}^{e*} and are as follows [14]:

$$V_{ds}^e = i_{ds}^e R_s + p\varphi_{ds}^e + j\omega_e\varphi_{ds}^e \tag{8}$$

$$V_{qs}^e = i_{qs}^e R_s + p\varphi_{qs}^e + j\omega_e\varphi_{qs}^e \tag{9}$$

$$0 = i_{dr}^e R_r + p\varphi_{dr}^e + j(\omega_e - \omega_r)\varphi_{dr}^e \tag{10}$$

$$0 = i_{qr}^e R_r + p\varphi_{qr}^e + j(\omega_e - \omega_r)\varphi_{qr}^e \tag{11}$$

$$\varphi_{ds}^e = L_s i_{ds}^e + L_m i_{dr}^e \tag{12}$$

$$\varphi_{qs}^e = L_s i_{qs}^e + L_m i_{qr}^e \tag{13}$$

$$\varphi_{dr}^e = L_r i_{dr}^e + L_m i_{ds}^e \tag{14}$$

$$\varphi_{qr}^e = L_r i_{qr}^e + L_m i_{qs}^e \tag{15}$$

$$T_e = \frac{3}{2}\frac{P}{2}\left(\frac{L_m}{L_r}\right)\left(\varphi_r^e \times i_s^e\right) \tag{16}$$

$$T_e = \frac{3}{2}\frac{P}{2}\left(\frac{L_m}{L_r}\right)\left(\varphi_{dr}^e i_{qs}^e - \varphi_{qr}^e i_{ds}^e\right) \tag{17}$$

From Equation (10):

$$i_{dr}^e = \frac{-p\varphi_{dr}^e - j(\omega_e - \omega_r)\varphi_{dr}^e}{R_r} \tag{18}$$

By substituting in Equation (14):

$$\varphi_{dr}^e = L_r \frac{-p\varphi_{dr}^e - j(\omega_e - \omega_r)\varphi_{dr}^e}{R_r} + L_m i_{ds}^e \tag{19}$$

Further after simplification,

$$p\varphi_{dr}^e + \{S_r + j(\omega_e - \omega_r)\}\varphi_{dr}^e = L_m S_r i_{ds}^e \tag{20}$$

Similarly,

$$p\varphi_{qr}^e + \{S_r + j(\omega_e - \omega_r)\}\varphi_{qr}^e = L_m S_r i_{qs}^e \tag{21}$$

From Equation (20) and Equation (21), the general equation is represented as:

$$p\varphi_r^e + \{S_r + j(\omega_e - \omega_r)\}\varphi_r^e = U i_s^e \tag{22}$$

where, $U = S_r L_m$; $S_r = \dfrac{R_r}{L_r}$

$$p\left(\varphi_{dr}^e + j\varphi_{qr}^e\right) + \{S_r + j(\omega_e - \omega_r)\}\left(\varphi_{dr}^e + j\varphi_{qr}^e\right) = U\left(i_{ds}^e + ji_{qs}^e\right) \tag{23}$$

Separating real and imaginary parts,

$$p\varphi_{dr}^e + S_r\varphi_{dr}^e - \omega_e\varphi_{qr}^e + \omega_r\varphi_{qr}^e = U i_{ds}^e \tag{24}$$

For constant flux operation, $p\varphi_{dr}^e = 0$ and $\varphi_{qr}^e = 0$ and i_{ds}^e is calculated as follows:

$$S_r\varphi_{dr}^{e*} = U i_{ds}^{*} = S_r L_m i_{ds}^{e*} \tag{25}$$

$$i_{ds}^{e*} = \frac{i_{ds}^{e*}}{L_m} \tag{26}$$

Torque developed in an induction motor

$$\tag{27}$$

$$T_e = \left(\frac{P}{2}\right)\frac{L_m}{L_r}\left(\varphi_r \times i_s\right)$$

$$T_e = \left(\frac{P}{2}\right)\frac{L_m}{L_r}\left(\varphi_{dr}^e i_{qs}^e - \varphi_{qr}^e i_{ds}^e\right) \tag{28}$$

i_{qs}^{e*} controls the average torque developed,

$$i_{qs}^{e*} = \frac{L_r}{\frac{P}{2} L_m \varphi_{dr}^*} T^* \tag{29}$$

$$i_{qs}^{e*} = \frac{L_r}{n_p L_m \varphi_{dr}^*} T^* \text{ where } n_p \text{ is the pole pair} \tag{30}$$

i_{ds}^{e*} is generated by comparing the actual flux with the set reference flux and the error is given to the PI controller which gives the desired value of i_{ds}^{e*}. In addition, it maintains the mean value of rotor flux as constant. The reference currents are transformed into stationary reference frame by rotor angle θ_e. The two phase dq-axes stator currents are transformed into three phase reference current by 2 to 3 conversion blocks (inverse Clarke's transformation).

3. FUZZY LOGIC CONTROLLER

Fuzzy logic can be described simply as "computing with words rather than numbers''; "control with sentences rather than equations''. A fuzzy controller includes empirical rules and is useful in operator controlled plants. Fuzzy control is preferred where robust control is desired, particularly with plant parameter variations and load disturbance effects. There is no design procedure in fuzzy control such as root-locus design, frequency response design, pole placement design or stability margins, because the rules are often nonlinear. Fuzzy controllers are being used in various control schemes. In this work, direct control is used, where the fuzzy controller is in the forward path in a feedback control system. The process output is compared with a reference, and if there is a deviation, the controller takes action according to the control strategy. Triangular memberfunctions were used in most of the literatures [15]-[16] whereas Gaussian membership functions are selected in this paper as they are smooth and nonzero at all points. The control signals are error (E) and change in error (CE). The fuzzy controller fuzzifies the input signals and generates the control signal through the evaluation of control rules and defuzzification. All the input and output signals use Gaussian membership functions. Mamdani type inference method and mean of maximum (This method disregards the shape of the fuzzy set, but the computational complexity is relatively good) defuzzification method are used. The linguistic membership functions are negative large (NL), negative small (NS), zero (Z), positive small (PS) and positive large (PL).

The rule matrix for fuzzy control is given in Table 1. As example, the control rules for E and CE are:

1. If E is Z and CE is Z then control signal is Z
2. If E is PS and CE is Z then control signal is PS
3. If E is Z and CE is NS then control signal is NS

The membership functions for the input variable error, change in error and the control signal are shown in Figure 3.

Table 1. Rule table for fuzzy control

CE E	NL	NS	Z	PS	PL
NL	NL	NL	NS	NS	Z
NS	NL	NS	NS	Z	PS
Z	NS	NS	Z	PS	PS
PS	NS	Z	PS	PS	PL
PL	Z	PS	PS	PL	PL

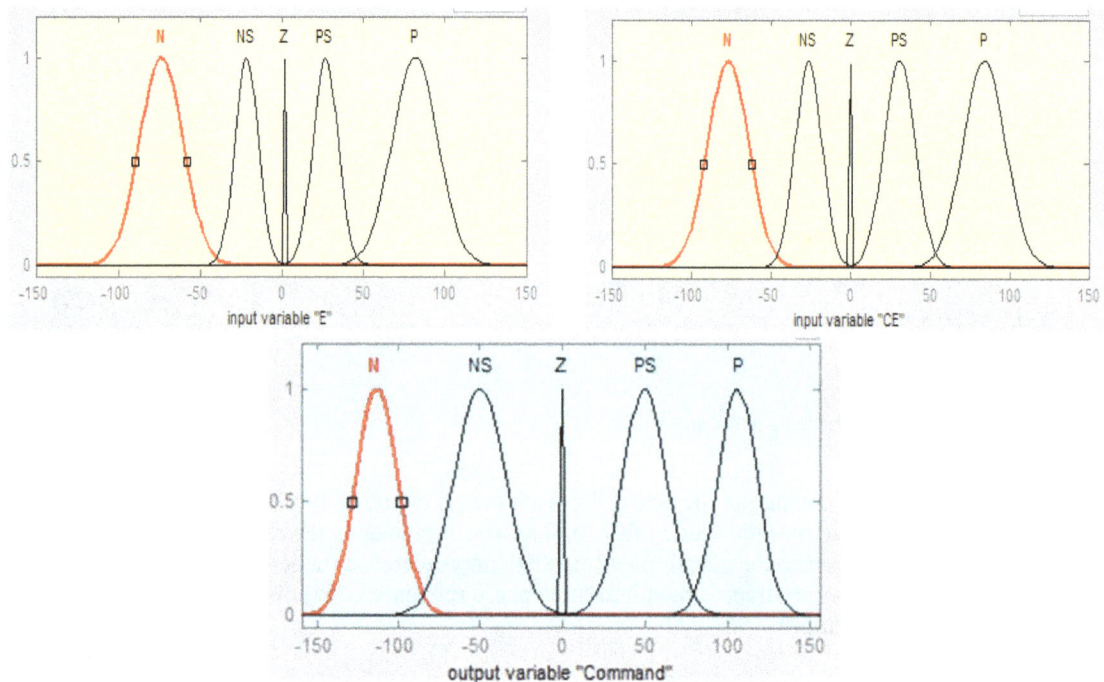

Figure 3. Fuzzy membership functions

4. SIMULATION RESULTS AND DISCUSSIONS

Simulations are done in MATLAB simulink atmosphere. The simulation blocks of sensorless vector control are constructed in MATLAB using power system blocksets and simulink libraries. Natural observer is used to estimate the speed, rotor fluxes and stator currents. Conventional PI controller and fuzzy controller are investigated. The simulation results are presented for different running conditions. The ratings and the parameters of the induction motor are given in Table 2. Direct field oriented sensorless vector control scheme is used and the rotor angle is determined from the estimated rotor fluxes. The torque adaptation gains are

K_p=0.08, K_I=0.2. Figure 4 and Figure 5 show the simulation diagram of sensorless vector control of induction motor drive with PI controller and fuzzy controller respectively. The induction motor and the natural observer are built with state space model and are constructed by MATLAB functions. In addition, various simple blocks available in simulink are used to construct the entire system. PI controller is constructed using PID block available in simulink libraries. The simulation blocks of fuzzy controller are constructed in MATLAB using fuzzy toolbox. Fuzzy controller is framed with 5 linguistic variables to generate the required torque reference signal from the speed error. It also reduces the computational burden in real time.

The simulation results of speed sensorless vector control of induction motor drive with PI controller and fuzzy speed controller for different running conditions are shown in Figure 6. The motor is at no load at the time of starting. The speed command is set at 500 rpm. At t = 1.5s, a step speed command is given to increase the speed from 500 rpm to 750 rpm. At t=3s, a load of 1.5 Nm is applied. It is observed from the Figure 6(a) that the estimated speed follows the actual speed. At steady state, the difference between estimated and actual speed is zero. During starting as well as change in speed, the peak magnitude of the actual torque is less in fuzzy controller than PI controller and also the transient torque at the time of change in speed is very high in PI controller and it is suppressed in fuzzy controller. The estimated and actual torque responses are shown in Figure 6(b). Simulations are also investigated at a speed of 1000 rpm and the results are described in Figure 7.

Table 2. Ratings and parameters of induction motor

Parameters	Ratings
Output	745.6 W
Poles	4
Speed	1415 rpm
Voltage	415 V
Current	1.8 A
R_s	19.355 Ω
R_r	8.43 Ω
L_s	0.715 H
L_r	0.715 H
L_m	0.689 H
f	50 Hz

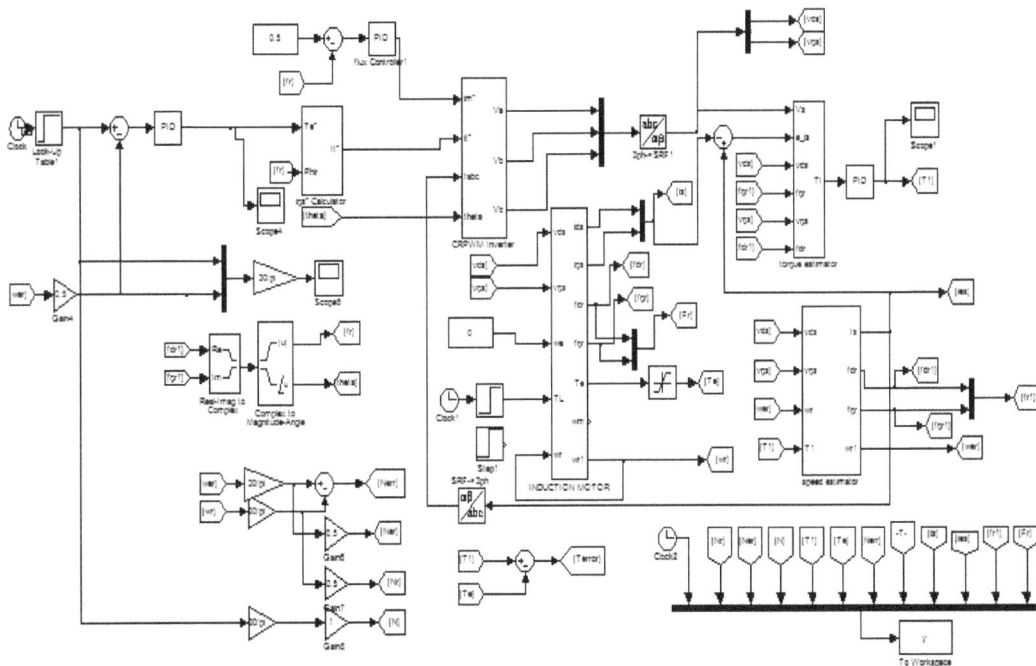

Figure 4. Simulation diagram of speed sensorless vector control of induction motor drive with PI controller

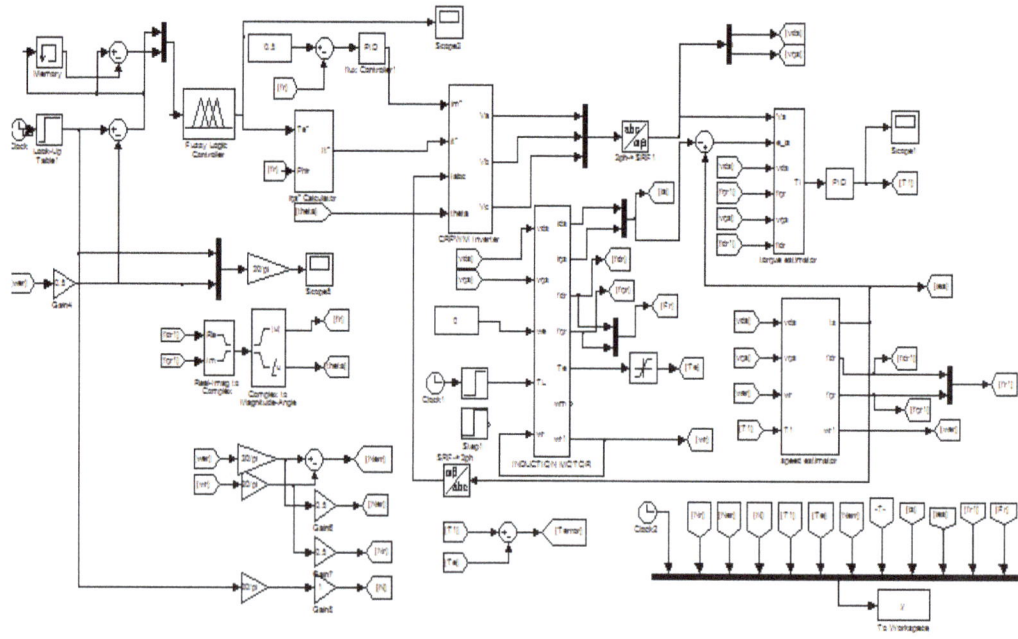

Figure 5. Simulation diagram of speed sensorless vector control of induction motor drive with fuzzy controller

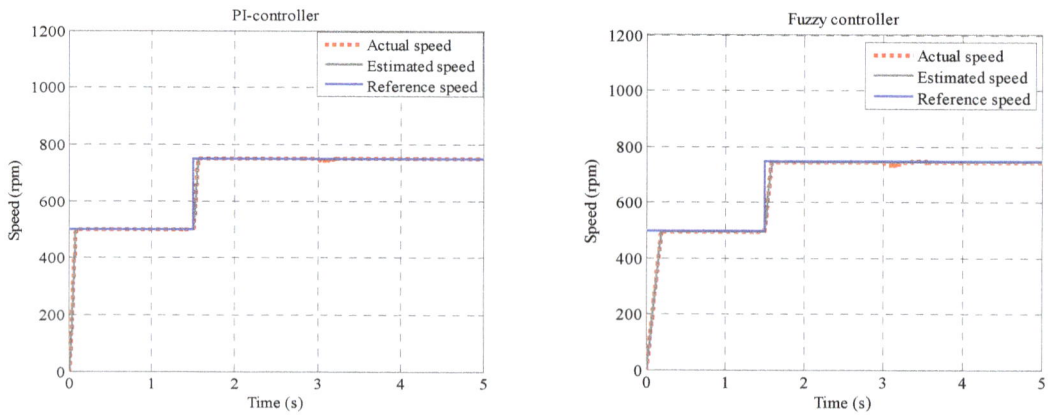

(a)   Estimated and actual speed response

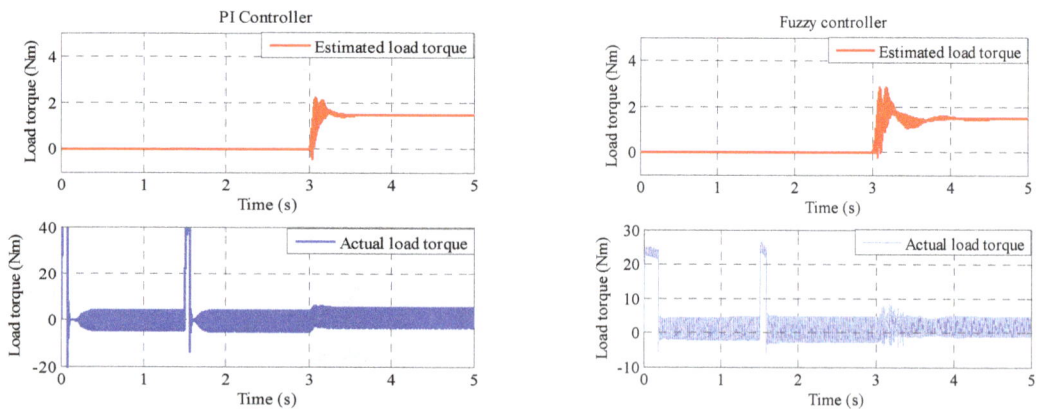

(b)   Estimated and actual torque response

Figure 6. Simulation waveform for a speed of 500 rpm and 750 rpm with 1.5 Nm load

(a) Estimated and actual speed response

(b) Estimated and actual torque response

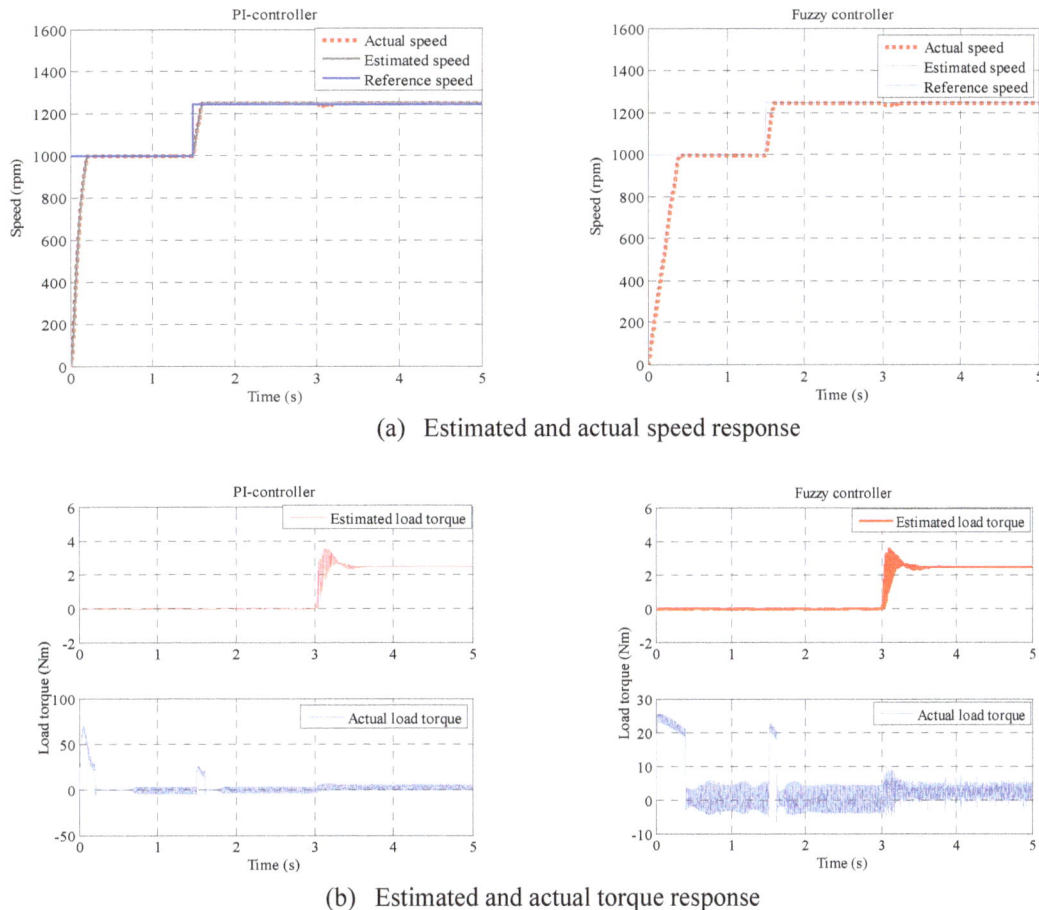

Figure 7. Simulation waveform for a speed of 1000 rpm and 1250 rpm with 2.5 Nm load

5. HARDWARE RESULTS AND DISCUSSIONS

Three-phase squirrel cage induction motor of 0.746 kW (1HP) is used for the experimental set up. Brake drum arrangements are provided for mechanical loading. The central processor unit is the TMS320F2812 DSP processor and it executes all the mathematical calculations. Various simulink blocks like natural observer and PI controllers are built in VISSIM. TMS320F2812 DSP processor supporting blocks are available in VISSIM. In VISSIM, the simulation blocks are converted into C- codes using the target support for TMS320F2812 and compiled using code composer studio internally and the output file is downloaded into the DSP processor through J-tag emulator. Three numbers of LEM current sensors and voltage sensors are used to measure the phase currents and terminal voltages of the induction motor respectively. The measured analog currents and voltages are converted into digital by on chip ADC with 12 bit resolution. The feedback signals are linked to DSP processor using 26 pin header and the processor estimates the stator current, rotor flux, load torque and speed. The processor also generates the required PWM pulses to enable the three phase IGBT inverter switches in the Intelligent Power Module (IPM). Highly effective over-current and short-circuit protection is realized through the use of advanced current sense IGBT chips that allow continuous monitoring of power device current. System reliability is further enhanced by the IPM's integrated over temperature and under voltage lock out protection.

The experimental results for a step change in speed of 1000 rpm to 1250 rpm for a load of 2.5 Nm are shown in Figure 8. The actual speed of the motor is measured by proximity sensor. The estimated and actual speed waveforms for step increase and decrease in speed are depicted in Figure 8(a) and Figure 8(b) respectively. It is observed that the estimated speed follows the actual speed and matches with the simulation waveform. The estimated speed response with respect to the reference speed of 1250 rpm (500 rpm/div) is presented in Figure 8(c) for a load of 2.5 Nm (1 Nm/div). It is inferred that drop in speed occurs at the time of applying the load and further the motor runs at a constant speed of 1250 rpm for a load of 2.5 Nm. The estimated load torque waveform is illustrated in Figure 8(d) and is equal is to the applied load. The experimental results are similar to the simulation results and the performance of natural observer is proved experimentally with PI controller.

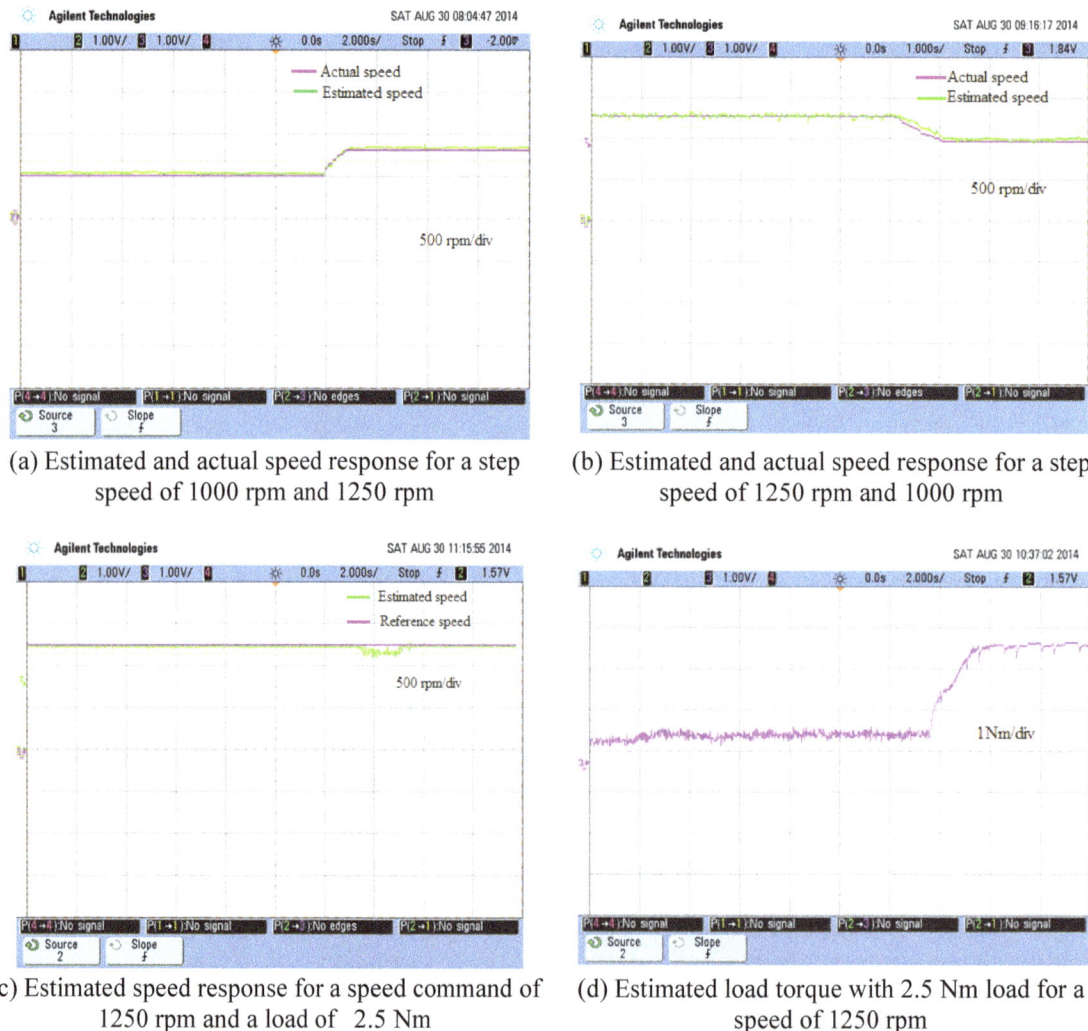

(a) Estimated and actual speed response for a step speed of 1000 rpm and 1250 rpm

(b) Estimated and actual speed response for a step speed of 1250 rpm and 1000 rpm

(c) Estimated speed response for a speed command of 1250 rpm and a load of 2.5 Nm

(d) Estimated load torque with 2.5 Nm load for a speed of 1250 rpm

Figure 8 Experimental results for a speed of 1250 rpm with 2.5 Nm load

5. CONCLUSION

The induction motor and the natural observer are modelled in MATLAB with state space and simulations have been carried out for different running conditions. It is concluded that fourth order induction motor model is used and the estimated parameters such as rotor speed and load torque follow the command value. PI controller and fuzzy controllers have been used in the speed control loop to generate the torque reference and their performances have been compared. It is validated that torque ripple in fuzzy controller is less than PI controller. The natural observer is simple and speedy and is a suitable estimator for sensorless vector control of induction motor drive. Mean value of the rotor flux has been maintained constant by employing rotor flux feedback. To validate the simulation, hardware results have been provided for different running conditions.

REFERENCES

[1] Salvatore L, Stasi S, Tarchioni L. A new EKF-based algorithm for flux estimation in induction machines. *IEEE Transactions on Industrial Electronics*. 1993; 40 (5): 496-504.

[2] Kim YR, Seung-Ki Sul, Park MH, Speed sensorless vector control of induction motor using Extended Kalman Filter. *IEEE Transactions on Industry Applications*. 1994; 30(5):1225-1233.

[3] Kim HW, Sul SK. A New motor Speed Estimator Using Kalman Filter in Low-Speed range. *IEEE Transactions on Industry Appications*. 1996; 43 (4):498-504.

[4] Shi KL, Chan TF, Wong YK, Ho SL. Speed estimation of an induction motor drive using an optimized Extended Kalman Filter. *IEEE Transactions on Industrial Electronics*. 2002; 49 (1): 124-133.

[5] Barut M, Bogosyan S, Gokasan M. Speed-sensorless estimation for induction motors using Extended Kalman Filters. *IEEE Transactions on Industrial Electronics*. 2007; 54 (1): 272-280.

[6] Barut M, Bogosyan S, Gokasan M. Experimental evaluation of Braided EKF for sensorless control of induction motors. *IEEE Transactions on Industrial Electronics*. 2008; 55 (2): 620-632.

[7] Kubota H, Matsuse K, Nakano T. *New adaptive flux observer of induction motor for wide speed range motor drives.* 16th IEEE industrial electronics society annual conference, 1990; 921-926.

[8] Kubota H, Matsuse K. Speed sensorless field-oriented control of induction motor with rotor resistance adaptation. *IEEE Transactions on Industry Appications*. 1994; 30 (5): 1219–1224.

[9] Yamada T, Matsuse K, Sasagawa K. *Sensorless control of direct field oriented induction motor operating at high efficiency using adaptive rotor flux observer.* 22nd IEEE international conference on industrial electronics, control and instrumentation, 1996; 1149-1154.

[10] Tsuji M, Umesaki Y, Nakayama R, Izumi K. *A Simplified MRAS based sensorless vector control method of induction motor.* IEEE international conference on power conversion, 2002; 3: 1090-1095.

[11] Bowes SR, Sevinc A, Holliday D. New natural observer applied to speed sensorless DC servo and induction motors. *IEEE Transactions on Industry Appications.* 2004; 51 (5), 1025–1032.

[12] Erenoglu I, Eksin I, Yesil E, Guzelkaya M, *An intelligent fuzzy hybrid fuzzy PID controller.* 20th European Conference on Modelling and Simulation Wolfgang Borutzky, Alessandra Orsoni, Richard Zobel, 2006.

[13] Abdullah I, Al-Odienat, Ayman A, Al-Lawama. The advantages of PID fuzzy controllers over the conventional types. *American Journal of Applied Sciences*, 2008; 6: 653-658.

[14] Subramanyam MV, Prasad PVN, Poornachandra Rao G. Fuzzy logic closed loop control of 5 level MLI driven three phase induction motor. *International Journal of Power Electronics and Drive System* (IJPEDS). 2013: 3 (2): 200-208.

[15] Khobaragade T, Barve A. Enhancement of power system stability using fuzzy logic controller. *International Journal of Power Electronics and Drive System* (IJPEDS), 2012: 2 (4): 389-401.

[16] Bose, BK. Modern Power Electronics and AC Drives, Prentice-Hall, Upper Saddle River, New Jersey, 2001.

20

Power Control of Wind Turbine Based on Fuzzy Sliding-Mode Control

Tahir Khalfallah*, Belfedal Cheikh*, Allaoui Tayeb*, Gerard Champenois**

*Laboratoire de Génie Energitique et Génie Informatique LGEGI, Université Ibn Khaldoun de Tiaret, Algérie
**University of Poitiers, Laboratoire d'Informatique et d'Automatique pour les Systèmes, Bâtiment B25, 2, rue Pierre Brousse, 86022 Poitiers, France

---

Keyword:	ABSTRACT
Fuzzy sliding mode control Maximum power point tracking Wind energy conversion system Wound field synchronous generator	This paper presents the study of a variable speed wind energy conversion system (WECS) using a Wound Field Synchronous Generator (WFSG) based on a Fuzzy sliding mode control (FSMC) applied to achieve control of active and reactive powers exchanged between the stator of the WFSG and the grid to ensure a Maximum Power Point Tracking (MPPT) of a wind energy conversion system. However the principal drawback of the sliding mode, is the chattering effect which characterized by torque ripple, this phenomena is undesirable and harmful for the machines, it generates noises and additional forces of torsion on the machine shaft. A direct fuzzy logic controller is designed and the sliding mode controller is added to compensate the fuzzy approximation errors. The simulation results clearly indicate the effectiveness and validity of the proposed method, in terms of convergence, time and precision.

This paper presents the study of a variable speed wind energy conversion system (WECS) using a Wound Field Synchronous Generator (WFSG) based on a Fuzzy sliding mode control (FSMC) applied to achieve control of active and reactive powers exchanged between the stator of the WFSG and the grid to ensure a Maximum Power Point Tracking (MPPT) of a wind energy conversion system. However the principal drawback of the sliding mode, is the chattering effect which characterized by torque ripple, this phenomena is undesirable and harmful for the machines, it generates noises and additional forces of torsion on the machine shaft. A direct fuzzy logic controller is designed and the sliding mode controller is added to compensate the fuzzy approximation errors. The simulation results clearly indicate the effectiveness and validity of the proposed method, in terms of convergence, time and precision.

Corresponding Author:

Tahir Khalfallah,
Departement of Electrical and Computer Engineering,
University Ibn Khaldun Tiaret, Algeria,
Email: tahir.commande@gmail.com

1. INTRODUCTION

Wind energy is becoming one of the most important renewable energy sources [1]. Recently, power converter control has mostly been studied and developed for WECS integration in the electrical grid.

In recent years, variable speed WECSs have become the industry standard because of their advantages over fixed speed ones such as improved energy capture, better power quality. They are capable of extracting optimal energy capture in addition to having reduced mechanical stress and aerodynamic noise. [2].

In terms of the generators for WECS, several types of electric generators are used such as Squired-Cage Induction Generator (SCIG), Synchronous Generator with external field excitation, Doubly Fed Induction Generator (DFIG) and Permanent Magnet Synchronous Generator (PMSG) with power electronic converter system [3]. Therefore, the study of synchronous generator has regained importance. The primary advantages of Wound Field Synchronous Generator are: The efficiency of this machine is usually high, because it employs the whole stator current for the electromagnetic torque production. The main benefit of the employment of wound field synchronous generator with salient pole is that it allows the direct control of the power factor of the machine, consequently the stator current may be minimized any operation circumstances [4].

The Sliding Mode Controller (SMC) is a particular type of variable structure control systems that is designed as a robust control to drive and then constrain the system to lie within of the switching function. However in the presence of large uncertainties or higher switching gain is required which produce higher amplitude of chattering.

Fuzzy logic has emerged as a powerful in control applications. It allows one to design a controller using linguistic rules without knowing the mathematical model of the plant.

In this paper our objective is to apply a fuzzy controller combined with sliding mode to overcome shattering of both sliding mode and fuzzy logic controllers and then to obtain a control system for a high performance for power system [5]. Simulation results are provided to show the effectiveness of the proposed overall WFSG control system.

2. WIND CONVERSION SYSTEM MODEL

The WECS described in this article includes the wind turbine, gearbox, WFSG, and back-to-back converters. The rotor winding of the WFSG is connected to the grid by DC/AC converter, whereas the stator winding is fed by back-to-back bidirectional PWM-VSC. In this system, the wind energy is transmitted through the turbine to the three-phase WFSG and generated in electrical form. This energy is transmitted directly through a bridge rectifier and inverter to the electrical network (Figure 1). We consider in this study that the rectifier is perfect. So semiconductors are ideal [6]. In this paper our study is limited to the generation of power in continuous form.

Figure 1 shows the equivalent diagram of the electrical portion of the string conversion of wind energy.

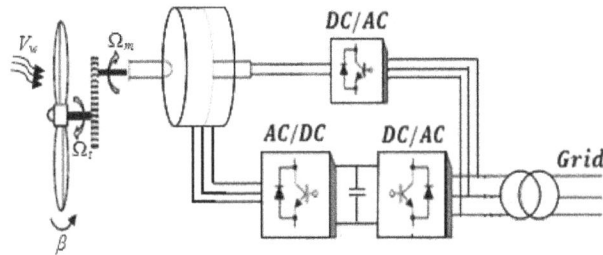

Figure 1. WFSG based wind energy conversion system

2.1. Modeling of the Wind Turbine and Gearbox

The turbine power and torque developed are given by the following relation [7]:

$$P_a = \frac{1}{2} \rho \pi R^2 V_w^3 C_p (\lambda, \beta)$$

(1)

$$T_a = \frac{P_a}{\Omega_t} = \frac{1}{2\lambda} \rho \pi R^3 V_w^2 C_p (\lambda, \beta)$$

(2)

Which λ presents the ratio between the turbine angular speed and the wind speed. This ratio called the tip speed ration and is defined as:

$$\lambda = \frac{\Omega_t R}{V_w}$$

(3)

Where ρ is the air density, R is the blade length, V_w is the wind speed, C_p is the power coefficient, Ω_t is the turbine angular speed.

The power coefficient (C_p) presents the aerodynamic efficiency of the turbine and depends on the specific speed λ and the angle of the blades. It is different from a turbine to another, and is usually provided by the manufacturer and can be used to define a mathematical approximation.

The wind turbine shaft is connected to the WFSG rotor through a gearbox which adapts the slow speed of the turbine to the WFSG speed. This gearbox is modeled by the following equations [8]:

$$\Omega_t = \frac{\Omega_m}{G} \; ; \; T_m = \frac{T_a}{G} \qquad (4)$$

From the dynamics fundamental relation, the turbine speed is determined as follows:

$$J \frac{d\Omega_m}{dt} = T_m - T_{em} - f\Omega_m \qquad (5)$$

J and f are the total moment of inertia and the viscous friction coefficient appearing at the generator side, T_m is the gearbox torque, T_{em} is the generator torque, and Ω_m is the mechanical generator speed.

Figure 2 represents the power coefficient C_p as a function of β and λ.

Figure 3 shows the mechanical power as a function of rotor speed of the turbine for different values of wind speed [9].

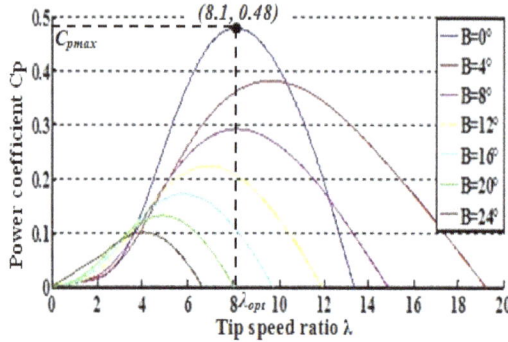

Figure 2. Power coefficient versus tip speed ratio

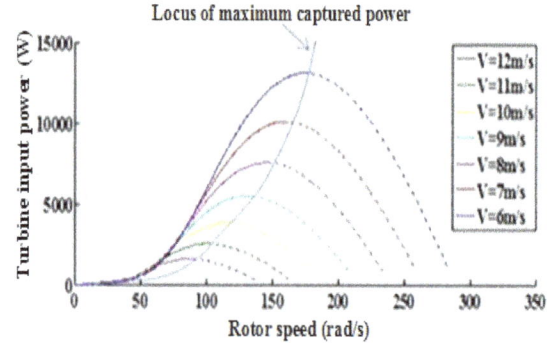

Figure 3. Rotor power versus rotational speed of generator

2.2. Modeling of the WFSG

In the synchronous d-q coordinates, the voltage equation of the WFSG is expressed as follows [10]:

$$v_{ds} = -r_s i_{ds} + \omega_e L_q i_{qs} - \omega_e m_{sQ} i_Q - L_d \frac{di_{ds}}{dt} + m_{sf} \frac{di_f}{dt} + m_{sD} \frac{di_D}{dt} \qquad (6)$$

$$v_{qs} = -r_s i_{qs} - \omega_e L_d i_{ds} + \omega_e m_{sf} i_f + \omega_e m_{sD} i_D - L_q \frac{di_{qs}}{dt} + m_{sQ} \frac{di_Q}{dt} \qquad (7)$$

$$\begin{cases} 0 = r_D i_D + \dfrac{d}{dt}\varphi_D \\[2mm] 0 = r_Q i_Q + \dfrac{d}{dt}\varphi_Q \end{cases} \begin{cases} \varphi_D = L_D i_D + m_{fD} i_f - m_{sD} i_{ds} \\[2mm] \varphi_Q = L_Q i_Q - m_{sQ} i_{qs} \end{cases} \qquad (8)$$

Where:

L_D , L_Q : inductances of the direct and quadrature damper windings.

L_f : inductance of the main field winding.

L_d , L_q : inductances of the d-axis stator winding and q-axis stator winding.

m_{sf} : mutual inductance between the field winding and the d-axis stator winding.

m_{sD} : mutual inductance between the d-axis stator winding and the d-axis damper winding.

m_{sQ} : mutual inductance between the q-axis stator winding and the q-axis damper winding.

m_{fD} : mutual inductance between the field winding and the d-axis damper winding.

ω_e : is the electrical angular speed, $\omega_e = p\Omega_m$

The electromagnetic torque is expressed by:

$$T_{em} = p(\varphi_d i_{qs} - \varphi_q i_{ds}).$$ (9)

3. SLIDING MODE CONTROL

To achieve the maximum power at below rated wind speed, sliding mode based torque control is proposed in [11]. The main objective of this controller is to track the reference rotor speed Ω_{m_ref} for maximum power extraction. In conventional sliding mode control, sliding surface generally depends on error, and derivative of the error signal is given in (10).

$$\sigma(x) = \left(\lambda_x + \frac{d}{dt}\right)^{n+1}(x_{ref} - x)$$ (10)

Where λ is the positive constant and n is the order of the uncontrolled system.

The speed error is defined by [12]:

$$e_{\Omega m} = \Omega_{m\_ref} - \Omega_m.$$ (11)

For $n = 1$, the position control manifold equation can be obtained from Equation (10) as follow:

$$\sigma(\Omega_m) = \Omega_{m\_ref} - \Omega_m.$$ (12)

The derivative of this surface is given by the expression:

$$\dot{\sigma}(\Omega) = c_2 \Omega_m - c_1 + \dot{\Omega}_{m\_ref} + c_3(m_{sf} i_f + m_{sD} i_D)i_{qs}.$$ (13)

During the sliding mode and in permanent regime, we have:

$$\sigma(\Omega_m) = 0, \dot{\sigma}(\Omega_m) = 0, i_{qs}^n = 0.$$ (14)

The current control i_{qs} is defined by:

$$i_{qs} = i_{qs}^{eq} - i_{qs}^n.$$ (15)

The control voltage i_{qs_ref} is defined by:

$$i_{qs\_ref} = \frac{-c_2 \Omega_m + c_1 - \Omega_{m\_ref}}{c_3(m_{sf} i_f + m_{sD} i_D)} + k_{\Omega_m} sat(\sigma(\Omega_m)).$$ (16)

The stator currents i_{qs} and i_{ds} are the images, respectively, of the P_s and the Q_s, which must follow their references.

3.1. Quadratic Rotor Current Control with SMC
The sliding surface representing the error between the measured and reference quadratic rotor current is given by:

$$\sigma(i_{qs}) = e_{i_{qs}} = i_{qs\_ref} - i_{qs} \tag{17}$$

$$\dot{\sigma}(i_{qs}) = \dot{i}_{qs\_ref} - \dot{i}_{qs} \tag{18}$$

Substituting the expression of $\dot{i}_{qs}$ Equation (7) in Equation (18), Equation (19) and Equation (20) can be obtained.

$$\dot{\sigma}(i_{qs}) = \dot{i}_{qs\_ref} + \frac{1}{L_q}\left(r_s i_{qs} + a_1 i_{ds} - a_2 i_f - a_3 i_D - m_{sQ} i_Q + v_{qs}\right) \tag{19}$$

And,

$$v_{qs} = v_{qs}^{eq} - v_{qs}^{n}. \tag{20}$$

During the sliding mode and in permanent regime, there is:

$$\sigma(i_{qs}) = 0, \dot{\sigma}(i_{qs}) = 0, v_q^n = 0 \tag{21}$$

Where the equivalent control is:

$$v_{qs}^{eq} = -L_q \dot{i}_{qs\_ref} - r_s i_{qs} - a_1 i_{ds} + a_2 i_f + a_3 i_D + m_{sQ} i_Q \tag{22}$$

Therefore, the correction factor is given by:

$$v_{qs}^{n} = K_{v_q} sat\left(\sigma(i_{qs})\right) \tag{23}$$

Where K_{v_q} is positive constant.

3.2. Direct Rotor Current Control with SMC
The sliding surface representing the error between the measured and reference direct rotor current is given by:

$$\sigma(i_{ds}) = e_{i_{ds}} = i_{ds\_ref} - i_{ds} \tag{24}$$

$$\dot{\sigma}(i_{ds}) = \dot{i}_{ds\_ref} - \dot{i}_{ds} \tag{25}$$

Substituting the expression of $\dot{i}_{ds}$ Equation (6) in Equation (25), there is:

$$\dot{\sigma}(i_{ds}) = \dot{i}_{ds\_ref} + \frac{1}{L_d}\left(r_s i_{ds} - b_1 i_{qs} + b_2 i_Q - m_{sf} \dot{i}_f - m_{sD} \dot{i}_D + v_{ds}\right)$$

(26)

And,

$$v_{ds} = v_{ds}^{eq} - v_{ds}^{n}.$$

(27)

During the sliding mode and in permanent regime, Equation (28) can be obtained.

$$\sigma(i_{ds}) = 0, \dot{\sigma}(i_{ds}) = 0, v_{ds}^{n} = 0$$

(28)

Where the equivalent control is:

$$v_{ds}^{eq} = -L_d \dot{i}_{ds\_ref} - r_s i_{ds} + b_1 i_{qs} - b_2 i_Q + m_{sf} \dot{i}_f + m_{sD} \dot{i}_D$$

(29)

Therefore, the correction factor is given by:

$$v_{ds}^{n} = K_{v_d} sat\left(\sigma(i_{ds})\right)$$

(30)

Where K_{v_d} is positive constant.

$$a_2 = m_{sf}\omega_e; \quad b_2 = m_{sQ}\omega_e; \quad c_2 = \frac{f}{J}; \quad c_3 = \frac{p}{J}; \quad a_1 = L_d\omega_e; \quad b_1 = L_q\omega_e; \quad a_3 = m_{sD}\omega_e;$$

$$c_1 = \frac{T_m}{J}$$

4.   FUZZY LOGIC CONTROLLER

Fuzzy-logic control has the capability to control nonlinear, uncertain and adaptive systems with parameter variation. Fuzzy control does not strictly need any mathematical model of the plant. Its control rule can be qualitatively expressed on the basis of logic-language variation and the fuzzy model of a plant is very easy to apply. In fact, fuzzy control is good adaptive control among the techniques discussed so far. In this paper, fuzzy-logic control is associated with sliding-mode control to generate the switching controller term $Ksat\left(\sigma(i_{dqs})\right)$, which ensures the precision and robustness of the control [12].

The general structure of a fuzzy-control system is shown in Figure 4. There are two input signals to the fuzzy controller, the error E and the change in error CE, which is related to the derivative DE/dt of error. The closed-loop error E and change in error CE signals are converted to the respective scale factors, $e = E/GE$ and $ce = CE/GC$. The output plant control signal DU is derived by multiplying by the scale factor GU, that is $DU = du * GU$, and then integrated to generate the U signal [13].

The scale factors can change the sensitivity of the controller without changing its structure. The fuzzy controller is composed of three blocks: fuzzification, rule bases, and defuzzification. The membership functions for inputs output variables are shown in Figure 5. The fuzzy subsets are as follows: GN (Grand negative), N (Negative), ZR (Zero), P (Positive), and GP (Grand positive). There are seven fuzzy subsets for each variable, which gives $5 \times 5 = 25$ possible rules. The fuzzy rules that produce these control actions are reported in Table 1.

The Defuzzification of the output control is accomplished using the method of center of gravity. When the error is below zero, the universe of the control value should be expanded by a contraction-expansion factor $F(x)$. When the error is above zero, the universe should be contracted. Therefore $F(x)$ is defined as $F(x) = M.x^{1/M}$ (M gain positive).

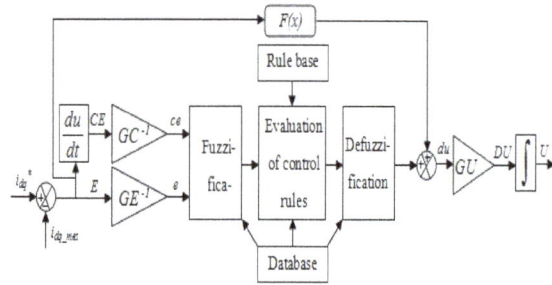

Figure 4. Structure of the fuzzy controller

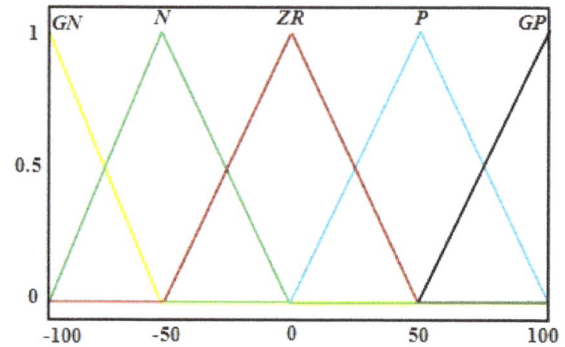

Figure 5. Membership functions of e, ce and DU

Table 1. Rules base

	GN	N	ZR	P	GP
GN	GN	GN	GN	N	ZR
N	GN	N	N	ZR	P
ZR	N	N	ZR	P	P
P	N	ZR	P	P	GP
GP	ZR	P	GP	GP	GP

5.   SIMULATION RESULTS AND DISCUSSION

To demonstrate the pertinence of the proposed WFSG Fuzzy-sliding-control approach (Figure 6), simulation has been performed for 7.5KW WFSG wind power system using Matlab/Simulink™. The wind profile used in our simulations is shown in Figure 7(a).

In addition, aerodynamic power is optimized with MPPT strategy and keeps at his nominal value when the wind speed exceeds the nominal value as shows in Figure 7(b), and the power coefficient C_p is the maximum around 0.48 as shown in Figure 7(c).

By applying the proposed control scheme, the optimal speed command is accurately tracked to extract the maximum power from the wind energy at any moment. In Figure 7(d) the generated torque reference follows the optimum mechanical torque of the turbine quite well. Figure 7(e) shows the speed tracking results of the WFSG. In terms of the actual wind speed, the optimal WFSG speed command is obtained by Eq. (3).

The decoupling effect of the between the direct and quadratic stator current of the WFSG is illustrated in Figure 7(f).

The stator current and voltage waveforms and these zoom of the WFSG are presented in Figire 7(g). As shown in this Figures, the stator currents are proportional to the wind speed. This is due to the reason that when the wind speed increases (not larger than 9.1 m/s), there is more power generated, thus yield more currents in the stator windings of the WFSG.

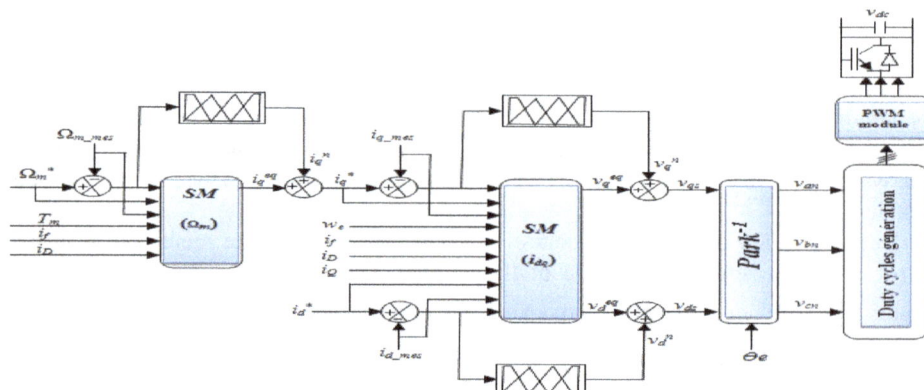

Figure 6. Global diagram of simulation and control of WFSG with Fuzzy-SMC

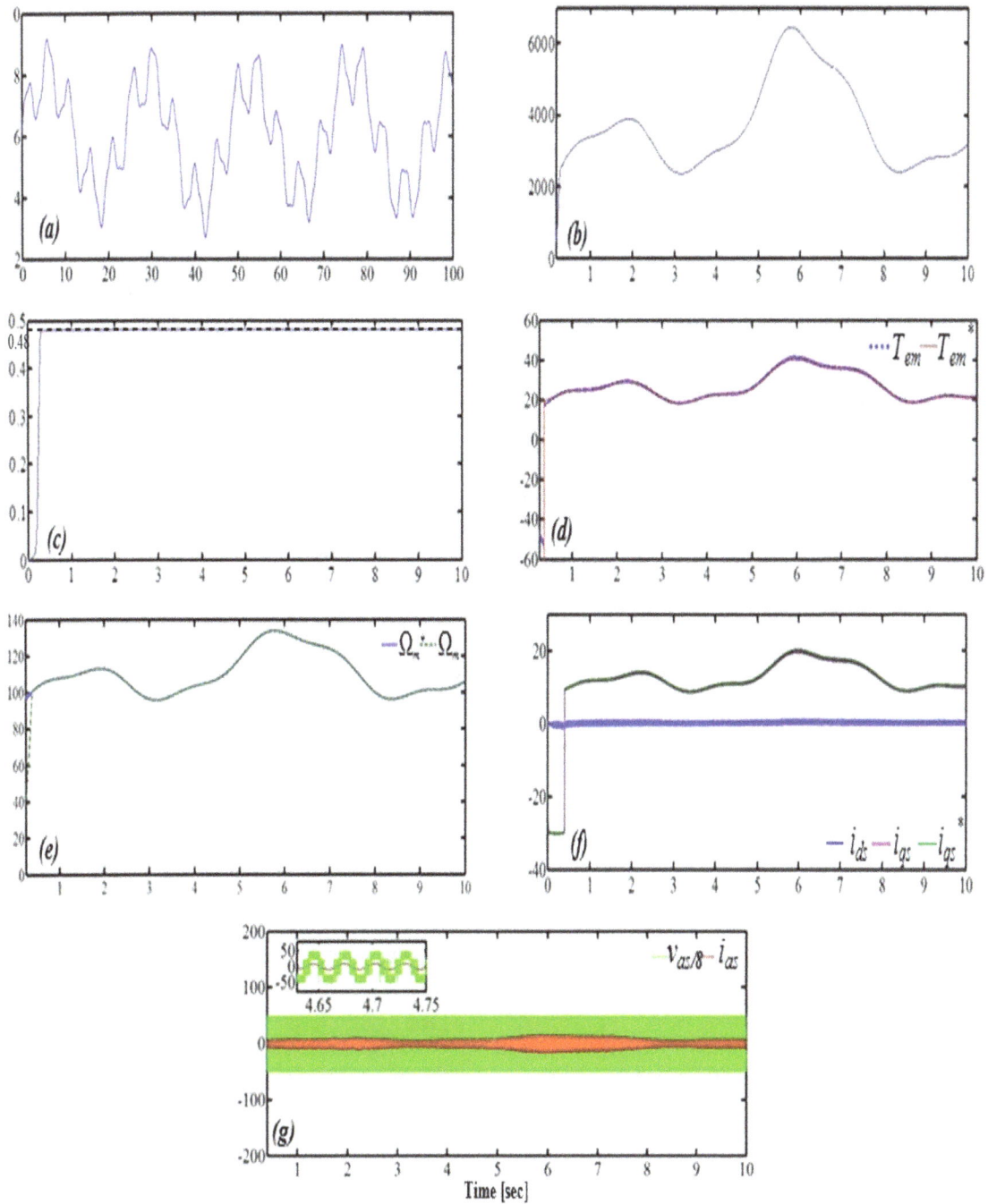

Figure 7. System performance under wind speed variation. (a) Wind speed [m/s]. (b) Aerodynamic power [W]. (c) Power coefficient. (d) Generated torque [N.m]. (e) Generator speed [rad/s]. (f) Direct and quadratic stator current [A]. (g) Stator current and voltage with zoom [A, V]

6.   CONCLUSION

In this paper, a fuzzy sliding mode controller is applied to control the power generated By the WECS based on wound field synchronous generator and to realize nonlinear control. We have established a model of the wind conversion chain, and design a control strategy based on vector control. This structure has been used for reference tracking of active and reactive powers exchanged between the stator and the grid by controlling the stator converter. A series of simulations are performed to test the effectiveness of this controller. The simulation results show that the proposed fuzzy-SMC is very good in dealing with the time-varying, nonlinear nature of WECS. The fuzzy-SMC was also proven more effective than the FLC and SM controller regarding the control performance and power capture.

ACKNOWLEDGEMENTS

The authors gratefully appreciate the support of Tiaret University, Algeria.

REFERENCES

[1]   Bouzid MA, Zine S, Allaoui T, Massoum A. Adaptive Fuzzy Logic Control of Wind Turbine Emulator. *International Journal of Power Electronics and Drive System (IJPEDS)*. 2014; 4(2); 233-240

[2]   Thongam JS, Bouchard P, Beguenane R, Okou AF. *Control of variable speed wind energy conversion system using a wind speed sensorless optimum speed MPPT control method.* [37]th Annual Conference on IEEE Industrial Electronics Society (IECON). 2011; 7-10.

[3]   Blaabjerg F, Iov F, Chen Z, Ma K. *Power electronics and controls for wind turbine systems.* IEEE International Energy Conference and Exhibition. 2010; 333-344 Dec.

[4]   E. Topal and L. Ergene, "designing a wind turbine with permanent magnet synchronous machine" *IU-JEEE* 2011.

[5]   Khaddouj BM, Faiza D, Ismail B. fuzzy sliding mode controller for power system SMIB. *Journal of Theoretical and Applied Information Technology*. 2013; 54 (2); 331-338.

[6]   Chunting Mi, Filippa M, Shen J, Natarajan N. Modeling and control of variable speed constant frequency synchronous generator with brushless exciter. IEEE Transactions on Industry Applications. 2004; 40(2); 565–73.

[7]   Camblong H. Minimizing the impact of wind turbine disturbances in the electricity generation by wind turbines with variable speed" (Minimisation de l'impact des perturbations d'origine éolienne dans la Génération d'électricité par des aérogénérateurs à vitesse variable). *PhD thesis; ENSAM; Bordeaux*; 2003.

[8]   Belabbas B, Tayeb A, Mohamed T, Ahmed S. Hybrid Fuzzy Sliding Mode Control of a DFIG Integrated into the Network. International Journal of Power Electronics and Drive System (IJPEDS). 2013; 3(4); 351-364.

[9]   Refoufi L, Al Zahawi BAT, Jack AG. Analysis and modelling of the steady state behavior of the static Kramer induction generator. *IEEE Transaction Energy Conversion*. 1999; 14(3); 333–339.

[10]   Abdallah B, Slim T, Gérard C, Emile M. Analysis of synchronous machine modeling for simulation and industrial applications. *Simulation Modelling Practice and Theory*. 2010; 18(3); 1382–1396.

[11]   Merabet A, Beguenane R, Thongam JS, Hussein I. *Adaptive sliding mode speed control for wind turbine systems.* In: Proceedings of the [37]th Annual Conf on IEEE Industrial Electronics Society. 2011; 2461– 2466.

[12]   Yongfeng Ren, Hanshan Li, Jie Zhou, Zhongquan An. *Dynamic Performance Analysis of Grid-connected DFIG Based on Fuzzy Logic Control*. International Conference on Mechatronics and Automation (ICMA). 2009; 719–723.

[13]   Zadeh LA. *Fuzzy setes, Information and Control*. 8 338–353. 1965.

21

A New Control Method for Grid-Connected PV System Based on Quasi-Z-Source Cascaded Multilevel Inverter Using Evolutionary Algorithm

Hamid Reza Mohammadi, Ali Akhavan
Department of Electrical Engineering, University of Kashan, Kashan, Iran

ABSTRACT

Keyword:

Cascaded multilevel inverter
Particle swarm optimization
Photovoltaic system
Quasi-Z-source inverter

In this paper, a new control method for quasi-Z-source cascaded multilevel inverter based grid-connected photovoltaic (PV) system is proposed. The proposed method is capable of boosting the PV array voltage to a higher level and solves the imbalance problem of DC-link voltage in traditional cascaded H-bridge inverters. The proposed control system adjusts the grid injected current in phase with the grid voltage and achieves independent maximum power point tracking (MPPT) for the separate PV arrays. To achieve these goals, the proportional-integral (PI) controllers are employed for each module. For achieving the best performance, this paper presents an optimum approach to design the controller parameters using particle swarm optimization (PSO). The primary design goal is to obtain good response by minimizing the integral absolute error. Also, the transient response is guaranteed by minimizing the overshoot, settling time and rise time of the system response. The effectiveness of the new proposed control method has been verified through simulation studies based on a seven level quasi-Z-Source cascaded multilevel inverter.

Corresponding Author:

Hamid Reza Mohammadi,
Departement of Electrical Engineering,
University of Kashan,
Kashan, Iran.
Email: mohammadi@kashanu.ac.ir

1. INTRODUCTION

Photovoltaic (PV) power generation has a great potential to serve as a clean and inexhaustible renewable energy source. However, output power of the PV arrays is greatly affected by environmental conditions such as stochastic changes of the temperature and solar irradiance. In PV systems, extracting the maximum power of the PV array andcurrent injection into the grid at unity power factor are necessary. In recent years, applying various multilevel inverter topologies to PV systems is getting more and more attention due to the large power-scale and high voltage demands. Among various topologies, cascaded H-bridge (CHB) inverter has unique advantages and has been identified as a suitable topology for transformer-less, grid-connected PV systems [1]. Applying CHB inverter in the PV systems has some advantages such as the independent maximum power point tracking (MPPT) of each array. However, the DC-link voltage in each inverter module is not constant, because PV array voltage varies due to the changes of environmental conditions such as temperature and solar irradiation or partial shadows. These cases will cause an imbalance DC-link voltage among different H-bridge modules. Furthermore, in the conventional cascaded multilevel inverter (CMI) based PV system, each module is a buck inverter because the first component of the output AC voltage, always is lower than the input DC voltage. Therefore, an additional DC-DC boost converter is necessary to obtain the desired output voltage, if the input voltage is lower than the desired output voltage and also, to balance the DC-link voltages. This DC-DC boost converter increases the complexity of the power and control circuit and reduces the efficiency [2].

In recent years, the Z-source inverter (ZSI) and quasi-Z-source inverter (QZSI) have been employed for PV power generation system due to some unique advantages and features. Unlike quasi Z-source inverter, ZSI has a discontinuous input current during the shoot-through state due to the blocking diode. Nowadays, quasi-Z-source cascaded multilevel inverter (QZS-CMI) based PV systems were proposed which inherits the advantages of traditional CMI while overcoming issues with imbalance DC-link voltages among independent modules and PV array voltage boost. References [3]-[4] present the various multi-carrier bipolar PWM techniques for QZS-CMI and [5] focused on the parameter design of the QZS-CMI. The phase shifted sinusoidal pulse width modulation (PS-SPWM) is used in [6] as a modulation scheme, but PV system has never been modeled in detail to design the controllers.

In this paper a new control method for a QZS-CMI based PV system is proposed. The control objectives are independent DC-link voltage control, independent MPPT control and current injection to the grid at unity power factor. The proportional-integral (PI) controllers are employed to control each QZS-CMI module. To achieve a high and fast performance, this paper presents an optimum approach to design PI parameters using particle swarm optimization (PSO) and also the PS-SPWM modulation scheme is used for the single phase QZS-CMI. This paper is organized as follows: Section 2 consists an overview of the system with proposed control strategy; Section 3 focuses on the system modeling and grid-connected control; design of the PI parameters using PSO is presented in section 4; the PS-SPWM modulation scheme is presented in section 5; the effectiveness of the proposed strategy is verified by simulation and comparison the results with the reference [2] in section 6; and finally, a conclusion is made in section 7.

2. QZS-CMI AND ITS CONTROL STRATEGY

The QZS-CMI based grid-connected PV system with the proposed control strategy is shown in Figure 1. Comparing to the conventional CMI module, an inductor-capacitor impedance network is included in the input stage of each module. Thisstructureis used to synthesize DC voltage sources to generate $2n + 1$ staircase output waveform where,n is the number of independent PV array. The individual PV source is an array composed of identical PV panels in series and parallel.

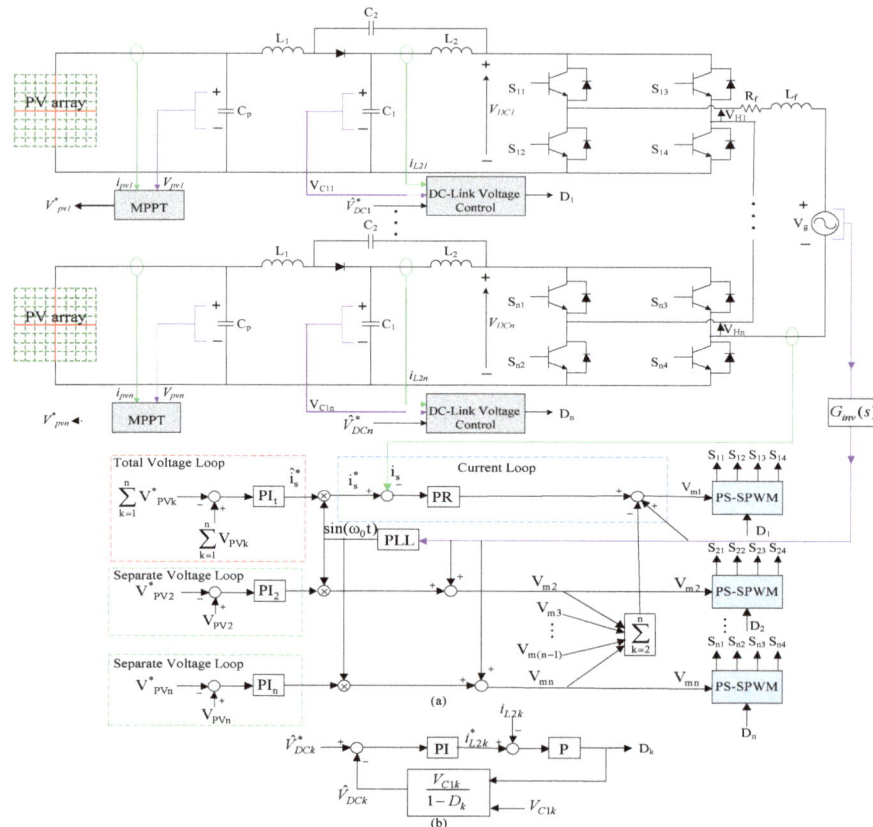

Figure 1. (a) qZS-CMI based grid-tie PV power system and (b) dc-link peak voltage control

2.1. Quasi-Z-source inverter operation

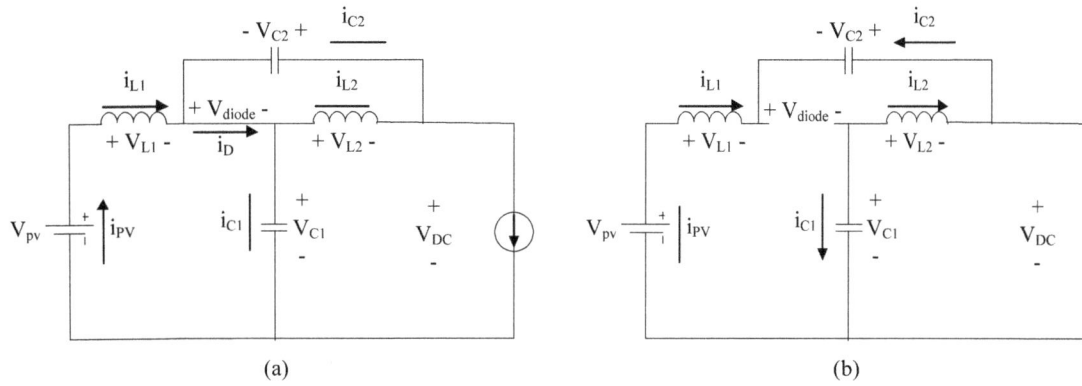

Figure 2. Equivalent circuit of the qZSI; (a) non-shoot-through state and (b) shoot-through state

The QZS-CMI combines the QZS network into each CMI module. The QZSI can be operated in two modes, i.e., the non-shoot-through and the shoot-through [6]. Figure 2 shows the QZSI equivalent circuits operating in the two modes and defines the polarities of all voltages and currents. If the switching period is T_s, the shoot-through period is T_{sh} and non-shoot-through period is T_{nsh}, where:

$$T_s = T_{sh} + T_{nsh} \tag{1}$$

Therefore, the shoot-through duty ratio is $D = T_{sh}/T_s$. When the kth module is in non-shoot-through state, the power is transmitted from theDC side to theAC side and inverter operates as a traditional CMI. In steady state, the following relations can be obtained.

$$\hat{V}_{DCk} = \frac{1}{1-2D_k} V_{PVk} = B_k V_{PVk} \quad , \quad V_{Hk} = S_k \hat{V}_{DCk} \tag{2}$$

While a shoot-through mode, there is no power transmission, because the DC-link voltage is zero. In this mode, there are:

$$\hat{V}_{DCk} = 0 \quad , \quad V_{Hk} = 0 \tag{3}$$

For the QZS-CMI, the synthesized voltage is:

$$V_H = \sum_{k=1}^{n} V_{Hk} = \sum_{k=1}^{n} S_k \hat{V}_{DCk} \tag{4}$$

In the above equations, $\hat{V}_{DCk}$ is the kth module DC-link peak voltage; V_{PVk} is the output voltage of the kth PV array; D_k and B_k are the shoot-through duty ratio and boost factor of the kth module, respectively; V_{Hk} is the output voltage of the kth module and $S_k \in \{-1, 0, 1\}$ is the switching function of the kth module.

2.2. Principle of Control Strategy

Each QZS-CMI module has two independent control commands: shoot-through duty ratio D_n and modulation signal V_m. D_n is used to adjust the DC-link voltage to a desired reference value. While, V_n is used to control the grid injected power. The main goals of the control system of QZS-CMI based grid-connected PV system are: 1) Independent MPPT for each module to ensure the maximum power extraction from each PV array; 2) Current injection into the grid at unity power factor and 3) Balance DC-link peak voltage for all QZS-CMI modules.

For achieving these goals, the PI controllers are employed. The total PV array voltage loop adjusts the sum of n PV array voltages using a PI controller, PI_t. Each PV array voltage reference is calculated by its MPPT block. Also, the current loop achieves a sinusoidal grid-injected current in phase with the grid voltage. A proportional-resonant (PR) controller makes the actual grid current to track the desired grid injected current [7]. The n-1 independent PV array voltage loops control the other n-1 PV array voltages to achieve their own MPPTs through the n-1 PI controllers, named as PI_2 to PI_n, respectively. Also, as shown in Figure 1 (b), DC-link peak voltage is controlled in terms of its shoot-through duty ratio for each QZS-CMI module. A PI controller is used for the DC-link voltage loop to make the DC-link peak voltage tracks its reference value. A proportional controller (P) is used to improve the dynamic of the response. Finally, the independent shoot-through duty ratio D_k and modulation signal V_{mk} are combined into the PS-SPWM modulation scheme for k^{th} module to achieve the desired goals.

3. SYSTEM MODELING

The block diagram of the QZS-CMI based grid-connected PV system is shown in Figure 3. The details will be explained as follows.

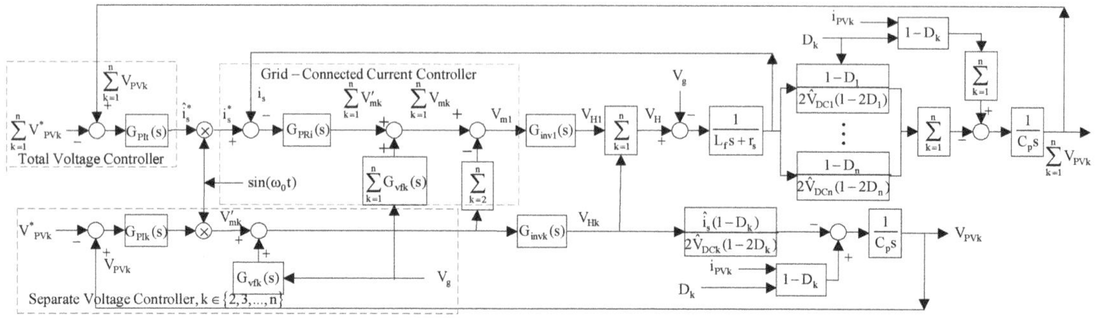

Figure 3. Block diagram of the proposed control grid-connected system for the QZS-CMI based PV system

3.1. Independent PV Voltage and Injected Current Control

In the kth QZS-CMI module the current of the inductor L_1 is:

$$i_{L1k} = i_{PVk} - C_p \frac{dV_{PVk}}{dt}$$

(5)

Where i_{L1k} is the current of inductor L_1 and also, i_{PVk} is the current of kth PV array; and C_p is the shunt capacitor with the PV array. The total output voltage of the QZS-CMI can be written as:

$$V_H = V_g + L_f \frac{di_s}{dt} + r_f i_s$$

(6)

Where V_g is the grid voltage and i_s is the grid injected current; r_f is parasitic resistance and L_f is the filter inductance. Consequently, the transfer function of the grid injected current can be obtained by:

$$G_f(s) = \frac{I_s(s)}{V_H(s) - V_g(s)} = \frac{1}{L_f s + r_f}$$

(7)

To make the actual grid injected current to track the desired reference, a PR controller $G_{PRi}(s) = k_{iP} + \dfrac{k_{iR}\omega_0}{s^2 + \omega_0^2}$ is used, where ω_0 is the resonant frequency i.e. 314 rad/s.

In the next step, a grid voltage feed forward control loop is applied. Therefore, the kth module has the following modulation signal:

$$V_{mk} = V'_{mk} + V_g(s)G_{vfk}(s)$$

(8)

In the above equation, V_{mk} is the kth module modulation signal; V'_{mk} is output of the PI controller in the kth module and $G_{vfk}(s)$ is:

$$G_{vfk}(s) = \frac{1}{nG_{invk}(s)} \quad , \quad G_{invk}(s) = \frac{V_{Hk}(s)}{V_{mk}(s)} = \hat{V}_{DCk}$$

(9)

Due to DC-link peak voltage balance control, the DC-link peak voltages are equal. Therefore:

$$G_{invk}(s) = G_{inv}(s) \quad , \quad k \in \{1, 2, ..., n\}$$

(10)

According to Figure 3, the closed-loop transfer function of the grid injected current control system can be written as:

$$G_{ic} = \frac{I_s(s)}{I_s^*(s)} = \frac{G_{PRi}(s)G_f(s)G_{inv}(s)}{1 + G_{PRi}(s)G_f(s)G_{inv}(s)} = \frac{\hat{V}_{DCk}(k_{iP}S^2 + k_{iP}\omega_0^2 + k_{iR}\omega_0)}{L_f s^3 + (r_f + \hat{V}_{DCk}k_{iP})s^2 + L_f\omega_0^2 s + \hat{V}_{DCk}k_{iP}\omega_0^2 + \hat{V}_{DCk}k_{iR}\omega_0 + r_f\omega_0^2}$$

(11)

According to Figure 2, each PV array voltage can be obtained by:

$$V_{PVk}(s) = \frac{1}{C_p s}[I_{PVk}(s) - I_{L1k}(s)]$$

(12)

In the non-shoot-through mode, the output power is equal to input power. Therefore:

$$\frac{\hat{i}_s \hat{v}_{Hk}}{2} = \hat{v}_{DCk}\bar{i}_{DCk} = v_{PVk}\bar{i}_{L1k\_nsh}$$

(13)

In the above equation, $\hat{v}_{Hk}$ is the output peak voltage of the kth module; $\bar{i}_{DCk}$ is the average current of the DC-link in the kth module; $\bar{i}_{L1k\_nsh}$ is the average current of inductor L_1 in non-shoot-through mode. Equation (13) can be rewritten using (2) as follows:

$$\bar{i}_{L1k\_nsh} = \frac{\hat{i}_s \hat{v}_{Hk}}{2\hat{v}_{DCk}(1 - 2D_k)}$$

(14)

Also, in the shoot-through mode, the average current of the inductor L_1 is:

$$\bar{i}_{L1k\_sh} = i_{PVk}$$

(15)

Therefore, the average current of the inductor L_1 in the one switching cycle can be obtained as follows:

$$\bar{i}_{L1k} = D_k\bar{i}_{L1k\_sh} + (1 - D_k)\bar{i}_{L1k\_nsh} = D_k i_{PVk} + \frac{\hat{i}_s(1 - D_k)\hat{v}_{Hk}}{2\hat{v}_{DCk}(1 - 2D_k)}$$

(16)

In addition, a PI controller $G_{PIt}(s) = k_{Pt} + \frac{k_{It}}{s}$ is used to track the total reference voltage coming from MPPT. The block diagram of the total PV array voltage loop is shown in Figure 4. Also, for modules 2 to n a PI controller $G_{PIk}(s) = k_{Pk} + \frac{k_{Ik}}{s}$ is applied to separate PV voltage to achieve their own MPPTs. The block diagram of the separate PV array voltage loop is shown in Figure 5.

3.2. Independent DC-link Voltage Control

The independent DC-link voltage control scheme is shown in Figure 1 (b). This control loop, adjust DC-link peak voltage using the capacitor-C_1 voltage and the inductor-L_2 current for each QZS-CMI module. Reference [8] presents the kth QZS-CMI module's transfer function from the shoot-through duty ratio to the DC-link peak voltage, $G_{Vdk}(s)$ and from the shoot-through duty ratio to the inductor-L_2 current, $G_{iLdk}(s)$ as follows:

$$G_{Vdk}(s) = \frac{(\frac{L}{1-2D_k}\hat{i}_s)s + (1-2D_k)(V_{ck1}+V_{ck2})}{LCs^2 + (R+r)Cs + (1-2D_k)^2}$$

(17)

$$G_{iLdK}(s) = \frac{LC(V_{ck1}+V_{ck2})s^2 + (R+r)(V_{ck1}+V_{ck2})Cs + (Ls+R+r)\dfrac{\hat{i}_s}{(1-2D_k)}}{(Ls+R+r)[LCs^2 + (R+r)Cs + (1-2D_k)^2]}$$

(18)

Where, L is the inductor and C is the capacitor of the impedance network. R is the series resistance of capacitors and r is the parasitic resistance of inductors; V_{ck1} and V_{ck2} are capacitor C_1 and C_2 voltages, respectively.

A proportional gain K_{dPk} is employed for the inductor current loop to improve the dynamic response. As shown in Figure 6, A PI controller with the transfer function of $G_{ViPIk}(s) = k_{VdPk} + \dfrac{k_{Vdlk}}{s}$ is cascaded to the inductor current loop to control the DC-link peak voltage.

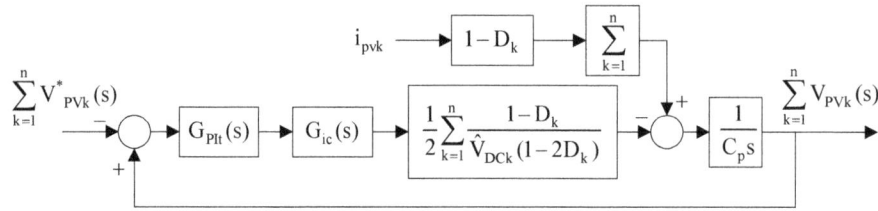

Figure 4. Block diagram of the total PV voltage loop

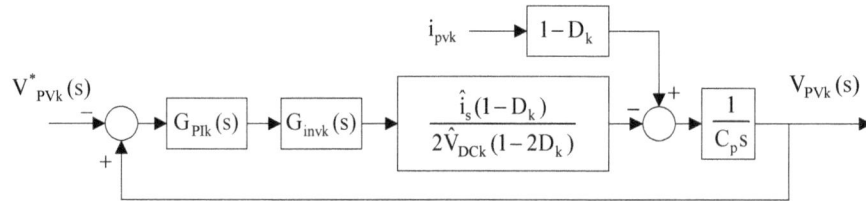

Figure 5. Block diagram of the separate PV voltage loop

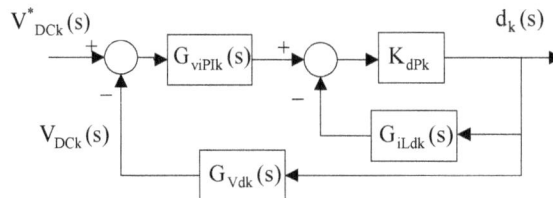

Figure 6. Block diagram of the DC-link peak voltage control of the kth module

4. DESIGN OF PI CONTROLLERS USING PARTICLE SWARM OPTIMIZATION

The PI controller is a well-known method for industrial control processes. This is due to its robust performance and simple structure in a wide range of operating conditions. Tuning of such a controller requires specification of two parameters: proportional gain K_p and integral gain K_i [9]. In the past, this problem has been handled by a trial and error technique. In this paper, the problem of the PI parameters tuning is formulated as an optimization problem. The problem formulation employs four performance indexes, i.e., the overshoot, the settling time, the rise time and the integral absolute error of the step response as the objective function to tuning the PI parameters for getting a well performance under a given plant. In this study, the primary design goal is to obtain good response by minimizing the integral absolute error. At the same time, the transient response is guaranteed by minimizing the overshoot, the settling time and the rise timeof the step response.Furthermore, we employ a solution algorithm based on particle swarm optimization.

The PSO is a stochastic optimization technique, which uses swarming behaviors observed in flock of birds. In fact, the PSO was inspired by the sociological behavior associated with swarms. PSO was developed by James Kennedy and Russell Eberhart in 1995 as a new heuristic method [10]. It uses a number of particles that constitute a swarm moving around in one N-dimensional search space looking for the best position. The individuals in the swarm are called particles. Each particle in the PSO algorithm is a potential solution for the optimizationproblem and keep track of its co-ordinates in the problem space and tries to search the best position through flying in a multidimensional space, which are associated with the best solution (called best fitness) it has achieved so far that called "*pbest*". Another "best" value called "*gbest*" that is tracked bythe global version of the particle swarm optimizer is the overall best value and its location obtained so far by each particle.

The transient response is very important, therefore both the amplitude and time duration of the transient response must be kept within tolerable limits. Hence, four indexes of the transient response are utilized to characterize the performance of PI control systems.

4.1. Response Parameters
Overshoot:

With the assumption of y as the step response, y_{max} is the maximumvalue and y_{ss} is the steady-state value of y. therefore the overshoot (f_O) is equal to:

$$f_O = y_{\max} - y_{ss} \qquad (19)$$

Rise Time:

The rise time (f_{RT}) is defined as the time required for the step response to rise from 10% to 90% of its final value. Hence:

$$f_{RT} = t_{y90\%} - t_{y10\%} \qquad (20)$$

Settling Time:

The settling time (f_{ST}) is defined as the time required for the step response to decrease and stay within a 5% of its final value.

$$f_{ST} = t_{st} \qquad (21)$$

Integral Absolute Error:
The integral absolute error (f_{IAE}) can be written as:

$$f_{IAE} = \int_0^{\infty} |e(t)| \, dt \qquad (22)$$

4.2. Objective Function
The objective function (f_{Totel}) for optimal design of PI controllers can be formulated as follows:

$$f_{Total} = f_O + f_{RT} + f_{ST} + f_{IAE} \qquad (23)$$

To apply PSO for tuning the PI controllers, the closed-loop transfer function of the total PV voltage loop, separate PV voltage loop and DC-link peak voltage is necessary. These closed-loop transfer functions are calculated using block diagram of Figure 4–6, respectively.

$$\frac{\sum V_{PVk}}{\sum V_{PVk}^*} = \frac{\left[k_{Pr}k_{iP}s^3 + k_{Ir}k_{iP}s^2 + k_{Pr}(k_{iP}\omega_0^2 + k_{iR}\omega_0)s + k_{Ir}(k_{iP}\omega_0^2 + k_{iR}\omega_0)\right] \times (\frac{1-D_1}{2(1-2D_1)} + \frac{1-D_2}{2(1-2D_2)} + \frac{1-D_3}{2(1-2D_3)})}{C_P L_f s^5 + C_P(\hat{V}_{DC}k_{iP} + r_f)s^4 + (C_P L_f \omega_0^2 + k_P k_{iP} \times (\frac{1-D_1}{2(1-2D_1)} + \frac{1-D_2}{2(1-2D_2)} + \frac{1-D_3}{2(1-2D_3)}))s^3 +}$$

$$+ (C_P(\hat{V}_{DC}k_{iP}\omega_0^2 + \hat{V}_{DC}k_{iR}\omega_0 + r_f \omega_0^2) + k_{Ir}k_{iP}(\frac{1-D_1}{2(1-2D_1)} + \frac{1-D_2}{2(1-2D_2)} + \frac{1-D_3}{2(1-2D_3)}))s^2 +$$

$$+ k_{Pr}(k_{iP}\omega_0^2 + k_{iR}\omega_0)(\frac{1-D_1}{2(1-2D_1)} + \frac{1-D_2}{2(1-2D_2)} + \frac{1-D_3}{2(1-2D_3)})s$$

$$+ k_{Ir}(k_{iP}\omega_0^2 + k_{iR}\omega_0)(\frac{1-D_1}{2(1-2D_1)} + \frac{1-D_2}{2(1-2D_2)} + \frac{1-D_3}{2(1-2D_3)})$$

(24)

$$\frac{V_{PVk}}{V_{PVk}^*} = \frac{k_{Pk}(\frac{1-D_k}{1-2D_k})\frac{i_s}{2}s + k_{Ik}(\frac{1-D_k}{1-2D_k})\frac{i_s}{2}}{C_P s^2 + k_{Pk}(\frac{1-D_k}{1-2D_k})\frac{i_s}{2}s + k_{Ik}(\frac{1-D_k}{1-2D_k})\frac{i_s}{2}}$$

(25)

$$\frac{d_k(s)}{V_{DCk}^*(s)} =$$

(26)

$$\frac{k_{VdPk}L^2 C s^5 + k_{dPk}k_{VdPk}(LC(R+r)+LC)s^4 + k_{VdIk}k_{dPk}L^2 C s^3 + (k_{dPk}k_{VdPk}(R+r)(1-2D_k)^2)s^2 + (Lk_{dPk}\frac{\hat{i}_s}{1-2D_k})s + k_{dPk}k_{VdPk}(R+r)}{L^2 C s^6 + LC(R+r)s^5 + (LCk_{dPk}\hat{V}_{DCk})s^4 + (L(1-2D_k)^2 + (R+r)^2)s^3 + (k_{dPk}(R+r)C\hat{V}_{DCk})s^2 + (R+r)^2 s + k_{dPk}k_{VdIk}(1-2D_k)^2}$$

5.   THE PS-SPWM FOR QZS-CMI

The modulation technique applied in the proposed system is a phase shifted sinusoidal pulse width modulation that shown in Figure 7. The shoot-through states are inserted with the simple boost control method. In this control method, two straight lines, which are denoted as $1-D_n$ and D_n-1, envelops equal to or greater than the peak value of the sinusoidal reference signals are used to control shoot-through duty ratio.

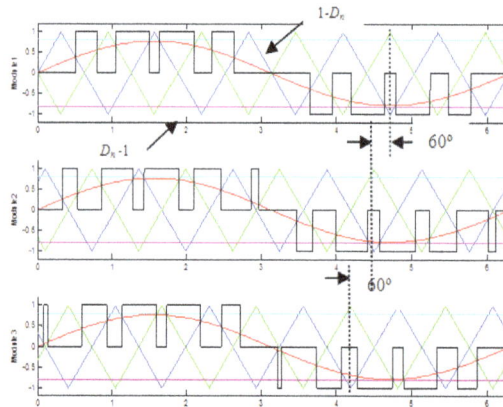

Figure 7. Modulation scheme for the proposed system

If the triangular carrier signal is smaller than D_n-1 or bigger than $1-D_n$ the two switches of one leg in H-bridge module are turned on simultaneously [6]. In PS-SPWM schemes, the number of triangular carrier waves is equal to $m-1$, where m is the level number. Also, the required phase shifts among different carriers is given as:

$$\varphi = \frac{360^\circ}{m-1}$$

(27)

Therefore, in this case ($m=7$) triangular carriers of different H-bridge modules are shifted 60° with respect to each other.

6. SIMULATION RESULTS

The performance evaluation of the proposed control strategy was carried out by different simulation in PSCAD/EMTDC. The parameters of the QZS-CMI are shown in Table 1. Also the PSO algorithm is programmed in MATLAB to obtain the best parameters for the PR and PI controllers. As mentioned earlier, the closed-loop transfer function of the total PV voltage control loop, the separate PV voltage control loop and the DC-link peak voltage control loop are used for optimization problem. The results of the PSO algorithm are shown in Table 2. The effectiveness of the proposed method is shown by comparing the results of the new proposed method for parameter tuning of PI controllers with the results of [2] in Table 3. In this table, the overshoot, the settling time and the rise time of the step response for these closed-loop transfer functions are presented.As shown in thistable, the overshoot, the settling time and the rise time of step response for all the closed-loop transfer functions obtained using PSO algorithm is smaller than the results of [2].

Table 1. QZS-CMI Parameters

QZS-CMI Parameters	Value
QZS inductance, L_1 and L_2	1.8 mH
QZS capacitance, C_1 and C_2	3300μF
PV array parallel capacitance, C_p	1100 μF
Filter inductance, L_f	1mH
Carrier frequency, f_c	5kHz
Grid voltage	220V/50Hz

Table 2. The Parameters of the Control System by Pso Algorithm

Parameters	Value	Parameters	Value
k_{iP}	0.00491	k_{iR}	-0.01673
k_{Pt}	1.1511	k_{It}	1.5348
k_{Pk}	0.0263	k_{Jk}	0.0017
k_{dPk}	0.0073		
k_{VdPk}	0.0313	k_{Vdlk}	2.8122

Table 3. Transfer Functions Characteristics

Transfer functions	Overshoot (%)		Settling time (sec)		Rise time (sec)	
	using PSO data	using [2] data	using PSO data	using [2] data	using PSO data	using [2] data
Total PV voltage loop	8.62	56.1	0.362	2.84	0.0395	0.177
Separate PV voltage loop	12.9	33.3	0.438	0.511	0.0578	0.0651
DC-link peak voltage loop	≈ 0	≈ 0	0.243	1.38	0.13	0.779

The P-V characteristic of the PV array is shown in Figure 8. The measured voltage and current of each PV array are used to calculate the MPPT search algorithm for the PV voltage reference at the MPP. Incremental conductance algorithm is used for tracking the maximum power point of a PV array in this paper [11].

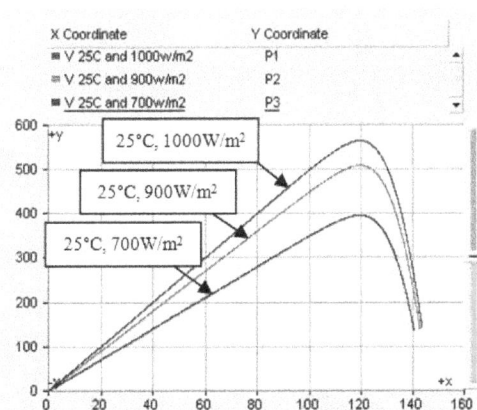

Figure 8. PV array power-voltage characteristic

At first, all modules are working at 1000 W/m^2 and 25°C condition and all the initial voltage references of MPPT algorithm are given at 120 V. The peak value of DC-link voltage of allmodules is controlled at 145V. After 1 second, the third module's irradiation decreases to 700 W/m^2. Hence, according to Figure 8, the reference of MPPT decreases to about 119.3 V. It should be noted that, the change of temperature mostly affects the voltage of maximum power point, so that, temperature rising causes the voltage of maximum power point to drop. While the change of solar irradiation affects the current injection.

The total PV voltage (sum of all three PV array voltages) and other PV array voltages are shown in Figure 9(a)-(d), respectively. It can be seen that during the change in the MPPT reference value due to change of environmental condition, the proposed control method have excellent tracking performance after a very short transient. It can be seen in Figure 9(d) that, after a change in the environmental condition of module 3, the controller of this module tracks the new reference. While, module 2 have notany transient because no change is happening in its condition. As shown in Figure 9(b-d), PV array voltage of module 1 is different with respect to modules 2 and 3. This is due to the fact that the modulation signal generation of module 1 is different from other modules.

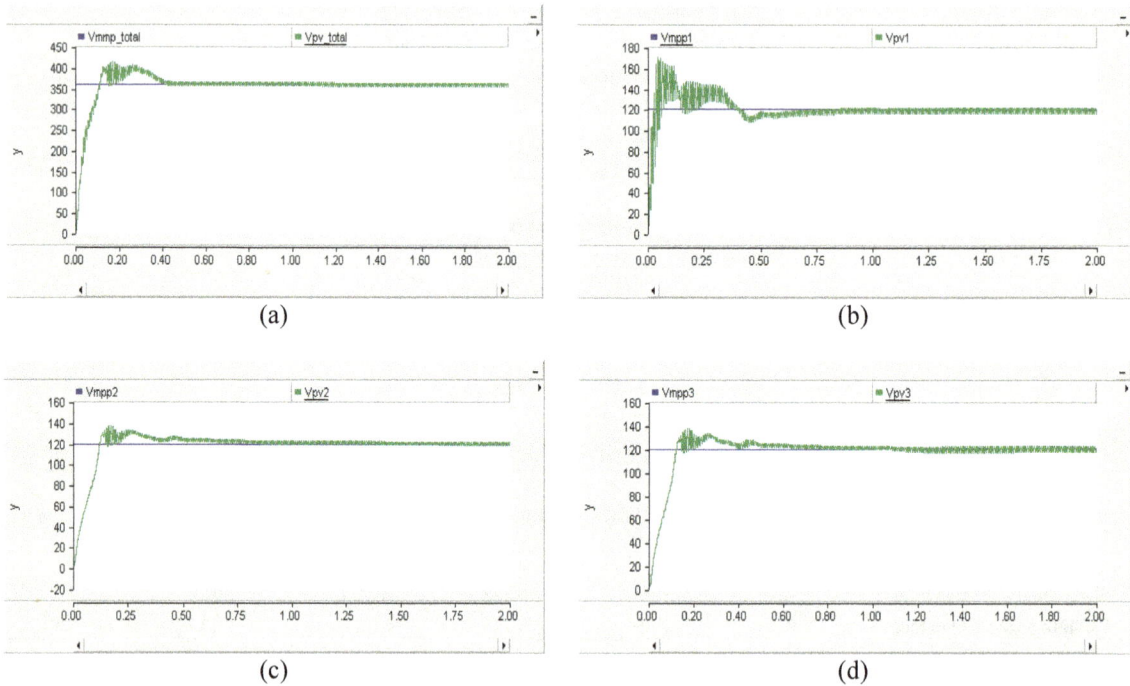

(a)   (b)

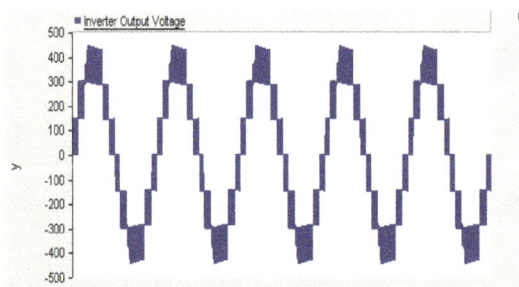

(c)   (d)

Figure 9. Simulation results (a) Total PV array voltage; (b) PV array voltage of module 1; (c) PV array voltage of module 2; (d) PV array voltage of module 3

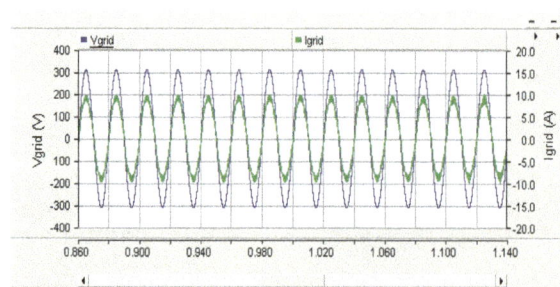

Figure 10. Inverter output voltage   Figure 11. Grid voltage and current

The inverter output voltage is shown in Figure 10. It can be indicated that the solar irradiation or temperature does not affect the seven-level staircase output voltage of the inverter due to shoot-through operation. It is no matter the temperature and the solar irradiation changes or not, the QZS-CMI always outputs a constant voltage, which verifies the voltage balancing capability.

Figure 11 shows the grid-injected current which is exactly in phase with the grid voltage and ensures unity power factor.It should be noted that, lower irradiation causes reduction of the grid-injected current as Figure 12, which the RMS value of the grid injected current is shown. As shown in this figure, the injected current reduces about 0.3A after reduction of irradiation.

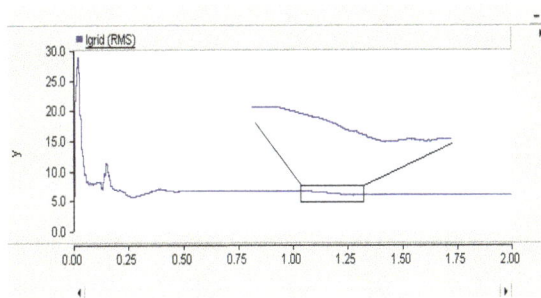

Figure 12. RMS value of the grid injected current   Figure 13. DC-link voltage of module 3

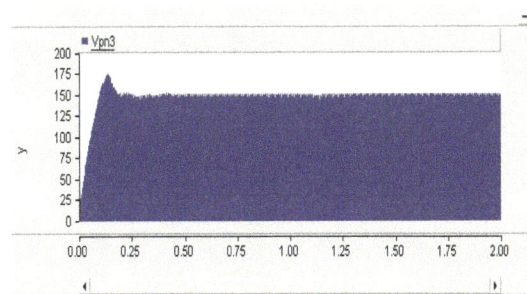

The DC-link voltage of module 3 is shown in Figure 13. It can be seen that, with the independent DC-link voltage control, DC-link peak voltage are kept at the reference value (145 V). It should be noted that, according to (2), after decrement of the MPPT reference value of module 3, longer shoot-through time interval is necessary for this module to fix the DC-link peak voltage at 145 V. As shown in Figure 13, DC-link peak voltage control loop, adjusts new shoot-through duty ratio after a very short transient, so that, third module's DC-link peak voltage have a very low distortion after a change in the solar irradiation.

7.   CONCLUSION

This paper proposed a new control method for QZS-CMI based single-phase grid-connected PV system. The proposed system is a combination of QZSI and CHB multilevel topology and has both advantages of them. This enables independent MPPT control even if some modules' PV array had different conditions.Moreover, independent DC-link voltage control enforced all QZS-HBI modules have the balanced voltage and also, grid-injected current was fulfilled at unity power factor.The control parameters designed using PSO algorithm and it was shown that the system has fast and accurate response. A PS-SPWMtechnique was proposed for modulation to synthesize the staircase voltage waveform of the single-phase QZS-CMI. A simulation was carried out on a seven level QZS-CMI in the variable environmental condition. The simulation results show the effectiveness of the proposed control strategy for a QZS-CMI based single-phase grid-connected PV system.

REFERENCES

[1]   J Sastry, P Bakas, H Kim, L Wang, A Marinopoulos. Evaluation of cascaded H-bridge inverter for utility-scale photovoltaic systems. *Renewable Energy*. 2014; 69: 208-218.

[2]   Y Liu, B Ge, H Abu-Rub, FZ Peng. An Effective Control Method for Quasi-Z-Source Cascade Multilevel Inverter-Based Grid-Tie Single-Phase Photovoltaic Power System. *IEEE Transactions on Industrial Informatics*. 2014; 10(1): 399-407.

[3]   I Colak, E Kabalci, R Bayindir. Review of multilevel voltage source inverter topologies and control schemes. *Energy Conversion and Management*. 2011; 52(2): 1114-1128.

[4]   Y Liu, H Abu-Rub, B Ge, FZ Peng. Phase-shifted pulse-width-amplitude modulation for quasi-Z-source cascade multilevel inverter based PV power system. *Energy Conversion Congress and Exposition (ECCE), IEEE*. 2013; 94-100.

[5]   D Sun, B Ge, X Yan, D Bi, H Zhang, Y Liu, H Abu-Rub, L BenBrahim, F Peng. Modeling, Impedance Design, and Efficiency Analysis of Quasi-Z Source Module in Cascade Multilevel Photovoltaic Power System. *IEEE Transactions onIndustrial Electronics*. 2014; 61(11): 6108-6117.

[6]   D Sun, B Ge, FZ Peng, AR Haitham, D Bi, Y Liu. A new grid-connected PV system based on cascaded H-bridge quasi-Z source inverter. *IEEE International Symposium onIndustrial Electronics (ISIE2012)*. 2012; 951,956.

[7]   DN Zmood, DG Holmes. Stationary frame current regulation of PWM inverters with zero steady-state error. *IEEE Transactions on Power Electronics*. 2003; 18(3): 814,822.

[8]   Y Liu, B Ge, FZ Peng, AR Haitham, AT de Almeida, Ferreira FJTE. Quasi-Z-Source inverter based PMSG wind power generation system. *Energy Conversion Congress and Exposition (ECCE), IEEE.* 2011; 291-297.

[9]   YT Hsiao, CL Chuang, CC Chien. Ant colony optimization for designing of PID controllers. *IEEE International Symposium onComputer Aided Control Systems Design.* 2004; 321-326.

[10] J Kennedy, R Eberhart. Particle swarm optimization. *IEEE International Conference onNeural Networks.* 1995; 1942-1948.

[11] V Salas, E Olías, A Barrado, A Lázaro. Review of the maximum power point tracking algorithms for stand-alone photovoltaic systems. *Solar Energy Materials and Solar Cells.* 2006; 90(11); 1555-1578.

Permissions

List of Contributors

Muhd Zharif Rifqi Zuber Ahmadi
Faculty of Electrical Engineering, Faculty of Electrical Engineering,Universiti Teknikal Malaysia Melaka Hang Tuah Jaya,76100 Durian Tunggal, Melaka Malaysia

Auzani Jidin
Faculty of Electrical Engineering, Faculty of Electrical Engineering,Universiti Teknikal Malaysia Melaka Hang Tuah Jaya,76100 Durian Tunggal, Melaka Malaysia

Maaspaliza Azri
Faculty of Electrical Engineering, Faculty of Electrical Engineering,Universiti Teknikal Malaysia Melaka Hang Tuah Jaya,76100 Durian Tunggal, Melaka Malaysia

Khairi Rahim
Faculty of Electrical Engineering, Faculty of Electrical Engineering,Universiti Teknikal Malaysia Melaka Hang Tuah Jaya,76100 Durian Tunggal, Melaka Malaysia

Tole Sutikno
Department of Electrical Engineering, Universitas Ahmad Dahlan, Yogyakarta, Indonesia

K. Viswanadha S Murthy
Department of Electrical and Electronics Engineering, KL University, Andhra Pradesh, India

M. Kirankumar
Department of Electrical and Electronics Engineering, KL University, Andhra Pradesh, India

G. R. K. Murthy
Department of Electrical and Electronics Engineering, KL University, Andhra Pradesh, India

A. Arivarasu
Department of Electrical and Electronics Engineering, SASTRA University, Thanjavur

R. Balasubramanium
Department of Electrical and Electronics Engineering, SASTRA University, Thanjavur

Mohammad Jannati
UTM-PROTON Future Drive Laboratory, Faculty of Electrical Engineering, Universiti Teknologi Malaysia, Johor Bahru, MALAYSIA

Seyed Hesam Asgari
UTM-PROTON Future Drive Laboratory, Faculty of Electrical Engineering, Universiti Teknologi Malaysia, Johor Bahru, MALAYSIA

Nik Rumzi Nik Idris
UTM-PROTON Future Drive Laboratory, Faculty of Electrical Engineering, Universiti Teknologi Malaysia, Johor Bahru, MALAYSIA

Mohd Junaidi Abdul Aziz
UTM-PROTON Future Drive Laboratory, Faculty of Electrical Engineering, Universiti Teknologi Malaysia, Johor Bahru, MALAYSIA

Souha Boukadida
Laboratory EµE of the FSM, University of Monastir, Tunisia

Soufien Gdaim
Laboratory EµE of the FSM, University of Monastir, Tunisia

Abdellatif Mtibaa
Laboratory EµE of the FSM, University of Monastir, Tunisia

M. Mohammadi
Department of Electrical Engineering, College of Engineering, Borujerd Branch, Islamic Azad University, Borujerd, Iran

A. Mohammadi Rozbahani
Department of Electrical Engineering, College of Engineering, Borujerd Branch, Islamic Azad University, Borujerd, Iran

S. Abasi Garavand
Department of Electrical Engineering, College of Engineering, Borujerd Branch, Islamic Azad University, Borujerd, Iran

M. Montazeri
Department of Electrical Engineering, College of Engineering, Borujerd Branch, Islamic Azad University, Borujerd, Iran

H. Memarinezhad
Department of Electrical Engineering, College of Engineering, Borujerd Branch, Islamic Azad University, Borujerd, Iran

Kodanda Ram R B P U S B
Department of Electrical and Electronics Engineering, K L University, Guntur, AP, INDIA

M. Venu Gopala Rao
Department of Electrical and Electronics Engineering, K L University, Guntur, AP, INDIA

K. Ramalingeswara Rao
Departement of Electrical and ElectronicsEngineering, K L University

K. S. Srikanth
Departement of Electrical and ElectronicsEngineering, K L University

Zerzouri Nora
Department of Electrical Engineering, Badji Mokthar University Annaba, Algeria

Labar Hocine
Department of Electrical Engineering, Badji Mokthar University Annaba, Algeria

G. Madhusudhana Rao
Professor, Departement of Electrical and Electronics Engineering, TKRCET

V. Anwesha Kumar
Research Scholar, Departement of Electrical and Electronics Engineering, JNTUH

B. V.Sanker Ram
Professor, Departement of Electrical and Electronics Engineering, TKRCET

Byamakesh Nayak
School of Electrical Engineering, KIIT University, Bhubaneswar

Saswati Swapna Dash
Department of Electrical Engineering, YMCA University of science and technology, Faridabad

Subrat Kumar
Department of operation and control (Electrical), Bharat Petroleum Corporation limited, MMBPL, Mathura

N. M. Nordin
Faculty of Electrical Engineering, Universiti Teknologi Malaysia, Johor Bahru, Malaysia

N. R. N. Idris
Faculty of Electrical Engineering, Universiti Teknologi Malaysia, Johor Bahru, Malaysia

N. A. Azli
Faculty of Electrical Engineering, Universiti Teknologi Malaysia, Johor Bahru, Malaysia

M. Z. Puteh
MIMOS Berhad, Technology Park Malaysia, Kuala Lumpur, Malaysia

T. Sutikno
Department of Electrical Engineering, Universitas Ahmad Dahlan, Yogyakarta, Indonesia

Othmane Boughazi
Faculty of Sciences and technology, BECHAR University B.P. 417 BECHAR, 08000

Abdelmadjid Boumedienne
Faculty of Sciences and technology, BECHAR University B.P. 417 BECHAR, 08000

Hachemi Glaoui
Faculty of Sciences and technology, BECHAR University B.P. 417 BECHAR, 08000

Mohamed Barara
Laboratory of Power Electronic and Control, Mohamed V University Agdal Mohammadia School of Engineering, Rabat Morocco

Abderrahim Bennassar
Laboratory of Power Electronic and Control, Mohamed V University Agdal Mohammadia School of Engineering, Rabat Morocco

Ahmed Abbou
Laboratory of Power Electronic and Control, Mohamed V University Agdal Mohammadia School of Engineering, Rabat Morocco

Mohamed Akherraz
Laboratory of Power Electronic and Control, Mohamed V University Agdal Mohammadia School of Engineering, Rabat Morocco

Badre Bossoufi
Laboratory of Electrical Engineering and Maintenance, Higher School of Technology, EST-Oujda, University of Mohammed I, Morocco

Saber Krim
Laboratory of Electronics and Microelectronics (EuE), Faculty of Sciences of Monastir University of Monastir, Tunisia

Soufien Gdaim
Laboratory of Electronics and Microelectronics (EuE), Faculty of Sciences of Monastir University of Monastir, Tunisia

Abdellatif Mtibaa
Laboratory of Electronics and Microelectronics (EuE), Faculty of Sciences of Monastir University of Monastir, Tunisia
Department of Electrical Engineering, National Engineering School of Monastir, University of Monastir, Tunisia

Mohamed Faouzi Mimouni
Research Unit of industrial systems Study and renewable energy (ESIER), University of Monastir, Tunisia.
Department of Electrical Engineering, National Engineering School of Monastir, University of Monastir, Tunisia

G. Venu Madhav
Department of Electrical and Electronics Engineering, Padmasri Dr. B. V. Raju Institute of Technology

Y. P. Obulesu
Department of Electrical and Electronics Engineering, LakiReddy BaliReddy College of Engineering

Goh Wee Yen
Department of Electrical Power Engineering, Universiti Teknologi Malaysia (UTM), Johor Bahru, Malaysia

Ali Monadi
Department of Electrical Power Engineering, Universiti Teknologi Malaysia (UTM), Johor Bahru, Malaysia

Nik Rumzi Nik Idris
Department of Electrical Power Engineering, Universiti Teknologi Malaysia (UTM), Johor Bahru, Malaysia

Auzani Jidin
Department of Power Electronics and Drives, Universiti Teknikal Malaysia Melaka (UTeM), Malacca, Malaysia

Tole Sutikno
Department of Power Electronics and Drives, Universiti Teknikal Malaysia Melaka (UTeM), Malacca, Malaysia

Md. Rashedul Islam
Department of Electrical and Electronic Engineering, Dhaka University of Engineering & Technology, Gazipur-1700, Bangladesh

Md. Maruful Islam
Department of Electrical and Electronic Engineering, Dhaka University of Engineering & Technology, Gazipur-1700, Bangladesh

Md. Kamal Hossain
Department of Electrical and Electronic Engineering, Dhaka University of Engineering & Technology, Gazipur-1700, Bangladesh

Pintu Kumar Sadhu
Service Operation Center, Grameenphone Limited, Bangladesh

R. Gunabalan
Departement of Electrical and Electronics Engineering, Dr. Sivanthi Aditanar College of Engineering, Tiruchendur

V. Subbiah
Departement of Electrical and Electronics Engineering, PSG College of Technology, Coimbatore

Tahir Khalfallah
Laboratoire de Génie Energitique et Génie Informatique LGEGI, Université Ibn Khaldoun de Tiaret, Algérie

Belfedal Cheikh
Laboratoire de Génie Energitique et Génie Informatique LGEGI, Université Ibn Khaldoun de Tiaret, Algérie

Allaoui Tayeb
Laboratoire de Génie Energitique et Génie Informatique LGEGI, Université Ibn Khaldoun de Tiaret, Algérie

Gerard Champenois
University of Poitiers, Laboratoire d'Informatique et d'Automatique pour les Systèmes, Bâtiment B25, 2, rue Pierre Brousse, 86022 Poitiers, France

Hamid Reza Mohammadi
Department of Electrical Engineering, University of Kashan, Kashan, Iran

Ali Akhavan
Department of Electrical Engineering, University of Kashan, Kashan, Iran